Theory of Lie Groups

Claude Chevalley

DOVER PUBLICATIONS
Garden City, New York

Bibliographical Note

This Dover edition, first published in 2018, is an unabridged republication of the work originally published by Princeton University Press, Princeton, New Jersey, in 1946.

Library of Congress Cataloging-in-Publication Data

Names: Chevalley, C. (Claude), 1909–1984, author.
Title: Theory of Lie groups / Claude Chevalley.
Other titles: Thâeorie des groupes de Lie. English
Description: Dover edition. | Garden City, New York : Dover Publications 2018. | Originally published: Princeton : Princeton University Press, 1946.
Identifiers: LCCN 2017046119| ISBN 9780486824536 | ISBN 0486824535
Subjects: LCSH: Lie groups. | Continuous groups.
Classification: LCC QA387 .C4613 2018 | DDC 512/.482—dc23
LC record available at https://lccn.loc.gov/2017046119

Manufactured in the United States of America
82453502 2023
www.doverpublications.com

To

ELIE CARTAN

and

HERMANN WEYL

INTRODUCTION

Expository books on the theory of Lie groups generally confine themselves to the local aspect of the theory. This limitation was probably necessary as long as general topology was not yet sufficiently well elaborated to provide a solid base for a theory in the large. These days are now passed, and we have thought that it would be useful to have a systematic treatment of the theory from a global point of view. The present volume introduces the main basic principles which govern the theory of Lie groups.

A Lie group is at the same time a group, a topological space and a manifold: it has therefore three kinds of "structures," which are interrelated with each other. The elementary properties of abstract groups are by now sufficiently well known to the general mathematical public to make it unnecessary for such a book as this one to contain a purely group-theoretic chapter. The theory of topological groups, however, has been included and is treated in Chapter II. The greatest part of this chapter is concerned with the theory of covering spaces and groups, which is developed independently from the theory of paths. Chapter III is concerned with the theory of (analytic) manifolds (independently of the notion of group). Our definition of a manifold is inspired by the definition of a Riemann surface given by H. Weyl in his book "Die Idee der Riemannschen Flache"; it has, compared with the definition by overlapping system of coordinates, the advantage of being intrinsic. The theory of involutive systems of differential equations on a manifold is treated not only from the local point of view but also in the large. In order to achieve this, a definition of the submanifolds of a manifold is given according to which a submanifold is not necessarily a topological subspace of the manifold in which it is imbedded.

The notions of topological group and manifold are combined together in Chapter IV to give the notions of analytic group and Lie group. An analytic group is a topological group which is given *a priori* as a manifold; a Lie group (at least when it is connected) is a topological group which can be endowed with a structure of manifold in such a way that it becomes an analytic group. It is shown that, if this is possible, the manifold-structure in question is uniquely determined, so that connected Lie groups and analytic groups are in reality the same things defined in different ways. We shall see however in the second volume that the difference becomes a real one when *complex* analytic groups are considered instead of the real ones which are treated here.

Chapter V contains an exposition of the theory of exterior differential forms of Cartan which plays an essential role in the general theory of Lie groups, as well in its topological as in its differential geometric aspects. This theory leads in particular to the construction of the invariant integral on a Lie group. In spite of the fact that this invariant integration can be defined on arbitrary locally compact groups, we have thought that it is more in the spirit of a treatise on Lie groups to derive it from the existence of left invariant differential forms.

Chapter VI is concerned with the general properties of *compact* Lie groups. The fundamental fact is of course contained in the statement of Peter-Weyl's theorem which guarantees the existence of faithful linear representations. We have also included a proof of the generalization by Tannaka of the Pontrjagin duality theorem. A slight modification of the original proof of Tannaka shows that a compact Lie group may be considered as the set of real points of an algebraic variety in a complex affine space, the whole variety being itself a Lie group on which complex coordinates can be introduced.

The second volume of this book, now in preparation, will be mainly concerned with the theory and classification of semi-simple Lie Groups.

In preparing this book, I have received many valuable suggestions from several of my friends, in particular from Warren Ambrose, Gerhardt Hochschild, Deane Montgomery and Hsiao Fu Tuan. I was helped in reading the proofs by John Coleman and Norman Hamilton. I have also received precious advice from Professor H. Weyl and Professor S. Lefschetz. To all of them I am glad to express here my deep gratitude.

C. C.

CONTENTS

Introduction v

I. The Classical Linear Groups 1

II. Topological Groups 25

III. Manifolds 68

IV. Analytic Groups. Lie Groups 99

V. The Differential Calculus of Cartan 139

VI. Compact Lie Groups and Their Representations 171

Index 215

Some Notations Used in This Book

I. We denote by ϕ the empty set, by $\{a\}$ the set composed of the single element a.

If f is a mapping of a set A into a set B, and if X is a sub-set of B, we denote by $\overset{-1}{f}(X)$ the set of the elements $a \epsilon A$ such that $f(a) \epsilon X$. If g is a mapping of B into a third set C, we denote by $g \circ f$ the mapping which assigns to every $a \epsilon A$ the element $g(f(a))$.

We use the signs $\cup$, $\cap$ to represent respectively the intersection and the union of sets. If E_α is a collection of sets, the index α running over a set A, we denote by $\bigcup_{\alpha \epsilon A} E_\alpha$ the union of all sets E_α and by $\bigcap_{\alpha \epsilon A} E_\alpha$ their intersection. We denote by δ_{ij} the Kronecker symbol, equal to 1 if $i = j$ and to 0 if $i \neq j$.

II. If G is a group, we call "neutral element" the element ϵ of G such that $\epsilon\sigma = \sigma$ for every $\sigma \epsilon G$.

We say that a sub-group H of G is "distinguished" if the conditions $\sigma \epsilon G$, $\tau \epsilon H$ imply $\tau\sigma\tau^{-1} \epsilon H$.

If $\sigma = (a_{ij})$ represents a matrix, the symbol $\boxed{\sigma} = \boxed{a_{ij}}$ stands for the determinant of the matrix; $\delta p \sigma$ stands for the trace of the matrix.

If $\mathfrak{M}$, $\mathfrak{N}$ are vector spaces over the same field K, we call *product* of $\mathfrak{M}$ and $\mathfrak{N}$, and denote by $\mathfrak{M} \times \mathfrak{N}$, the set of the pairs $(\mathbf{e}, \mathbf{f})$ with $\mathbf{e} \epsilon \mathfrak{M}$, $\mathbf{f} \epsilon \mathfrak{N}$, this set being turned in a vector space by the conventions

$$(\mathbf{e}, \mathbf{f}) + (\mathbf{e}', \mathbf{f}') = (\mathbf{e} + \mathbf{e}', \mathbf{f} + \mathbf{f}')$$
$$a(\mathbf{e}, \mathbf{f}) = (a\mathbf{e}, a\mathbf{f}) \qquad \text{for } a \epsilon K.$$

III. *Topology.* We call topological spaces only the spaces in which Hausdorf separation axiom is satisfied.

A neighbourhood of a point p in space $\mathfrak{V}$ is understood to be a set N such that there exists an open set U such that $p \varepsilon U \subset N$; N need not be open itself.

The adherence $\bar{A}$ of a set A in a topological space is the set of those points p such that every neighbourhood of p meets A. Every point of $\bar{A}$ is said to be adherent to A. We shall make use of the possibility of defining the topology in a space by the operation $A \rightarrow \bar{A}$ of adherence (cf. Alexandroff-Hopf, *Topologie*, Kap. 1).

Intervals. If a and b are real numbers such that $a \leqslant b$, we denote by $]a, b[$ the open interval of extremities a and b. We set $]a, b] =]a, b[\cup \{b\}$, $[a, b[=]a, b[\cup \{a\}$, $[a, b] =]a, b[\cup \{a\} \cup \{b\}$.

CHAPTER I

The Classical Linear Groups

Summary. Chapter I introduces the classical linear groups whose study is one of the main objects of Lie group theory. The unitary and orthogonal groups are defined in §I, together with a series of other groups. Their fundamental property of being compact is established.

Section II is concerned with the study of the exponential of a matrix. The property for a matrix of being orthogonal or unitary is defined by a system of non-linear relationships between its coefficients; the exponential mapping gives a parametric representation of the set of unitary (or orthogonal) matrices by matrices whose coefficients satisfy *linear* relations (Cf. Proposition 5, §II, p. 8). The reader may observe that the spaces M^s, M^{sh}, M^S, M^R which are introduced on p. 8 all contain $YX - XY$ whenever they contain X and Y. Although we could have given here an elementary explanation of this fact, we have not done so, on account of the fact that the full importance of this result can only be grasped much later (in Chapter IV). In the cases of the orthogonal and unitary group, the linearization can also be accomplished by the Cayley parametrization (which we have not introduced); however, the exponential mapping is more advantageous from our point of view because it preserves some properties of the ordinary exponential function (Cf. Proposition 3, §IV, p. 13).

Sections III and IV are preliminary to the result which will be proved in Section V (Proposition 1, p. 14). Hermitian matrices are defined in terms of the unitary geometry in a complex vector space (unitary geometry is defined by the notion of hermitian product of two vectors, just as euclidean geometry can be defined in terms of the scalar product). Proposition 2, §III, p. 10 shows that the unitary matrices are the isometric transformations of a unitary geometry.

The proposition which asserts that the full linear group can be decomposed topologically into the product of the unitary group and the space of positive definite hermitian matrices (Proposition 1, §V, p. 14) is the prototype of the theorems which allow us to derive topological properties of general Lie groups from the properties of compact groups. A similar decomposition is given for the complex orthogonal group (Proposition 2, §V, p. 15).

Sections VI and VII are preliminary to the definition of the symplectic groups. The symplectic group is defined to be the group of isometric transformations of a symplectic geometry (Definition 1, §VII, p. 20). In §IX, we construct a representation of $Sp(n)$ by complex matrices of degree $2n$. The consideration of the conditions which the matrices of this representation must satisfy leads to the introduction of a new group, the complex symplectic group $Sp(n, C)$. It can be seen easily that $Sp(n, C)$ stands in the same relation to $Sp(n)$ as $GL(n, C)$ to $U(n)$ or as $O(n, C)$ to $O(n)$. A proposition of the type of Proposition 1, §V, p. 14 could be derived without much difficulty for $Sp(n, C)$. However, we have not found it necessary to state this proposition, which is contained as a special case of a theorem proved later (Corollary to Theorem 5, Chapter VI, §XII, p. 211).

§I. THE FULL LINEAR GROUP AND SOME OF ITS SUBGROUPS

The n-dimensional complex cartesian space C^n may be considered as a vector space of dimension n over the field C of complex numbers. Let $\mathbf{e}_i$ be the element of C^n whose i-th coordinate is 1 and whose other coordinates are 0. The elements $\mathbf{e}_1, \cdots, \mathbf{e}_n$ form a base of C^n over C.

A linear endomorphism α of C^n is determined when the elements $\alpha\mathbf{e}_i = \Sigma_{j=1}^n a_{ji}\mathbf{e}_j$ are given. There corresponds to this endomorphism a matrix (a_{ij}) of degree n; we shall denote this matrix by the same letter α as the endomorphism itself. Conversely, to any matrix of degree n with complex coefficients, there corresponds an endomorphism of C^n.

Let α and β be two endomorphisms of C^n, and let (a_{ij}) and (b_{ij}) be the corresponding matrices. Then $\alpha \circ \beta$ is again an endomorphism, whose matrix (c_{ij}) is the product of the matrices (a_{ij}) and (b_{ij}); i.e.

$$c_{ij} = \Sigma_{k=1}^n a_{ik}b_{kj} \tag{1}$$

We shall denote by $\mathfrak{M}_n(C)$ the set of all matrices of degree n with coefficients in C. If $(a_{ij}) \varepsilon \mathfrak{M}_n(C)$, we set $b_{i+(j-1)n} = a_{ij}$ and we associate with the matrix (a_{ij}) the point of coordinates $b_1, \cdots, b_{n^2}$ in C^{n^2}. We obtain in this way a one-to-one correspondence between $\mathfrak{M}_n(C)$ and C^{n^2}. Since C^{n^2} is a topological space, we can define a topology in $\mathfrak{M}_n(C)$ by the requirement that our correspondence shall be a homeomorphism between $\mathfrak{M}_n(C)$ and C^{n^2}.

Let $\mathfrak{E}$ be any topological space, and let φ be a mapping of $\mathfrak{E}$ into $\mathfrak{M}_n(C)$. If $t \varepsilon \mathfrak{E}$, let $a_{ij}(t)$ be the coefficients of the matrix $\varphi(t)$. It is clear that φ will be continuous if and only if each function $a_{ij}(t)$ is continuous.

It follows immediately from this remark and from the formulas (1) that the product $\sigma\tau$ of two matrices σ, τ is a continuous function of the pair (σ, τ), considered as a point of the space $\mathfrak{M}_n(C) \times \mathfrak{M}_n(C)$.

If $\alpha = (a_{ij})$, we shall denote by ${}^t\alpha$ the *transpose* of α, i.e. the matrix (a'_{ij}), with $a'_{ij} = a_{ji}$. We shall denote by $\bar{\alpha}$ the complex conjugate of α, i.e. the matrix $\bar{\alpha} = (\bar{a}_{ij})$. It is clear that the mappings $\alpha \to {}^t\alpha$, $\alpha \to \bar{\alpha}$ are homeomorphisms of order 2 of $\mathfrak{M}_n(C)$ with itself. If α and β are any two matrices, we have

$${}^t(\alpha\beta) = {}^t\beta\,{}^t\alpha \qquad \overline{\alpha\beta} = \bar{\alpha}\bar{\beta}$$

A matrix σ will be called *regular* if it has an inverse, i.e., if there exists a matrix σ^{-1} such that $\sigma\sigma^{-1} = \sigma^{-1}\sigma = \epsilon$, where ϵ is the unit matrix of degree n. A necessary and sufficient condition for a matrix σ to be regular is that its determinant $\boxed{\sigma}$ be $\neq 0$.

If an endomorphism σ of C^n maps C^n onto itself (and not onto some subspace of lower dimension), the corresponding matrix σ is regular and σ has a reciprocal endomorphism σ^{-1}.

If σ is a regular matrix, we have

$$^t(\sigma^{-1}) = (^t\sigma)^{-1} \qquad \bar{\sigma}^{-1} = \overline{(\sigma^{-1})}$$

If σ and τ are regular matrices, $\sigma\tau$ is also regular and we have

$$(\sigma\tau)^{-1} = \tau^{-1}\sigma^{-1}$$

It follows that the regular matrices form a group with respect to the operation of multiplication.

Definition 1. *The group of all regular matrices of degree n with complex coefficients is called the general linear group. We shall denote it by $GL(n, C)$.*

Since the determinant of a matrix is obviously a continuous function of the matrix, $GL(n, C)$ is an open subset of $\mathfrak{M}_n(C)$. We may consider the elements of $GL(n, C)$ as points of a topological space, which is a subspace of $\mathfrak{M}_n(C)$.

If $\sigma = (a_{ij})$ is a regular matrix, the coefficients b_{ij} of σ^{-1} are given by expressions of the form

$$b_{ij} = A_{ij}\boxed{\sigma}^{-1}$$

where the A_{ij}'s are polynomials in the coefficients of σ. It follows that the mapping $\sigma \rightarrow \sigma^{-1}$ of $GL(n, C)$ onto itself is continuous. Since this mapping coincides with its reciprocal mapping, it is a homeomorphism of order 2 of $GL(n, C)$ with itself.

The mappings $\sigma \rightarrow \bar{\sigma}$ and $\sigma \rightarrow {}^t\sigma$ are homeomorphisms of $GL(n, C)$ with itself. The first but not the second is also an automorphism of the group $GL(n, C)$.

If $\sigma \varepsilon GL(n, C)$, we shall denote by σ^* the matrix defined by the formula

$$\sigma^* = {}^t\sigma^{-1}$$

We have

$$(\sigma\tau)^* = \sigma^*\tau^* \qquad (\sigma^*)^{-1} = (\sigma^{-1})^*$$

Hence, the mapping $\sigma \rightarrow \sigma^*$ is a homeomorphism and an automorphism of order 2 of $GL(n, C)$.

Definition 2. *A matrix σ is said to be orthogonal if $\sigma = \bar{\sigma} = \sigma^*$. The set of all orthogonal matrices of degree n will be denoted by $O(n)$. If only $\sigma = \sigma^*$, σ is said to be complex orthogonal; the set of these matrices will be denoted by $O(n, C)$. If only $\bar{\sigma} = \sigma^*$, σ is said to be unitary. The set of all unitary matrices will be denoted by $U(n)$.*

Since the mappings $\sigma \to \bar{\sigma}$ and $\sigma \to \sigma^*$ are continuous, the sets $O(n)$, $O(n, C)$ and $U(n)$ are closed subsets of $GL(n, C)$. Because these mappings are automorphisms, $O(n)$, $O(n, C)$ and $U(n)$ are subgroups of $GL(n, C)$. We have clearly

$$O(n) = O(n, C) \cap U(n).$$

Definition 3. *We shall say that the matrix σ is real if its coefficients are real,* i.e. *if $\sigma = \bar{\sigma}$. The set of all real matrices of degree n will be denoted by $\mathfrak{M}_n(R)$. The set $\mathfrak{M}_n(R) \cap GL(n, C)$ will be denoted by $GL(n, R)$.*

Therefore, we have also

$$O(n) = GL(n, R) \cap O(n, C)$$

The determinant of the product of two matrices being the product of the determinants of these matrices, it follows that the matrices of determinant 1 form a subgroup of $GL(n, C)$.

Definition 4. *The group of all matrices of determinant* 1 *in $GL(n, C)$ is called the special linear group. This group is denoted by $SL(n, C)$. We set $SL(n, R) = SL(n, C) \cap GL(n, R)$; $SO(n) = SL(n, C) \cap O(n)$; $SU(n) = SL(n, C) \cap U(n)$.*

It is clear that $SL(n, C)$, $SL(n, R)$, $SO(n)$, $SU(n)$ are subgroups and closed subsets of $GL(n, C)$. They may be considered as subspaces of $GL(n, C)$.

Theorem 1. *The spaces $U(n)$, $O(n)$, $SU(n)$ and $SO(n)$ are compact.*

Since $O(n)$, $SU(n)$ and $SO(n)$ are closed subsets of $U(n)$, it is sufficient to prove that $U(n)$ is compact. A matrix σ is unitary if and only if ${}^t\sigma\bar{\sigma} = \epsilon$, where ϵ is the unit matrix (in fact, this condition implies that σ is regular and that $\sigma^* = \bar{\sigma}$). If $\sigma = (a_{ij})$, the equation ${}^t\sigma\bar{\sigma} = \epsilon$ is equivalent to the conditions

$$\Sigma_j a_{ji}\bar{a}_{jk} = \delta_{ik}$$

Since the left sides of these equations are continuous functions of σ, $U(n)$ is not only a closed subset of $GL(n, C)$ but also of $\mathfrak{M}_n(C)$. Moreover, the conditions $\Sigma_j a_{ji}\bar{a}_{ji} = 1$ imply $|a_{ij}| \leqslant 1 (1 \leqslant i, j \leqslant n)$. It follows that the coefficients of a matrix $\sigma \varepsilon U(n)$ are bounded. If we take into account the homeomorphism established between $\mathfrak{M}_n(C)$

and C^{n^2}, we see that $U(n)$ is homeomorphic to a closed bounded subset of C^{n^2}. Theorem 1 is thereby proved.

§II. The Exponential of a Matrix

Let α be any matrix of degree n, and let μ be an upper bound for the absolute values of the coefficients $x_{ij}(\alpha)$ of α. Let $x_{ij}^{(p)}(\alpha)$ be the coefficients of $\alpha^p (0 \leqslant p < \infty$; we set $\alpha^0 = \epsilon =$ the unit matrix). We assert that $|x_{ij}^{(p)}(\alpha)| \leqslant (n\mu)^p$. This is true for $p = 0$. Assume that our inequality holds for some integer $p \geqslant 0$; then

$$|x_{ij}^{(p+1)}(\alpha)| = \left|\Sigma_k x_{ik}^{(p)}(\alpha) x_{kj}(\alpha)\right| \leqslant n(n\mu)^p\mu = (n\mu)^{p+1}$$

which proves that the inequality holds for $p + 1$.

It follows that each of the n^2 series $\Sigma_{p=0}^{\infty} \frac{1}{p!} x_{ij}^{(p)}(\alpha)$ converges uniformly on the set of all α such that $|x_{ij}(\alpha)| \leqslant \mu$. In other words, the series $\epsilon + \frac{\alpha}{1} + \frac{\alpha^2}{2!} + \cdots + \frac{\alpha^p}{p!} + \cdots$ is always convergent, and uniformly so when α remains in a bounded region of the set $\mathfrak{M}_n(C)$.

Definition 1. *We denote by* exp α *the sum of the series* $\Sigma_0^{\infty} \frac{1}{p!} \alpha^p$.

The function exp α is thus defined and continuous on $\mathfrak{M}_n(C)$ and maps $\mathfrak{M}_n(C)$ into itself.

Proposition 1. *If σ is a regular matrix of degree n, then*

$$\exp(\sigma\alpha\sigma^{-1}) = \sigma(\exp \alpha)\sigma^{-1}$$

In fact, we have $\sigma\alpha^p\sigma^{-1} = (\sigma\alpha\sigma^{-1})^p$, and hence $\exp(\sigma\alpha\sigma^{-1}) = \Sigma_0^{\infty} \frac{1}{p!} (\sigma\alpha\sigma^{-1})^p = \Sigma_0^{\infty}\sigma\left(\frac{1}{p!}\alpha^p\right)\sigma^{-1} = \sigma\left(\Sigma_0^{\infty}\frac{1}{p!}\alpha^p\right)\sigma^{-1} = \sigma(\exp\alpha)\sigma^{-1}$.

Proposition 2. *If $\lambda_1, \cdots, \lambda_n$ are the characteristic roots of α, each occurring a number of times equal to its multiplicity, the characteristic roots of* exp α *are* exp $\lambda_1, \cdots,$ exp λ_n.

We shall prove this by induction on n. It is obvious for $n = 1$, because then α is a complex number. Now, assume that $n > 1$ and that the proposition holds for matrices of degree $n - 1$.

Let λ_1 be a characteristic root of α; then there is an element $\mathbf{a} \neq 0$ in C^n such that $\alpha\mathbf{a} = \lambda_1\mathbf{a}$. Let $\mathbf{e}_1$ be the point whose coordinates are $1, 0, \cdots, 0$. Because $\mathbf{a} \neq 0$, there exists a regular matrix σ such that $\sigma\mathbf{a} = \mathbf{e}_1$. Then $\sigma\alpha\sigma^{-1}\mathbf{e}_1 = \lambda\mathbf{e}_1$; in other words,

$$\sigma\alpha\sigma^{-1} = \begin{pmatrix} \lambda_1 & * \cdots\cdots * \\ 0 & \\ \cdot & (\tilde{\alpha}) \\ \cdot & \\ 0 & \end{pmatrix}$$

where the *'s indicate complex numbers and $\tilde{\alpha}$ is a matrix of degree $n - 1$. We have

$$\sigma\alpha^p\sigma^{-1} = \begin{pmatrix} \lambda_1^p & * \cdots\cdots * \\ 0 & \\ \cdot & (\tilde{\alpha}^p) \\ \cdot & \\ \cdot & \\ 0 & \end{pmatrix}$$

and therefore

$$\exp(\sigma\alpha\sigma^{-1}) = \begin{pmatrix} \exp \lambda_1 & * \cdots\cdots * \\ 0 & \\ \cdot & \\ \cdot & (\exp \tilde{\alpha}) \\ \cdot & \\ \cdot & \\ 0 & \end{pmatrix}$$

If $\lambda_2, \cdots, \lambda_n$ are the characteristic roots of $\tilde{\alpha}$, those of α, which are the same as those of $\sigma\alpha\sigma^{-1}$, are $\lambda_1, \lambda_2, \cdots, \lambda_n$. The proposition being true for matrices of degree $n - 1$, it follows that the characteristic roots of $\exp \tilde{\alpha}$ are $\exp \lambda_2, \cdots, \exp \lambda_n$, and those of $\exp(\sigma\alpha\sigma^{-1})$ are $\exp \lambda_1, \exp \lambda_2, \cdots, \exp \lambda_n$. But these are also the characteristic roots of $\sigma(\exp \alpha)\sigma^{-1}$ (Cf. Proposition 1) and hence of $\exp \alpha$. Proposition 2 is thereby proved.

Corollary 1. *The determinant of the matrix* $\exp \alpha$ *is* $\exp \operatorname{Sp} \alpha$.

This follows at once from the facts that the trace and the determinant of a matrix are respectively the sum and the product of the characteristic roots.

Corollary 2. *The exponential of any matrix is a regular matrix.*

Proposition 3. *If α and β are permutable matrices* (i.e. *if* $\alpha\beta = \beta\alpha$) *then* $\exp(\alpha + \beta) = (\exp \alpha)(\exp \beta)$.

Since α and β are permutable, we can expand $(\alpha + \beta)^p$ by the binomial formula:

$$\frac{1}{P!}(\alpha + \beta)^P = \Sigma_0^P \frac{\alpha^k}{k!}\frac{\beta^{p-k}}{(p-k)!}$$

Therefore, for any integer P, we have

$$\Sigma_0^{2P} \frac{(\alpha+\beta)^p}{p!} = \left(\Sigma_0^P \frac{\alpha^p}{p!}\right)\left(\Sigma_0^P \frac{\beta^p}{p!}\right) + R_p$$

where R_P is the sum $\Sigma_{(k,l)}\, \alpha^k/k!\, \beta^l/l!$, extended over all combinations (k, l) such that $\max(k, l) > P$, $k + l \leqslant 2P$. The number of these combinations of indices is $P(P + 1)$. On the other hand, if μ is an upper bound for the coefficients of α and β, the absolute value of any coefficient of $\alpha^k/k!\, \beta^l/l!$ is at most $n(n\mu)^k/k!(n\mu)^l/l! \leqslant (n\mu_0)^{2P}/P!$, where μ_0 is some number > 0. It follows that the coefficients of R_P are smaller than $P(P + 1)(n\mu_0)^{2P}/P!$ in absolute value and that R_P tends to 0 as P increases indefinitely. The formula to be proved is an immediate consequence of this fact.

Corollary. *If t is a real variable and α a fixed matrix, the mapping $t \rightarrow \exp t\alpha$ is a continuous homomorphism of the additive group of real numbers into $GL(n, C)$.*

If α is any matrix, we have clearly

$$\exp\,({}^t\alpha) = {}^t(\exp \alpha);\ \exp \bar{\alpha} = \overline{\exp \alpha}$$

It follows from the Corollary to Proposition 3 that we have also

$$\exp\,(-\alpha) = (\exp \alpha)^{-1}.$$

Proposition 4. *There exists a neighbourhood U of O in $\mathfrak{M}_n(C)$ which is mapped topologically onto a neighbourhood of ϵ in $GL(n, C)$ by the mapping $\alpha \rightarrow \exp \alpha$.*

We represent a matrix $\alpha \varepsilon \mathfrak{M}_n(C)$ by the point of C^{n^2} whose coordinates are the coefficients $x_{ij}(\alpha)$ of α (these coefficients being arranged in some fixed order). From the uniform convergence of the series $\Sigma_0^\infty \alpha^P/P!$ it follows that the coefficients $y_{ij}(\alpha)$ of $\exp \alpha$ are integral analytic functions $F_{ij}(\cdot\cdot\cdot, x_{kl}(\alpha), \cdot\cdot\cdot)$ of the coefficients of α. It is clear that the terms of degrees < 2 in the Maclaurin expansion of $F_{ij}(\cdot\cdot\cdot, x_{kl}, \cdot\cdot\cdot)$ are $\delta_{ij} + x_{ij}$. It follows immediately that the Jacobian of the n^2 functions F_{ij} with respect to their n^2 arguments is equal to 1 when $x_{kl} = 0 (1 \leqslant k, l \leqslant n)$. By the theorem on implicit functions, we know that the mapping of C^{n^2} into itself which assigns to the point of coordinates x_{ij} the point of coordinates $F_{ij}(\cdot\cdot\cdot, x_{kl}, \cdot\cdot\cdot)$ maps topologically some neighbourhood of the origin onto a neighbourhood of the point of coordinates $y_{ij} = \delta_{ij}$. Proposition 4 follows immediately.

Definition 2. *A matrix α is said to be skew symmetric if ${}^t\alpha + \alpha = 0$, skew hermitian if ${}^t\alpha + \bar{\alpha} = 0$.*

We shall denote by M^s the set of skew symmetric matrices, by M^{sh} the set of skew hermitian matrices, by M^S the set of matrices of trace O and by M^R the set of real matrices.

Lemma 1. *We can find a neighbourhood U of O in $\mathfrak{M}_n(C)$ which satisfies the following conditions:* 1) *it is mapped topologically onto a neighbourhood of ϵ in $GL(n, C)$ by the mapping $\alpha \to \exp \alpha$;* 2) *the trace of any $\alpha \in U$ is smaller than 2π in absolute value;* 3) *the condition $\alpha \in U$ implies $-\alpha \in U$, ${}^t\alpha \in U$, $\bar{\alpha} \in U$.*

Let U_1 be a neighbourhood of O which satisfies the first and second conditions; we denote by $-U_1$ the set of matrices $-\alpha$ with $\alpha \in U_1$, and we define similarly ${}^tU_1$, $\bar{U}_1$. The set $U = U_1 \cap (-U_1) \cap ({}^tU_1) \cap \bar{U}_1$ satisfies the conditions of Lemma 1.

Proposition 5. *Let U be a neighbourhood of O in $\mathfrak{M}_n(C)$ which satisfies the conditions of Lemma* 1. *The sets $M^S \cap U$, $M^{sh} \cap U$, $M^S \cap M^{sh} \cap U$, $M^R \cap U$, $M^R \cap M^S \cap U$, $M^R \cap M^{sh} \cap U$, $M^R \cap M^S \cap M^{sh} \cap U$, $M^s \cap U$ are mapped topologically under the mapping $\alpha \to \exp \alpha$ onto neighbourhoods of ϵ in the following groups: $SL(n, C)$, $U(n)$, $SU(n)$, $GL(n, R)$, $SL(n, R)$, $O(n)$, $SO(n)$, $O(n, C)$.*

We know that the mapping $\alpha \to \exp \alpha$ maps every subset of U topologically. If $\alpha \in M^S$, we have $\exp \alpha \in SL(n,C)$ by Corollary 1 to Proposition 2 above. If $\alpha \in M^s$, we have ${}^t(\exp \alpha) = \exp({}^t\alpha) = \exp(-\alpha) = (\exp \alpha)^{-1}$, which proves that $\exp \alpha$ is complex orthogonal. In a similar way, we prove that, if $\alpha \in M^{sh}$, then $\exp \alpha$ is unitary. Conversely, if $\exp \alpha \in SL(n, C)$, $\alpha \in U$, the conditions $\exp(Sp\ \alpha) = 1$, $|Sp\ \alpha| < 2\pi$ imply $Sp\ \alpha = 0$, whence $\alpha \in M^S$. If $\exp \alpha \in O(n, C)$, $\alpha \in U$, we have ${}^t\alpha \in U$, $-\alpha \in U$ and $\exp({}^t\alpha) = \exp(-\alpha)$, whence ${}^t\alpha = -\alpha$ and $\alpha \in M^s$. In a similar way, we see that if $\alpha \in U$ and $\exp \alpha$ is unitary, then $\alpha \in M^{sh}$. If α is real, $\exp \alpha$ is also real; conversely, if $\alpha \in U$ is such that $\exp \alpha$ is real, we have $\exp \alpha = \exp \bar{\alpha}$ whence $\alpha = \bar{\alpha}$. Proposition 5 follows immediately from these facts.

The sets M^S, M^{sh}, M^R, $M^S \cap M^{sh}$, $M^R \cap M^S$, $M^R \cap M^{sh}$, $M^R \cap M^S \cap M^{sh}$, M^s may all be considered as vector spaces over the field R of real numbers; as such, their dimensions are $2n^2 - 2$, n^2, n^2, $n^2 - 1$, $n^2 - 1$, $n(n - 1)/2$, $n(n - 1)/2$ and $n(n - 1)$ respectively. We have therefore proved:

Proposition 6. *In each of the groups $GL(n, C)$, $SL(n, C)$, $U(n)$, $SU(n)$, $GL(n, R)$, $SL(n, R)$, $O(n)$, $SO(n)$, $O(n, C)$ there exists a neighbourhood of the neutral element which is homeomorphic to an open set in a real cartesian space of suitable dimension. These dimensions are:*

$2n^2$ *for* $GL(n, C)$, $2n^2 - 2$ *for* $SL(n, C)$, n^2 *for* $U(n)$, $n^2 - 1$ *for* $SU(n)$, n^2 *for* $GL(n, R)$, $n^2 - 1$ *for* $SL(n, R)$, $n(n - 1)/2$ *for* $O(n)$ *and* $SO(n)$, $n(n - 1)$ *for* $O(n, C)$.

§III. Hermitian Product

As we have already observed, the space C^n may be considered as a vector space of dimension n over C, with the base $\{\mathbf{e}_1, \cdots, \mathbf{e}_n\}$ introduced in §I, p. 2. In this section, we shall use the notation $\mathbf{a}z$ (instead of $z\mathbf{a}$) for the product of a vector $\mathbf{a}$ by a number z; this notation will be preferable when we come to quaternions.

Definition 1. *Let* $\mathbf{a} = \Sigma_1^n \mathbf{e}_i z_i$ *and* $\mathbf{b} = \Sigma_1^n \mathbf{e}_i u_i$ *be vectors in* C^n. *We define their hermitian product* $\mathbf{a} \cdot \mathbf{b}$ *by*

$$\mathbf{a} \cdot \mathbf{b} = \Sigma_1^n \bar{z}_i u_i$$

We define the length of $\mathbf{a}$ *to be the number* $||\mathbf{a}|| = (\mathbf{a} \cdot \mathbf{a})^{\frac{1}{2}} = (\Sigma_1^n z_i \bar{z}_i)^{\frac{1}{2}}$.

We see immediately that $||\mathbf{a}|| \geqslant 0$, and that $||\mathbf{a}|| = 0$ implies $\mathbf{a} = 0$.

The number $\mathbf{a} \cdot \mathbf{b}$ is, for $\mathbf{a}$ fixed, a linear function of $\mathbf{b}$; i.e.

$$\mathbf{a} \cdot (\mathbf{b}_1 u_1 + \mathbf{b}_2 u_2) = (\mathbf{a} \cdot \mathbf{b}_1) u_1 + (\mathbf{a} \cdot \mathbf{b}_2) u_2$$

However, if $\mathbf{b}$ is fixed, $\mathbf{a} \cdot \mathbf{b}$ is *not* a linear function of $\mathbf{a}$, for we have

$$\mathbf{b} \cdot \mathbf{a} = \overline{(\mathbf{a} \cdot \mathbf{b})}$$

whence

$$(\mathbf{a}_1 z_1 + \mathbf{a}_2 z_2) \cdot \mathbf{b} = (\mathbf{a}_1 \cdot \mathbf{b}) \bar{z}_1 + (\mathbf{a}_2 \cdot \mathbf{b}) \bar{z}_2$$

Definition 2. *A vector* $\mathbf{a}$ *is called a unit vector if* $||\mathbf{a}|| = 1$. *Two vectors* $\mathbf{a}$ *and* $\mathbf{b}$ *are said to be othogonal if* $\mathbf{a} \cdot \mathbf{b} = 0$. *A set of vectors is said to be orthonormal if every vector of the set is a unit vector and any two different vectors of the set are orthogonal.*

Proposition 1. *Let* $\mathbf{a}_1, \cdots, \mathbf{a}_m$ *be* m *linearly independent vectors. Then there exists an orthonormal set* $\{\mathbf{b}_1, \cdots, \mathbf{b}_m\}$ *such that, for each* k $(1 \leqslant k \leqslant m)$, *the sets* $\{\mathbf{a}_1, \cdots, \mathbf{a}_k\}$ *and* $\{\mathbf{b}_1, \cdots, \mathbf{b}_k\}$ *span the same subspace of* C^n.

We proceed by induction on m. Proposition 4 holds for $m = 1$; in fact, we have $\mathbf{a}_1 \neq 0$, and we may define $\mathbf{b}_1$ to be $\mathbf{a}_1 ||\mathbf{a}_1||^{-1}$. Assume that $m > 1$ and that Proposition 1 holds for systems of $m - 1$ vectors. Then, we can find vectors $\mathbf{b}_1, \cdots, \mathbf{b}_{m-1}$ such that, for every $k \leqslant m - 1$, the sets $\{\mathbf{a}_1, \cdots, \mathbf{a}_k\}$ and $\{\mathbf{b}_1, \cdots, \mathbf{b}_k\}$ span the same subspace of C^n. Now let us consider the vector $\mathbf{c} = \mathbf{a}_m - \Sigma_{i=1}^{m-1} \mathbf{b}_i (\mathbf{b}_i \cdot \mathbf{a}_m)$. Because $\mathbf{a}_m$ is linearly independent of $\mathbf{a}_1, \cdots, \mathbf{a}_{m-1}$, $\mathbf{c}$ does not lie in the space spanned by $\mathbf{a}_1, \cdots, \mathbf{a}_{m-1}$. We define $\mathbf{b}_m$ to be $\mathbf{c}||\mathbf{c}||^{-1}$. Obviously, $||\mathbf{b}_m|| = 1$ and (using the orthogonality of $\mathbf{b}_1, \cdots, \mathbf{b}_{m-1}$),

$$\mathbf{b}_m \cdot \mathbf{b}_k = (\mathbf{a}_m \cdot \mathbf{b}_k - \mathbf{a}_m \cdot \mathbf{b}_k)||\mathbf{c}||^{-1} = 0$$

which shows that $\mathbf{b}_m$ is orthogonal to $\mathbf{b}_1, \cdots, \mathbf{b}_{m-1}$. Then $\{\mathbf{b}_1, \cdots, \mathbf{b}_m\}$ is an orthonormal set which spans the same space as $\{\mathbf{a}_1, \cdots, \mathbf{a}_m\}$: Proposition 1 is proved for systems of m vectors.

Corollary 1. *Any vector subspace of C^n has an orthonormal base.*

Corollary 2. *Any unit vector* $\mathbf{a}$ *of C^n belongs to an orthonormal base of C^n.*

In fact, $\mathbf{a}$ may be taken as the first element of a base of C^n. If we apply to this base the construction of the proof of Proposition 1, we obtain an orthonormal base of C^n whose first element is $\mathbf{a}$.

We shall now consider the matrices of degree n as endomorphisms of C^n, in the way which was explained in §I.

Proposition 2. *A necessary and sufficient condition that a matrix σ be unitary is that $||\sigma\mathbf{a}|| = ||\mathbf{a}||$ for all $\mathbf{a} \varepsilon C^n$. This condition implies that $\sigma\mathbf{a} \cdot \sigma\mathbf{b} = \mathbf{a} \cdot \mathbf{b}$ for any two vectors $\mathbf{a}$ and $\mathbf{b}$ in C^n.*

First, let $\alpha = (a_{ij})$ be any matrix. We have $\alpha\mathbf{e}_i = \Sigma_j \mathbf{e}_j a_{ji}$, whence $\bar{a}_{ji} = (\alpha\mathbf{e}_i) \cdot \mathbf{e}_j$. We have also ${}^t\alpha\mathbf{e}_j = \Sigma_i \mathbf{e}_i a_{ji}$, whence $a_{ji} = \mathbf{e}_i \cdot ({}^t\alpha\mathbf{e}_j)$ and $(\alpha\mathbf{e}_i) \cdot \mathbf{e}_j = \mathbf{e}_i \cdot ({}^t\bar{\alpha}\mathbf{e}_j)$. It follows easily that

$$(\alpha\mathbf{a}) \cdot \mathbf{b} = \mathbf{a} \cdot ({}^t\bar{\alpha}\mathbf{b}) \tag{1}$$

for any two vectors $\mathbf{a} = \Sigma_i \mathbf{e}_i a_i$ and $\mathbf{b} = \Sigma_j \mathbf{e}_j b_j$.

If σ is a unitary matrix, we have $\sigma\mathbf{a} \cdot \sigma\mathbf{b} = \mathbf{a} \cdot ({}^t\bar{\sigma}\sigma\mathbf{b}) = \mathbf{a} \cdot \mathbf{b}$, and, in particular, $||\sigma\mathbf{a}||^2 = ||\mathbf{a}||^2$, whence $||\sigma\mathbf{a}|| = ||\mathbf{a}||$.

Conversely, assuming that this condition is satisfied for every $\mathbf{a}$, we have

$$(\sigma\mathbf{a} + \sigma\mathbf{b}) \cdot (\sigma\mathbf{a} + \sigma\mathbf{b}) = (\mathbf{a} + \mathbf{b}) \cdot (\mathbf{a} + \mathbf{b})$$

whence

$$\sigma\mathbf{a} \cdot \sigma\mathbf{b} + \sigma\mathbf{b} \cdot \sigma\mathbf{a} = \mathbf{a} \cdot \mathbf{b} + \mathbf{b} \cdot \mathbf{a}.$$

Replacing $\mathbf{b}$ by $\sqrt{-1}\,\mathbf{b}$, we have also

$$\sigma\mathbf{a} \cdot \sigma\mathbf{b} - \sigma\mathbf{b} \cdot \sigma\mathbf{a} = \mathbf{a} \cdot \mathbf{b} - \mathbf{b} \cdot \mathbf{a}$$

whence $\sigma\mathbf{a} \cdot \sigma\mathbf{b} = \mathbf{a} \cdot \mathbf{b} = \mathbf{a} \cdot {}^t\bar{\sigma}\sigma\mathbf{b}$. We have therefore $\mathbf{a} \cdot (\mathbf{b} - {}^t\bar{\sigma}\sigma\mathbf{b}) = 0$ for every $\mathbf{a}$, whence $\mathbf{b} = {}^t\bar{\sigma}\sigma\mathbf{b}$ (we may for instance take $\mathbf{a} = \mathbf{b} - {}^t\bar{\sigma}\sigma\mathbf{b}$). The formula $\mathbf{b} = {}^t\bar{\sigma}\sigma\mathbf{b}$ being true for every $\mathbf{b}$, ${}^t\bar{\sigma}\sigma$ is the unit matrix, which proves that σ is unitary.

Because the set $\{\mathbf{e}_1, \cdots, \mathbf{e}_n\}$ is orthonormal, it follows that the set $\{\sigma\mathbf{e}_1, \cdots, \sigma\mathbf{e}_n\}$ is orthonormal for every unitary σ. Conversely, let $\{\mathbf{a}_1, \cdots, \mathbf{a}_n\}$ be any orthonormal set; then there exists a matrix $\sigma = (a_{ij})$ such that $\sigma\mathbf{e}_i = \mathbf{a}_i (1 \leqslant i \leqslant n)$. Since

$$\sigma\mathbf{e}_i \cdot \sigma\mathbf{e}_k = \Sigma_j \bar{a}_{ji} a_{jk} = \mathbf{a}_i \cdot \mathbf{a}_k = \delta_{ik}$$

we see that σ is unitary. In particular, we obtain

Proposition 3. *If* $\mathbf{a}$ *is a unit vector, there exists a unitary matrix* σ *such that* $\sigma\mathbf{e}_1 = \mathbf{a}$.

We shall say that a vector $\mathbf{a} = \Sigma_i \mathbf{e}_i x_i$ is *real* if its coordinates $x_1, x_2, \cdots, x_n$ are real. If $\mathbf{a}$ and $\mathbf{b}$ are real vectors, the number $\mathbf{a} \cdot \mathbf{b}$ is also real.

Proposition 4. *A matrix* σ *is orthogonal if and only if the two following conditions are satisfied:* 1) $\sigma\mathbf{a} \cdot \sigma\mathbf{b} = \mathbf{a} \cdot \mathbf{b}$ *for any two real vectors* $\mathbf{a}$ *and* $\mathbf{b}$; 2) *if* $\mathbf{a}$ *is any real vector,* $\sigma\mathbf{a}$ *is also real.*

These conditions are certainly satisfied if σ is orthogonal, since in that case σ is unitary and real. Conversely, let us assume that the conditions are satisfied. Let $\mathbf{a} = \Sigma_i \mathbf{e}_i x_i$ and $\mathbf{b} = \Sigma_j \mathbf{e}_j y_j$ be any two complex vectors. Since $\mathbf{e}_1, \cdots, \mathbf{e}_n$ are real vectors, we have

$$\sigma\mathbf{a} \cdot \sigma\mathbf{b} = \Sigma_{ij} \bar{x}_i y_j (\sigma\mathbf{e}_i \cdot \sigma\mathbf{e}_j) = \Sigma_{ij} \bar{x}_i y_j (\mathbf{e}_i \cdot \mathbf{e}_j) = \mathbf{a} \cdot \mathbf{b}$$

and hence σ is unitary. Since σ is also real, it is orthogonal.

The process of orthonormalisation which was used in proving Proposition 1, if applied to a system of real vectors, leads to real vectors. Hence:

Corollary 2a to Proposition 1. *Any real unit vector belongs to an orthonormal base of* C^n *composed of real vectors.*

In the same way that we proved Proposition 3, we derive:

Proposition 3a. *If* $\mathbf{a}$ *is a real unit vector, there exists an orthogonal matrix* σ *such that* $\sigma\mathbf{e}_1 = \mathbf{a}$.

§IV. Hermitian Matrices

Definition 1. *A matrix* α *is called hermitian if* ${}^t\alpha = \bar{\alpha}$.

The reader will observe that the mapping $\alpha \to {}^t\bar{\alpha}$ is not an automorphism of $GL(n, C)$ and that the hermitian matrices do not form a subgroup of $GL(n, C)$.

Proposition 1. *A matrix* α *is hermitian if and only if* $\alpha\mathbf{a} \cdot \mathbf{b} = \mathbf{a} \cdot \alpha\mathbf{b}$ *for any two vectors* $\mathbf{a}$ *and* $\mathbf{b}$ *in* C^n.

In fact, if α is hermitian, the result follows immediately from Formula (1), §III, p. 10. Conversely, if the condition is satisfied, and if $\mathbf{b}$ is any vector in C^n, we have $\mathbf{a} \cdot \alpha\mathbf{b} = \mathbf{a} \cdot {}^t\bar{\alpha}\mathbf{b}$ for all $\mathbf{a} \varepsilon C^n$, whence $\alpha\mathbf{b} = {}^t\bar{\alpha}\mathbf{b}$ and $\alpha = {}^t\bar{\alpha}$.

Proposition 2. *If α is a hermitian matrix and σ is unitary, $\sigma\alpha\sigma^{-1}$ is again hermitian. Moreover, there exists a unitary matrix σ_0 such that $\sigma_0\alpha\sigma_0^{-1}$ is a diagonal matrix. If α is real, σ_0 may be assumed to be orthogonal.*

The first part of the proposition follows at once from the fact that ${}^t(\sigma\alpha\sigma^{-1}) = {}^t(\sigma^{-1})^t\alpha^t\sigma = \sigma^*\bar{\alpha}^t\sigma = \bar{\sigma}\bar{\alpha}\bar{\sigma}^{-1} = \overline{\sigma\alpha\sigma^{-1}}$.

We shall prove the second part by induction on the degree n of the matrix α. It is obvious for $n = 1$. Assume that $n > 1$ and that our assertion holds for matrices of degree $n - 1$.

Let λ_1 be a characteristic root of α. Then there exists a vector $\mathbf{a}_1 \neq 0$ in C^n such that $\alpha\mathbf{a}_1 = \lambda_1\mathbf{a}_1$; multiplying $\mathbf{a}_1$ by a number $\neq 0$, we may assume that $||\mathbf{a}_1|| = 1$. Hence, there exists a unitary matrix σ_1 such that $\sigma_1\mathbf{a}_1 = \mathbf{e}_1$ (Proposition 3, §III, p. 11). Set $\alpha_1 = \sigma_1\alpha\sigma_1^{-1}$; then α_1 is hermitian and moreover we have $\alpha_1\mathbf{e}_1 = \lambda_1\mathbf{e}_1$. Suppose that $\alpha_1\mathbf{e}_i = \Sigma_j\mathbf{e}_ja_{ji}$ $(1 \leqslant i \leqslant n)$. We have $a_{11} = \lambda_1$, $a_{j1} = 0$ $(2 \leqslant j \leqslant n)$; since α_1 is hermitian, we have $a_{ij} = \bar{a}_{ji}$. It follows that λ_1 is real and that $a_{1j} = 0$ $(2 \leqslant j \leqslant n)$. Knowing that λ_1 is real, we see that, if α is real, we may assume $\mathbf{a}_1$ to be real (the coordinates of $\mathbf{a}_1$ have to satisfy a system of linear equations with real coefficients); we may therefore assume in this case that σ_1 is orthogonal (Proposition 3*a*, §III, p. 11).

The matrix α_1 has the form

$$\alpha_1 = \begin{pmatrix} \lambda_1 & 0 \ldots\ldots 0 \\ 0 & \\ \cdot & \tilde{\alpha}_1 \\ \cdot & \\ \cdot & \\ 0 & \end{pmatrix}$$

where $\tilde{\alpha}_1$ is a hermitian matrix of degree $n - 1$ and is real if α is real. By our induction assumption, there exists a unitary matrix $\tilde{\sigma}_2$ of degree $n - 1$ such that $\tilde{\sigma}_2\tilde{\alpha}_1\tilde{\sigma}_2^{-1}$ is a diagonal matrix; if α is real, $\tilde{\sigma}_2$ may be assumed to be orthogonal. We denote by σ_2 the matrix

$$\sigma_2 = \begin{pmatrix} 1 & 0 \ldots\ldots 0 \\ 0 & \\ \cdot & \\ \cdot & \tilde{\sigma}_2 \\ \cdot & \\ \cdot & \\ 0 & \end{pmatrix}$$

which is obviously unitary. If $\sigma_0 = \sigma_2\sigma_1$, σ_0 is also a unitary matrix and is orthogonal if α is real. Since $\sigma_0\alpha\sigma_0^{-1}$ is a diagonal matrix, we see that Proposition 2 holds for matrices of degree n. Moreover, our proof contains the following result:

Proposition 3. *The characteristic roots of a hermitian matrix are real numbers.*

We shall say that a vector $\mathbf{a}$ is an *eigenvector* of a matrix α if $\mathbf{a}$ is a unit vector and if $\alpha\mathbf{a} = \lambda\mathbf{a}$, where λ is a number; λ is necessarily a characteristic root of α. We shall say that $\mathbf{a}$ *belongs* to the root λ. If α is a diagonal matrix, the vectors $\mathbf{e}_1, \cdots, \mathbf{e}_n$ are eigenvectors of α, and conversely. If $\mathbf{a}$ is an eigenvector of α, and if σ is any regular matrix, $\sigma\mathbf{a}$ is an eigenvector of $\sigma\alpha\sigma^{-1}$. Hence, an equivalent formulation of Proposition 2 is the following:

Proposition 4. *If α is a hermitian matrix, the space C^n has an orthonormal base composed of eigenvectors of α.*

Definition 2. *A hermitian matrix α is called positive (semi-definite) if its characteristic roots are all $\geqslant 0$; if none of these roots is equal to 0, α is called positive definite.*

If α is a hermitian matrix, $\exp \alpha$ is also hermitian, because we have ${}^t(\exp \alpha) = \exp {}^t\alpha = \exp \bar{\alpha} = \overline{(\exp \alpha)}$. Moreover, each characteristic root of $\exp \alpha$ is of the form $\exp \lambda$, where λ is a characteristic root of α, and hence real (Proposition 2, §II, p. 5). It follows that $\exp \alpha$ is a positive definite hermitian matrix.

Conversely, let β be any positive definite hermitian matrix. We know that there exists a unitary matrix σ such that, if we set $\mathbf{a}_i = \sigma\mathbf{e}_i$ $(1 \leqslant i \leqslant n)$, we have $\beta\mathbf{a}_i = \mu_i\mathbf{a}_i$ with μ_i real > 0 $(1 \leqslant i \leqslant n)$. We set $\lambda_i = \log \mu_i$ $(1 \leqslant i \leqslant n)$ and define a matrix α by $\alpha\mathbf{a}_i = \lambda_i\mathbf{a}_i$ $(1 \leqslant i \leqslant n)$. We have $\sigma^{-1}\alpha\sigma\mathbf{e}_i = \lambda_i\mathbf{e}_i$, which shows that $\sigma^{-1}\alpha\sigma$ is a real diagonal matrix, hence hermitian; then $\alpha = \sigma(\sigma^{-1}\alpha\sigma)\sigma^{-1}$ is hermitian. Moreover, we have $(\exp \alpha)\mathbf{a}_i = (\exp \lambda_i)\mathbf{a}_i = \mu_i\mathbf{a}_i (1 \leqslant i \leqslant n)$, whence $\exp \alpha = \beta$.

We assert furthermore that the representation of β as the exponential of a hermitian matrix is unique. In fact, let α' be any hermitian matrix such that $\exp \alpha' = \beta$. Let $\mathbf{a}' = \Sigma_i\mathbf{a}_i x_i$ be any eigenvector of α', belonging to a characteristic root λ' of α'. Then we have $\beta\mathbf{a}' = (\exp \alpha')\mathbf{a}' = (\exp \lambda')\mathbf{a}' = \Sigma_i\mathbf{a}_i(\mu_i x_i)$, which proves that $x_i = 0$ if $\mu_i \neq \exp \lambda'$. Let i_0 be an index such that $x_{i_0} \neq 0$; then we have $\mu_{i_0} = \exp \lambda' = \exp \lambda_{i_0}$, whence $\lambda' = \lambda_{i_0}$ since λ' and λ_{i_0} are both real. On the other hand, we have $\alpha\mathbf{a}' = \Sigma\mathbf{a}_i(x_i \log \mu_i) = \lambda'\mathbf{a}' = \alpha'\mathbf{a}'$. Since α' produces the same effect as α on each of its eigenvectors, it follows from Proposition 4 that $\alpha = \alpha'$. We have proved:

Proposition 5. *The mapping $\alpha \to \exp \alpha$ maps in a univalent way the set of all hermitian matrices onto the set of all positive definite hermitian matrices.*

This mapping is clearly continuous. We shall prove that it is topological. In fact, let $(\beta_1, \cdots, \beta_p, \cdots)$ be a sequence of positive definite hermitian matrices which converges to a positive definite hermitian matrix β. The characteristic polynomial of β_p converges to the characteristic polynomial of β; it follows that the characteristic roots $\mu_{1,p}, \cdots, \mu_{n,p}$ of β_p (arranged in a suitable order) converge to the characteristic roots $\mu_1, \cdots, \mu_n$ of β. Since $\mu_i > 0$ $(1 \leqslant i \leqslant n)$, the numbers $\log \mu_{i,p}$ remain bounded when p increases indefinitely. It follows that, if α_p is the hermitian matrix such that $\exp \alpha_p = \beta_p$, the characteristic roots of α_p remain bounded. For each p, there exists a unitary matrix σ_p such that $\sigma_p\alpha_p\sigma_p^{-1} = \delta_p$ is a diagonal matrix, whose diagonal coefficients are the characteristic roots of α_p; hence the coefficients of δ_p remain bounded as p increases indefinitely. Since σ_p is unitary, any coefficient of σ_p is $\leqslant 1$ in absolute value. Therefore the coefficients of α_p remain bounded and the sequence $(\alpha_1, \cdots, \alpha_p, \cdots)$ belongs to a bounded i.e. compact, subset of $\mathfrak{M}_n(C)$. It follows that we can extract from the sequence $(\alpha_1, \cdots, \alpha_p, \cdots)$ a subsequence which converges to a matrix α. Since ${}^t\alpha_p = \bar{\alpha}_p$, we have also ${}^t\alpha = \bar{\alpha}$ and α is hermitian. The exponential mapping being continuous, $\exp \alpha$ is the limit of a subsequence extracted from $(\beta_1, \cdots, \beta_p, \cdots)$, whence $\exp \alpha = \beta$. But we know that the representation of β as the exponential of a hermitian matrix is unique; therefore all convergent subsequences of the sequence $(\alpha_1, \cdots, \alpha_p, \cdots)$ have the same limit α. It follows immediately that $\lim_{p\to\infty} \alpha_p = \alpha$. This proves that the mapping $\alpha \to \exp \alpha$ is topological.

A hermitian matrix $\alpha = (a_{ij})$ is obviously determined by the knowledge of the coefficients a_{ii} (which must be real) and a_{ij} for $i < j$ (which may be arbitrary complex numbers). Hence the set of all hermitian matrices is homeomorphic to $R^n \times C^{n(n-1)/2}$, i.e. also to R^{n^2}. We have proved

Proposition 6. *The set of all hermitian matrices of degree n and the set of all positive definite hermitian matrices of degree n are both homeomorphic to R^{n^2}. The mapping $\alpha \to \exp \alpha$ is a homeomorphism of the first of these sets onto the second.*

§V. Representation of GL(n,C) as a Product Space

Proposition 1. *Any regular matrix τ may be written in one and only one way as the product $\tau = \sigma\alpha$ of a unitary matrix σ and a positive definite hermitian matrix α.*

We consider τ as a linear endomorphism of the vector space C^n. If $\mathbf{a}$ is any vector in C^n, we have (Cf. Formula (1), §III, p. 10)

$$\mathbf{a} \cdot ({}^t\bar{\tau}\tau)\mathbf{a} = \tau\mathbf{a} \cdot \tau\mathbf{a} = ({}^t\bar{\tau}\tau)\mathbf{a} \cdot \mathbf{a} \geqslant 0$$

By Proposition 1, §IV, p. 11, it follows that the matrix $\alpha_1 = {}^t\bar{\tau}\tau$ is hermitian. Moreover, if $\mathbf{a}$ is an eigenvector of α_1, corresponding to a characteristic root μ, we have $\mu(\mathbf{a} \cdot \mathbf{a}) = \tau\mathbf{a} \cdot \tau\mathbf{a}$, whence $\mu \geqslant 0$. Since ${}^t\bar{\tau}\tau$ is regular, it is a positive definite hermitian matrix.

By Proposition 2, §IV, p. 12, there exists a unitary matrix ν such that $\nu\alpha_1\nu^{-1}$ is a diagonal matrix δ_1. Since the coefficients of the diagonal of δ_1 are real positive numbers, there exists a real diagonal matrix δ such that $\delta^2 = \delta_1$; moreover, the diagonal coefficients of δ may be taken to be positive. It follows that $\alpha = \nu^{-1}\delta\nu$ is a positive definite hermitian matrix and that $\alpha^2 = \alpha_1$.

We set $\sigma = \tau\alpha^{-1}$, whence $\sigma^* = \tau^*(\alpha^{-1})^* = {}^t\tau^{-1}{}^t\alpha = {}^t\tau^{-1}\bar{\alpha}$. We have $\bar{\alpha}^2 = \bar{\alpha}_1 = {}^t\tau\bar{\tau}$, whence ${}^t\tau^{-1}\bar{\alpha} = \bar{\tau}\bar{\alpha}^{-1} = \bar{\sigma}$; σ is unitary, and we have $\tau = \sigma\alpha$.

Suppose now that $\sigma_1\alpha_1 = \sigma_2\alpha_2$, with σ_1 and σ_2 unitary, α_1 and α_2 positive definite hermitian. We set $\sigma_3 = \sigma_2^{-1}\sigma_1$; then σ_3 is unitary and we have $\sigma_3\alpha_1 = \alpha_2$. It follows that $\alpha_2 = {}^t\bar{\alpha}_2 = {}^t\bar{\alpha}_1{}^t\bar{\sigma}_3 = \alpha_1\sigma_3^{-1}$ and $\alpha_2^2 = \alpha_1\sigma_3^{-1}\sigma_3\alpha_1 = \alpha_1^2$. By Proposition 5, §IV, p. 14, we have $\alpha_1 = \exp \beta_1$, $\alpha_2 = \exp \beta_2$, where β_1 and β_2 are hermitian matrices, and consequently $\exp 2\beta_1 = \alpha_1^2 = \alpha_2^2 = \exp 2\beta_2$. By Proposition 5, §IV, p. 14, it follows that $2\beta_1 = 2\beta_2$, $\beta_1 = \beta_2$ and $\alpha_1 = \alpha_2$. Hence σ_3 is the unit matrix and $\sigma_1 = \sigma_2$, which completes the proof of Proposition 1.

Remark. It follows easily from Proposition 1 that a regular matrix τ can also be written in one and only one way in the form $\tau = \alpha\sigma$, where σ is unitary and α positive definite hermitian.

Proposition 2. *Any complex orthogonal matrix ρ may be written in one and only one way in the form σ (exp $\sqrt{-1}\,\beta$) where σ is a real orthogonal matrix and where β is a real skew symmetric matrix.*

By Proposition 1, we have $\rho = \sigma\alpha$, where σ is unitary and α hermitian positive definite. The orthogonality condition ${}^t\rho\rho = \epsilon$ gives ${}^t\alpha{}^t\sigma\sigma = \alpha^{-1}$. We know that α can be represented in the form $\exp \beta_1$, with a hermitian β_1; hence $\alpha^{-1} = \exp(-\beta_1)$ is again hermitian. The matrix ${}^t\sigma$, and therefore also ${}^t\sigma\sigma$, is unitary. Since ${}^t\alpha = \exp({}^t\beta_1)$ is hermitian, the uniqueness assertion in the remark which follows Proposition 1 gives ${}^t\sigma\sigma = \epsilon$, ${}^t\alpha = \alpha^{-1}$. Since ${}^t\sigma\sigma = \epsilon$, $\sigma^{-1}\bar{\sigma} = \epsilon$ we have $\sigma = \bar{\sigma}$, which proves that σ is real orthogonal.

The equality ${}^t\alpha = \alpha^{-1}$ gives $\exp({}^t\beta_1) = \exp(-\beta_1)$. Making use of Proposition 5, §IV, p. 14, we see that ${}^t\beta_1 = -\beta_1$, i.e., β_1 is skew symmetric. Since also ${}^t\beta_1 = \bar{\beta}_1$, we have $\bar{\beta}_1 = -\beta_1$. If we set $\beta_1 = \sqrt{-1}\,\beta$, β is real skew symmetric.

Conversely, any real orthogonal matrix is also unitary, and, if β is real skew symmetric, $\sqrt{-1}\,\beta$ is hermitian and $\exp \sqrt{-1}\,\beta$ is positive hermitian. Hence the uniqueness assertion of Proposition 2 follows from Proposition 1.

The factors σ, α of the decomposition of τ given in Proposition 1 *are continuous functions of τ.* In fact, let $(\tau_1, \cdots, \tau_p, \cdots)$ be a sequence of regular matrices which converges to a regular matrix τ, and suppose that $\tau_p = \sigma_p\alpha_p$, $\sigma_p \epsilon U(n)$, α_p positive hermitian. Because $U(n)$ is compact, the sequence $(\sigma_1, \cdots, \sigma_p, \cdots)$ has a subsequence which converges to a limit $\sigma \epsilon U(n)$. The corresponding matrices $\alpha_p = \sigma_p^{-1}\tau_p$ clearly converge to the limit $\alpha = \sigma^{-1}\tau$. Since the set of positive hermitian matrices is obviously closed, α is positive hermitian. Since σ and τ are regular, α is positive definite. But there exists only one decomposition of τ in the product of a unitary matrix and a positive definite hermitian matrix. It follows that all convergent subsequences of the sequence $(\sigma_1, \cdots, \sigma_p, \cdots)$ have the same limit σ, which proves that $\lim_{p\to\infty} \sigma_p = \sigma$ and $\lim_{p\to\infty} \alpha_p = \alpha$. Our assertion is thereby proved. It follows easily that the matrices σ and β of Proposition 2 are continuous functions of the complex orthogonal matrix ρ.

The set of all positive definite hermitian matrices of degree n is homeomorphic to R^{n^2} (Proposition 5, §IV, p. 14). The set of all skew symmetric real matrices of degree n is obviously homeomorphic to $R^{n(n-1)/2}$. Hence we obtain the following results:

Proposition 3. *The space $GL(n, C)$ is homeomorphic to the topological product of the spaces $U(n)$ and R^{n^2}. The space $O(n, C)$ is homeomorphic to the product of the spaces $O(n)$ and $R^{n(n-1)/2}$.*

§VI. QUATERNIONS

The algebra Q of quaternions is an algebra of dimension 4 over the field R of real numbers, with a base composed of 4 elements e_0, e_1, e_2, e_3 whose multiplication table is given by the following formulas:

$$(1) \qquad e_0e_i = e_ie_0 = e_i; \qquad e_i^2 = -e_0; \qquad e_ie_j = -e_je_i = e_k$$

($1 \leqslant i, j, k \leqslant 3$; the mapping $1 \to i$, $2 \to j$, $3 \to k$ is assumed to be an even permutation of the set $\{1, 2, 3\}$). Therefore, a quaternion q may be expressed in the form $\Sigma_{i=0}^3 a_ie_i$, with real coefficients a_0, a_1, a_2, a_3. Addition and multiplication are defined by the usual distributivity laws and the formulas (1).

From these formulas we see at once that e_0 is the unit element of Q. Moreover, it is easy to check that $(e_ie_j)e_k = e_i(e_je_k)$ $(0 \leqslant i, j, k \leqslant 3)$; hence Q is an associative but not commutative algebra.

If $q = a_0e_0 + \Sigma_{i=1}^3 a_ie_i$ is a quaternion, we denote by q^ι the quaternion

$$q^\iota = a_0e_0 - \Sigma_{i=1}^3 a_ie_i$$

and call q^ι the conjugate of q.

We have $qq^\iota = q^\iota q = (\Sigma_0^3 a_i^2)e_0$. The number $\Sigma_0^3 a_i^2$ is a non negative real number which can be equal to 0 only if $q = 0$. This number is called the *norm* of q; it is denoted by $N(q)$.

If q' is another quaternion, it is easy to see that

$$(aq + bq')^\iota = aq^\iota + b(q')^\iota; \qquad (qq')^\iota = (q')^\iota q^\iota$$
$$(q^\iota)^\iota = q.$$

We express these facts by saying that the conjugation $q \to q^\iota$ is an involutory anti-automorphism of the algebra Q. We have

$$N(qq')e_0 = qq'(qq')^\iota = q(N(q')e_0)q^\iota = N(q')e_0N(q)e_0$$

whence

$$N(qq') = N(q)N(q')$$

From the existence of the norm, we deduce that Q is a *division algebra*, i.e. that every quaternion $q \neq 0$ has an inverse q^{-1} such that $qq^{-1} = q^{-1}q = e_0$, namely $q^{-1} = (N(q))^{-1}q^\iota$.

Let C_1 be the set of quaternions of the form $a_0e_0 + a_1e_1$ (with $a_0, a_1 \varepsilon R$). The sums and products of elements of C_1 are in C_1; if $q \varepsilon C_1$ and $a \varepsilon R$, then $aq \varepsilon C_1$. Hence C_1 is a subalgebra of Q. If we assign to every element $a_0e_0 + a_1e_1 \varepsilon C_1$ the complex number $a_0 + a_1\sqrt{-1}$, we clearly obtain an isomorphism of C_1 with the field C of complex numbers.

Since Q contains a field isomorphic with C, it can be considered as a vector space over C. To be more specific, we shall make the following definition: if $q \epsilon Q$ and $x = a_0 + a_1\sqrt{-1} \varepsilon C$, qx will represent the quaternion $q(a_0e_0 + a_1e_1)$. We have

$$(q + q')x = qx + q'x; \qquad q(x + x') = qx + qx';$$
$$q(xx') = (qx)x' = (qx')x$$

and these formulas show that Q may be considered as a vector space over C (we write the multiplier x to the right of q for convenience). Any quaternion $q = \Sigma_0^3 a_ie_i$ may be written in the form

$$q = e_0(a_0 + a_1\sqrt{-1}) + e_2(a_2 - a_3\sqrt{-1})$$

It follows immediately that Q is of dimension 2 over C.

To every quaternion q there is associated a mapping T_q of Q into itself, defined by the formula $T_q(q') = qq'$. We have $T_q(q'_1 + q'_2) = T_q(q'_1) + T_q(q'_2)$; furthermore, we have clearly $(qq')x = q(q'x)$ (where $x \varepsilon C$); it follows that $T_q(q'x) = T_q(q')x$, which shows that T_q is an endomorphism of Q, considered as a vector space over C. As such T_q can be represented by a matrix of degree 2, whose coefficients are defined by the formulas

$$T_q(e_0) = e_0x_{11} + e_2x_{21}$$
$$T_q(e_2) = e_0x_{12} + e_2x_{22}$$

In particular, we have

$$T_{e_0} = \begin{pmatrix} 1 & 0 \\ 0 & 1 \end{pmatrix} \qquad T_{e_1} = \begin{pmatrix} \sqrt{-1} & 0 \\ 0 & -\sqrt{-1} \end{pmatrix}$$
$$T_{e_2} = \begin{pmatrix} 0 & -1 \\ 1 & 0 \end{pmatrix} \qquad T_{e_3} = \begin{pmatrix} 0 & -\sqrt{-1} \\ -\sqrt{-1} & 0 \end{pmatrix}$$

whence

$$T_{e_0x+e_2y} = \begin{pmatrix} x & -\bar{y} \\ y & \bar{x} \end{pmatrix}$$

On the other hand, we have $T_{q_1q_2} = T_{q_1} \circ T_{q_2}$. It follows that the mapping $q \to T_q$ is a representation of the algebra Q by matrices of degree 2 with coefficient in C.

Finally, we observe that

(2) $$(e_0x + e_2y)^\iota = e_0\bar{x} - e_2y$$

whence $T_{q^\iota} = {}^t\bar{T}_q$.

§VII. SYMPLECTIC GEOMETRY

Let Q be the algebra of quaternions; n being some integer > 0, we denote by Q^n the product of n sets identical with Q. An element $(a_1, \cdots, a_n) \varepsilon Q^n$ ($a_i \varepsilon Q$, $1 \leqslant i \leqslant n$) will be called a (quaternionic) *vector*; $a_1, \cdots, a_n$ will be called the *coordinates* of this vector. The addition of vectors is defined by the addition of the corresponding coordinates; if $\mathbf{a} = (a_1, \cdots, a_n)$ is a vector and $q \varepsilon Q$, we denote by $\mathbf{a}q$ the vector $(a_1q, \cdots, a_nq)$. The vectors obviously form an additive group with respect to addition. Moreover, we have

$$(\mathbf{a}_1 + \mathbf{a}_2)q = \mathbf{a}_1q + \mathbf{a}_2q; \qquad \mathbf{a}(q_1 + q_2) = \mathbf{a}q_1 + \mathbf{a}q_2$$
$$\mathbf{a}(q_1q_2) = (\mathbf{a}q_1)q_2$$

where $\mathbf{a}$, $\mathbf{a}_1$, $\mathbf{a}_2$ are vectors and q, q_1, q_2 are quaternions. (We could also define the left multiplication $(q, \mathbf{a}) \to q\mathbf{a}$; however, we shall not use this type of multiplication.)

If $\mathbf{a} = (a_1, \cdots, a_n)$ and $\mathbf{b} = (b_1, \cdots, b_n)$ are vectors, we shall define their *symplectic product* $\mathbf{a} \cdot \mathbf{b}$ to be the quaternion

$$\mathbf{a} \cdot \mathbf{b} = \Sigma_{i=1}^n a_i^\iota b_i$$

This product has properties which are similar to those of the hermitian product introduced in §III. We have

$$(\mathbf{a}_1 + \mathbf{a}_2) \cdot \mathbf{b} = \mathbf{a}_1 \cdot \mathbf{b} + \mathbf{a}_2 \cdot \mathbf{b};$$
$$\mathbf{a} \cdot (\mathbf{b}_1 + \mathbf{b}_2) = \mathbf{a} \cdot \mathbf{b}_1 + \mathbf{a} \cdot \mathbf{b}_2; \qquad \mathbf{a} \cdot (\mathbf{b}q) = (\mathbf{a} \cdot \mathbf{b})q$$
$$(\mathbf{a}q) \cdot \mathbf{b} = q^\iota(\mathbf{a} \cdot \mathbf{b})$$

We have $\mathbf{a} \cdot \mathbf{a} = \Sigma_{i=1}^n a_i^\iota a_i = ||\mathbf{a}||e_0$, where $||\mathbf{a}||$ is a real number $\geqslant 0$ which is called the *length* of $\mathbf{a}$. This length is always $\neq 0$ if $\mathbf{a} \neq 0$.

A vector subspace of Q^n is a subset $\mathfrak{M}$ such that the conditions $\mathbf{a} \varepsilon \mathfrak{M}$, $\mathbf{b} \varepsilon \mathfrak{M}$ and $q \varepsilon Q$ imply $\mathbf{a} + \mathbf{b} \varepsilon \mathfrak{M}$ and $\mathbf{a}q \varepsilon \mathfrak{M}$. If $\mathbf{a}_1, \cdots, \mathbf{a}_h$ are a finite number of vectors, the set of all vectors of the form $\mathbf{a}_1 q_1 + \cdots + \mathbf{a}_h q_h (q_1, \cdots, q_h \varepsilon Q)$ is a vector subspace of Q^n which is said to be spanned by $\mathbf{a}_1, \cdots, \mathbf{a}_h$. If moreover the condition $\Sigma_1^h \mathbf{a}_i q_i = 0$ implies $q_1 = \cdots = q_h = 0$, we say that $\mathbf{a}_1, \cdots, \mathbf{a}_h$ are linearly independent.

In particular, the space Q^n itself is spanned by the n linearly independent vectors $\mathbf{e}_1, \cdots, \mathbf{e}_n$, where $\mathbf{e}_i$ is the vector whose j-coordinate is $\delta_{ij} e_0$.

Exactly as in the usual case, it can be proved that:

1) every vector subspace $\mathfrak{M}$ of Q^n can be spanned by a finite set of linearly independent vectors; such a set is called a *base* of the space $\mathfrak{M}$;

2) all bases of $\mathfrak{M}$ have the same number of elements, called the *dimension* of $\mathfrak{M}$;

3) if $\mathfrak{M}'$ is a vector subspace of $\mathfrak{M}$, the equality $\dim \mathfrak{M}' = \dim \mathfrak{M}$ implies $\mathfrak{M}' = \mathfrak{M}$.

An endomorphism of Q^n is a mapping σ of Q^n into itself such that

$$\sigma(\mathbf{a} + \mathbf{b}) = \sigma\mathbf{a} + \sigma\mathbf{b}; \qquad \sigma(\mathbf{a}q) = (\sigma\mathbf{a})q$$

for any $\mathbf{a}$, $\mathbf{b} \varepsilon Q^n$ and $q \varepsilon Q$. Such a mapping is obviously entirely determined when the vectors

$$\sigma\mathbf{e}_i = \Sigma_{j=1}^n \mathbf{e}_j q_{ji}$$

are given. Hence, the linear endomorphisms of Q^n are in a one-to-one correspondence with the matrices (q_{ij}) with coefficients in Q. We shall denote by the same letter σ the endomorphism itself and the corresponding matrix. If $\sigma = (q_{ij})$ and $\tau = (r_{ij})$ are two of these matrices, we denote (as usual) by $\sigma\tau$ the matrix (s_{ij}) with

$$s_{ij} = \Sigma_{k=1}^{n} q_{ik} r_{kj}$$

and then we have $\sigma\tau = \sigma \circ \tau$, i.e. $(\sigma\tau)\mathbf{a} = \sigma(\tau\mathbf{a})$ for every vector $\mathbf{a}$.

If σ is any matrix, either there exists a vector $\mathbf{a} \neq 0$ such that $\sigma\mathbf{a} = 0$, or σ has an inverse matrix σ^{-1} (i.e. $\sigma\sigma^{-1} = \sigma^{-1}\sigma = \epsilon$, where ϵ is the diagonal matrix of degree n whose diagonal coefficients are all equal to e_0). In fact, if the condition $\mathbf{a} \neq 0$ implies $\sigma\mathbf{a} \neq 0$, σ is an isomorphism of Q^n with a subspace $\mathfrak{M}$ of Q^n; hence $\dim \mathfrak{M} = \dim Q^n$, whence $\mathfrak{M} = Q^n$. It follows that σ has a reciprocal mapping, which is obviously also an endomorphism σ^{-1}.

The set of all matrices of degree n with coefficients in Q will be denoted by $\mathfrak{M}_n(Q)$.

A matrix $\sigma \varepsilon \mathfrak{M}_n(Q)$ is said to be *symplectic* if $||\sigma\mathbf{a}|| = ||\mathbf{a}||$ for every $\mathbf{a} \varepsilon Q^n$. Exactly as in §III, p. 9, we can prove the following facts:

1) if σ is symplectic, then $\sigma\mathbf{a} \cdot \sigma\mathbf{b} = \mathbf{a} \cdot \mathbf{b}$ for any two vectors $\mathbf{a}$, $\mathbf{b}$ in Q^n;

2) a necessary and sufficient condition for σ to be symplectic is that ${}^t\sigma^\iota \cdot \sigma = \epsilon$, where ϵ is the unit matrix defined above.

Therefore, if σ is symplectic, the condition $\sigma\mathbf{a} = 0$ implies $\mathbf{a} = 0$; σ has an inverse, which is obviously ${}^t\sigma^\iota$. We have also $\sigma \cdot {}^t\sigma^\iota = \epsilon$, which shows that σ^{-1} is also symplectic.

It follows easily that the symplectic matrices form a group.

Definition 1. *We define the symplectic group for the dimension n (denoted by $Sp(n)$) as the group of all matrices $\sigma \varepsilon \mathfrak{M}_n(Q)$ such that*

$$\sigma\mathbf{a} \cdot \sigma\mathbf{b} = \mathbf{a} \cdot \mathbf{b}$$

holds for any two vectors $\mathbf{a}$ *and* $\mathbf{b}$ *in* Q^n.

Definition 2. *A vector* $\mathbf{a}$ *is called a unit vector if the length of* $\mathbf{a}$ *is* 1. *The vectors* $\mathbf{a}$ *and* $\mathbf{b}$ *are said to be orthogonal if* $\mathbf{a} \cdot \mathbf{b} = 0$. *The set of vectors* $\{\mathbf{a}_1, \cdots, \mathbf{a}_h\}$ *is said to be orthonormal if we have* $\mathbf{a}_i \cdot \mathbf{a}_j = \delta_{ij} e_0$.

For instance, the n basic vectors $\mathbf{e}_1, \cdots, \mathbf{e}_n$ form an orthonormal set.

Proposition 1. *Let* $\mathbf{a}_1, \cdots, \mathbf{a}_m$ *be m linearly independent vectors in* Q^n. *There exists an orthonormal set* $\{\mathbf{b}_1, \cdots, \mathbf{b}_m\}$ *such that, for each k* $(1 \leqslant k \leqslant m)$, *the set of vectors* $\{\mathbf{b}_1, \cdots, \mathbf{b}_k\}$ *spans the same subspace as* $\{\mathbf{a}_1, \cdots, \mathbf{a}_k\}$.

The proof is entirely similar to the proof of Proposition 1, §III, p. 9.

On the basis of Proposition 1, we can obtain the following result by an argument similar to one used in §III:

Proposition 2. *If* **a** *is any unit vector in* Q^n, *there exists a symplectic matrix* σ *such that* $\sigma \mathbf{e}_1 = \mathbf{a}$.

§VIII. The Linear Symplectic Groups

Let us again consider the vector space Q^n which was introduced in the previous section. If $\mathbf{a} = (a_1, \cdots, a_n) \varepsilon Q^n$, we may represent a_i in the form $a_i = e_0 x_i + e_2 x_{n+i}$, where x_i and x_{n+i} are complex numbers ($1 \leqslant i \leqslant n$; Cf. §VI, p. 16). We assign to **a** the vector $\mathbf{a}'$ of C^{2n} whose coordinates are $x_1, \cdots, x_n, \cdots, x_{2n}$. This correspondence preserves addition. Moreover, if **a** corresponds to the vector $\mathbf{a}' \varepsilon C^{2n}$, then $\mathbf{a}(ue_0 + ve_1)$ corresponds to the vector $\mathbf{a}'(u + v\sqrt{-1})$ (where u and v are real numbers), because we have

$$a_i(ue_0 + ve_1) = a_i(u + v\sqrt{-1}) = e_0 x_i(u + v\sqrt{-1}) + e_2 x_{n+i}(u + v\sqrt{-1})$$

It follows immediately that there corresponds to every endomorphism σ of Q^n an endomorphism σ' of C^{2n} such that

$$\sigma \mathbf{a} \to \sigma' \mathbf{a}' \quad \text{if} \quad \mathbf{a} \to \mathbf{a}'$$

Moreover, to the product $\sigma\tau$ of two endomorphisms σ and τ of Q^n there corresponds the product $\sigma'\tau'$ of the corresponding endomorphisms of C^{2n}.

The correspondence $\sigma \to \sigma'$ gives an isomorphism of $Sp(n)$ with a subgroup of $GL(2n, C)$. The latter group will be called the *linear symplectic group.* The set of elements of the linear symplectic group, being a subset of $GL(2n, C)$, may be considered as the set of points of a subspace of $GL(2n, C)$. Hence, we may introduce in $Sp(n)$ a topology such that the mapping $\sigma \to \sigma'$ is a homeomorphism. When we speak of $Sp(n)$ as a topological space, we shall always have this topology in mind.

We shall now determine algebraically the linear symplectic group. Let $\sigma = (q_{ij})$ be any matrix in $Sp(n)$, and let $\sigma' = (r_{kl})$ ($1 \leqslant k, l \leqslant 2n$) be the corresponding matrix in $GL(2n, C)$. If $\mathbf{a}' = (x_1, \cdots, x_{2n})$ is any vector in C^{2n}, $\sigma'\mathbf{a}'$ is the vector $(\tilde{x}_1, \cdots, \tilde{x}_{2n})$ with

(1) $$\tilde{x}_k = \Sigma_{l=1}^{2n} r_{kl} x_l$$

Suppose that $\mathbf{a}'$ is the vector which corresponds to a vector $\mathbf{a} \varepsilon Q^n$. Let **b** be another vector in Q^n, and let $\mathbf{b}' = (y_1, \cdots, y_{2n})$ be the corresponding vector in C^{2n}. Then, we have, by an easy computation,

$$\begin{aligned}\mathbf{a}\cdot\mathbf{b} &= e_0\Sigma_1^{2n}\bar{x}_iy_i + e_2\Sigma_{i=1}^{n}(x_iy_{i+n} - x_{i+n}y_i)\\ &= e_0(\mathbf{a}'\cdot\mathbf{b}') + e_2\Sigma_{i=1}^{n}(x_iy_{i+n} - x_{i+n}y_i)\end{aligned}$$

Moreover, we have $\sigma'\mathbf{b}' = (\tilde{y}_1, \cdots, \tilde{y}_{2n})$, with

$$\tilde{y}_k = \Sigma_{l=1}^{2n} r_{kl}y_l \tag{2}$$

Since $\sigma\mathbf{a}\cdot\sigma\mathbf{b} = \mathbf{a}\cdot\mathbf{b}$, we see that the linear substitutions (1) and (2), performed on the variables x and y, leave unchanged the quantities $\Sigma_1^{2n}\bar{x}_iy_i$ and $\Sigma_1^n(x_iy_{i+n} - x_{i+n}y_i)$. The first property shows that σ' belongs to $U(2n)$; we shall express the second by saying that σ' leaves invariant the bilinear form $\Sigma_1^n(x_iy_{i+n} - x_{i+n}y_i)$.

Conversely, let σ' be a unitary matrix which leaves invariant the expression $\Sigma_1^n(x_iy_{i+n} - x_{i+n}y_i)$. We may assign to σ' a mapping σ of Q^n into itself such that $\sigma'\mathbf{a}'$ corresponds to $\sigma\mathbf{a}$ if $\mathbf{a}'$ corresponds to $\mathbf{a}$ (under the correspondence which was established above between vectors of Q^n and of C^{2n}). We have $\sigma(\mathbf{a}+\mathbf{b}) = \sigma\mathbf{a} + \sigma\mathbf{b}$ and $\sigma\mathbf{a}\cdot\sigma\mathbf{b} = \mathbf{a}\cdot\mathbf{b}$ (where $\mathbf{a}$ and $\mathbf{b}$ are arbitrary vectors in Q^n). This does not yet prove that σ is an endomorphism of Q^n; we still have to show that $(\sigma\mathbf{a})q = \sigma(\mathbf{a}q)$, where q is any quaternion. It will be sufficient to prove that $\mathbf{b}\cdot((\sigma\mathbf{a})q - \sigma(\mathbf{a}q)) = 0$ for any $\mathbf{b}\in Q^n$. We have

$$(\sigma\mathbf{c})\cdot((\sigma\mathbf{a})q - \sigma(\mathbf{a}q)) = ((\sigma\mathbf{c})\cdot(\sigma\mathbf{a}))q - (\sigma\mathbf{c})\cdot\sigma(\mathbf{a}q) = (\mathbf{c}\cdot\mathbf{a})q - (\mathbf{c}\cdot\mathbf{a}q) = 0$$

which proves our assertion. We have proved

Proposition 1. *The linear symplectic group is the group of all unitary matrices in $GL(2n, C)$ which leave invariant the bilinear form*

$$\Sigma_1^n(x_iy_{i+n} - x_{i+n}y_i).$$

This group is obviously a closed subset of $GL(2n, C)$. Hence:

Theorem 1a. *The space $Sp(n)$ is compact.*

Let us now consider the set $Sp(n, C)$ of matrices $\sigma'\in\mathfrak{M}_{2n}(C)$ which are not necessarily unitary, but which leave invariant the bilinear form $\Sigma_1^n(x_iy_{i+n} - x_{i+n}y_i)$. Now, the matrix J of the coefficients of this bilinear form is

$$J = \begin{pmatrix}0 & \epsilon_n\\ -\epsilon_n & 0\end{pmatrix}$$

where ϵ_n is the unit matrix of degree n. Thus our condition may be expressed by the equality

$$^t\sigma'J\sigma' = J \tag{3}$$

Since the determinant of J is $\neq 0$, it follows immediately that any matrix $\sigma' \varepsilon Sp(n, C)$ is regular. Moreover, the product of two matrices of $Sp(n, C)$ and the inverse of a matrix of $Sp(n, C)$ again belong to $Sp(n, C)$. Hence $Sp(n, C)$ is a subgroup of $GL(2n, C)$.

Definition 1. *The subgroup of $GL(2n, C)$ composed of the matrices σ' which satisfy condition* (3) *is called the complex symplectic group and is denoted by $Sp(n, C)$.*

We have seen in §II that the matrices of a suitable neighbourhood of ϵ in $GL(2n, C)$ can be represented in the form $\exp X$, with $X \varepsilon \mathfrak{M}_{2n}(C)$. We shall now investigate under which condition we have $\exp X \varepsilon Sp(n, C)$.

The condition ${}^t\sigma' J \sigma' = J$ may be written in the form ${}^t\sigma' = J\sigma'^{-1}J^{-1}$. Since $J(\exp X)^{-1}J^{-1} = \exp(-JXJ^{-1})$ and $\exp {}^tX = {}^t(\exp X)$, we see that we shall certainly have $\exp X \varepsilon Sp(n, C)$ if ${}^tX = -JXJ^{-1}$, or $JX + {}^tXJ = 0$. Conversely, let U be a neighbourhood of 0 in $\mathfrak{M}_{2n}(C)$ which satisfies the conditions of Lemma 1, §II, p. 8, and let U_1 be the neighbourhood $U \cap JUJ^{-1}$. If we observe that $J^2 = -\epsilon$ and that ${}^tJ = -J$, we see that U_1 also satisfies the conditions of the lemma, and that furthermore $JU_1J^{-1} = U_1$. It is clear that the conditions $X \varepsilon U_1$, $\exp X \varepsilon Sp(n, C)$ imply $JX + {}^tXJ = 0$.

The matrices X for which $JX + {}^tXJ = 0$ form a vector subspace $\mathfrak{S}$ of $\mathfrak{M}_{2n}(C)$. If we write X in the form

$$X = \begin{pmatrix} X_1 & X_2 \\ X_3 & X_4 \end{pmatrix}$$

where X_1, X_2, X_3, X_4 are matrices of degree n, the condition $JX + {}^tXJ = 0$ gives

$$X_4 = -{}^tX_1 \qquad X_3 = {}^tX_3 \qquad X_2 = {}^tX_2$$

It follows that $\mathfrak{S}$ is of dimension $n^2 + 2n(n + 1)/2 = 2n^2 + n$ over C; $\mathfrak{S}$ may also be considered as a vector space of dimension $2(2n^2 + n)$ over R.

The matrices $X \epsilon U_1$ for which $\exp X \epsilon Sp(n)$ are the matrices of $\mathfrak{S}$ which satisfy the supplementary condition ${}^tX + \bar{X} = 0$. They form a vector space of dimension $2n^2 + n$ over the field of real numbers. Therefore we have proved

Proposition 2. *In each of the groups $Sp(n)$, $Sp(n, C)$ there exists a neighbourhood of the neutral element which is homeomorphic with an open subset of a cartesian space of suitable dimension. These dimensions are $2n^2 + n$ for $Sp(n)$ and $2(2n^2 + n)$ for $Sp(n, C)$.*

Now, let $Y = (y_{ij})$ be any matrix in $\mathfrak{M}_n(Q)$; set

$$\begin{aligned} y_{ij} &= y_{ij}^{(0)}e_0 + y_{ij}^{(1)}e_1 + y_{ij}^{(2)}e_2 + y_{ij}^{(3)}e_3 (y_{ij}^{(0)}, y_{ij}^{(1)}, y_{ij}^{(2)}, y_{ij}^{(3)} \varepsilon R) \\ &= e_0(y_{ij}^{(0)} + y_{ij}^{(1)}\sqrt{-1}) + e_2(y_{ij}^{(2)} - y_{ij}^{(3)}\sqrt{-1}). \qquad (1 \leqslant i, j \leqslant n) \end{aligned}$$

Let $X = (x_{ij})$ $(1 \leqslant i, j \leqslant 2n)$ be the corresponding matrix in $\mathfrak{M}_{2n}(C)$. If $\mathbf{a}' = (x_1, \cdots, x_{2n})$ is the vector in C^{2n} which corresponds to the vector $\mathbf{a} = (a_1, \cdots, a_n)$ in Q^n, then $X\mathbf{a}'$ is the vector which corresponds to $Y\mathbf{a}$. Set $Y\mathbf{a} = (\tilde{a}_1, \cdots, \tilde{a}_n)$, $X\mathbf{a}' = (\tilde{x}_1, \cdots, \tilde{x}_{2n})$; we have

$$\begin{aligned} a_i &= e_0x_i + e_2x_{n+i} & \tilde{a}_i &= e_0\tilde{x}_i + e_2\tilde{x}_{n+i} \\ \tilde{x}_i &= \Sigma_{j=1}^{2n} x_{ij}x_j & \tilde{a}_i &= \Sigma_{j=1}^{n} y_{ij}a_j \end{aligned}$$

It follows by an easy computation that

$$\begin{aligned} x_{ij} &= y_{ij}^{(0)} + y_{ij}^{(1)}\sqrt{-1}; & x_{i,n+j} &= -y_{ij}^{(2)} - y_{ij}^{(3)}\sqrt{-1} \\ x_{n+i,j} &= y_{ij}^{(2)} - y_{ij}^{(3)}\sqrt{-1}; & x_{n+i,n+j} &= y_{ij}^{(0)} - y_{ij}^{(1)}\sqrt{-1} \qquad (1 \leqslant i, j \leqslant n) \end{aligned}$$

It follows immediately from these formulas that, if a matrix $X \varepsilon \mathfrak{M}_{2n}(C)$ satisfies the conditions ${}^tXJ + JX = 0$, ${}^tX + \bar{X} = 0$, it is the corresponding matrix of a matrix $Y \epsilon \mathfrak{M}_n(Q)$ which satisfies the condition ${}^tY + Y^\iota = 0$.

The mapping which assigns to every matrix in $\mathfrak{M}_n(Q)$ the corresponding matrix in $\mathfrak{M}_{2n}(C)$ is an isomorphism of the ring $\mathfrak{M}_n(Q)$ with a subring of $\mathfrak{M}_{2n}(C)$. On the other hand, this mapping may be used to define a topology in $\mathfrak{M}_n(Q)$ (we require our mapping to be a homeomorphism). If $Y \epsilon \mathfrak{M}_n(Q)$ and if X is the corresponding matrix in $\mathfrak{M}_{2n}(C)$, the convergence of the series $\Sigma_0^\infty \frac{1}{p!} X^p$ implies the convergence of the series $\Sigma_0^\infty \frac{1}{p!} Y^p$. We set $\exp Y = \Sigma_0^\infty \frac{1}{p!} Y^p$. Let $\mathfrak{Y}$ be the set of matrices $Y \epsilon \mathfrak{M}_n(Q)$ such that ${}^tY + Y^\iota = 0$; it follows from what we have said that *the mapping* $Y \to \exp Y$ *maps some neighbourhood of* 0 *in* $\mathfrak{Y}$ *topologically onto a neighbourhood of the neutral element in* $Sp(n)$.

CHAPTER II

Topological Groups

Summary. Chapter II is concerned with the properties of groups which result from the existence in these groups of a topology. Section I contains the definition of a topological group, of a topological subgroup and of products of topological groups. It turns out that, in many cases, the study of a topological group in a neighbourhood of the neutral element will give valuable information on the whole group (Cf. for instance Proposition 5, §III, p. 35; Theorem 1, §IV, p. 35; Theorem 3, §VII, p. 49 and Proposition 2, Chapter IV, §XIV, p. 134). In view of this, it is important to characterize *locally* the topological structure of a topological group: this is accomplished in §II.

If $\mathfrak{H}$ is a closed subgroup of a topological group $\mathfrak{G}$, the cosets modulo $\mathfrak{H}$ form a topological space $\mathfrak{G}/\mathfrak{H}$. The spaces which can be obtained in this way are called *homogeneous spaces*. The definition of such spaces is the object of §III. One of the reasons of the importance of homogeneous spaces is that they provide the most general representations of groups as transitive groups of transformations (satisfying certain topological conditions). In particular, the spheres are homogeneous spaces relative to the linear groups which were introduced in Chapter I. A good part of the knowledge we have of the topology of these groups is derived from this fact (Cf. Proposition 3, §IV, p. 33 and Proposition 5, §X, p. 59).

If $\mathfrak{H}$ happens to be a distinguished subgroup of $\mathfrak{G}$, then $\mathfrak{G}/\mathfrak{H}$ is not only a space, but also a topological group. These factor groups are also considered in Section III.

Section IV is concerned with the connectedness properties of topological groups. The essential fact there is contained in the statement of Theorem 1, p. 35 which allows us in many cases to pass from the local to the global in the study of a topological group.

If $\mathfrak{G}$ is a connected group, the elements of a neighbourhood V of the neutral element form a set of generators of $\mathfrak{G}$. Following the idea of studying a group locally, it is natural to inquire whether it is possible to construct all relations between these generators when the law of composition of the group is known only in V. Section V introduces the study of this question and correlates it with the notion of local isomorphism of topological groups. Examples are given which show that, in general, the answer to our question is negative.

A more profound study of the problem requires the development of a number of purely topological considerations centering around the notion of covering space. Sections VI, VII, VIII and IX are concerned with the elaboration of this notion and its applications to group theory. Following an idea of H. Cartan, we have departed from the usual method of using closed curves in order to define simple-connectedness. We feel that the notion of covering space (as introduced in Definition 3, §VI, p. 40) is the essential notion; roughly speaking, we define a simply connected space as a space which cannot be covered any more (Definition 1, §VII, p. 44). It seems to us that the main property of simply connected spaces is what we call the principle of monodromy (Theorem 2, §VII, p. 46). Theorem 3, §VII, p. 49 states the main property of simply connected groups; it is again a principle of extension from the local to the global. It should be noted that the proof of Theorem 3 gives a typical example of the method of application of the principle of monodromy.

In §VIII, we define the notion of Poincaré group for spaces which admit a simply connected covering space. The Poincaré group is the group of automorphisms of the simply connected covering space, playing therefore a role similar to that played by the Galois group of an algebraic extension. It is shown that the Poincaré group of a topological group is always abelian and may be identified with a subgroup of the center of the simply connected covering group. In section IX, we prove the existence of simply connected covering spaces for a large class of spaces.

In section X, we determine the Poincaré groups of some of the classical groups. The method consists in using the fact that the groups in question operate on spheres. Proposition 5 then yields an algorithm which allows us to proceed to an inductive determination of the Poincaré groups. However, the case of $SO(n)$ cannot easily be completely settled by this method. Although it is possible to prove by purely topological methods (making use of the notion of the second homotopy group of Hurewicz) that the Poincaré group of $SO(n)$ is of order 2 for $n \geqslant 3$, we have prefered to follow an algebraic approach by actually constructing the simply connected covering group of $SO(n)$. This is accomplished in Section XI by making use of the algebra of Clifford numbers.

The algebraic properties (center and ideals) of Clifford numbers are established by an elegant method which was communicated orally to me by V. Bargmann. We then define the spinor group (Definition 1, p. 65) and we prove that this group is the simply connected covering group of $SO(n)$.

§I. DEFINITION OF A TOPOLOGICAL GROUP

A topological group $\mathfrak{G}$ *is the composite object formed by a group* G *and a topological space* $\mathfrak{V}$ *which satisfy the following conditions:* 1) *the set of points of* $\mathfrak{V}$ *is the same as the set of elements of* G*;* 2) *the mapping* $(\sigma, \tau) \rightarrow \sigma\tau^{-1}$ *of* $\mathfrak{V} \times \mathfrak{V}$ *into* $\mathfrak{V}$ *is continuous. The group* G *is called the underlying group of the topological group* $\mathfrak{G}$*, and the space* $\mathfrak{V}$ *is called its underlying space.*

Any notion which is either group-theoretical (such as being abelian, for example), or topological (such as being connected, or compact, etc.) has a meaning when applied to a topological group.

It is clear that the mapping $\tau \rightarrow \tau^{-1}$ of a topological group $\mathfrak{G}$ onto itself is a homeomorphism and that the mapping $(\sigma, \tau) \rightarrow \sigma\tau$ of $\mathfrak{G} \times \mathfrak{G}$ onto $\mathfrak{G}$ is continuous. Conversely, these two conditions together imply condition 2) above.

Examples of topological groups

1) The additive group of real numbers is the underlying group of a topological group, whose underlying space is the usual space of real

numbers R; in fact, the difference $x - y$ is a continuous function of the pair (x, y) of real numbers.

2) The group $GL(n, C)$, with the topology which has been defined on it in §I, Chapt. I, p. 2, gives a topological group, which we shall denote also by $GL(n, C)$.

3) Let G be any group, and let $\mathfrak{B}$ be the discrete topological space whose set of points is the set of elements of G. The group G and the space $\mathfrak{B}$ together make up a topological group. Such a group is called *discrete*.

We shall now indicate how we can construct new topological groups, starting with those which we have already defined.

Subgroups of a topological group

Let H be a subgroup of a topological group $\mathfrak{G}$. The set of elements of H is also the set of points of a subspace of $\mathfrak{G}$, and this subspace forms with H a topological group $\mathfrak{H}$, which will be called a ***topological subgroup*** of $\mathfrak{G}$.

For instance, the groups $GL(n, R)$, $O(n)$, $O(n, C)$, $U(n)$, $SL(n, C)$, $SL(n, R)$, $SO(n)$, $SU(n)$ (which were defined in §I, Chapt. I) are the underlying groups of topological subgroups of $GL(n, C)$. The groups $Sp(n)$, $Sp(n, C)$ (which were defined in §VII, Chapt. I) are the underlying groups of topological subgroups of $GL(2n, C)$.

Products of topological groups

Let $(\mathfrak{G}_\alpha)$ be a family of topological groups, α running over some set of indices. Let G_α and $\mathfrak{B}_\alpha$ be the underlying group and the underlying space of $\mathfrak{G}_\alpha$. The direct product $G = \Pi_\alpha G_\alpha$ is a group, and the product $\mathfrak{B} = \Pi_\alpha \mathfrak{B}_\alpha$ is a space. The group G and the space $\mathfrak{B}$ have the same set of elements. The mapping $(\sigma, \tau) \to \sigma\tau^{-1}$ of $\mathfrak{B} \times \mathfrak{B}$ into $\mathfrak{B}$ is continuous. In fact, the α-coordinate of $\sigma\tau^{-1}$ is $\sigma_\alpha\tau_\alpha^{-1}$, if σ_α and τ_α are the α-coordinates of σ and τ; $\sigma_\alpha\tau_\alpha^{-1}$ is a continuous function of the pair $(\sigma_\alpha, \tau_\alpha)$, which is itself a continuous function of the pair (σ, τ). Therefore, every coordinate of $\sigma\tau^{-1}$ is a continuous function of the pair (σ, τ), which proves our assertion.

It follows that the group G and the space $\mathfrak{B}$ give rise to a topological group $\mathfrak{G}$, which is called the product of the groups $\mathfrak{G}_\alpha$, and which is denoted by $\Pi_\alpha\mathfrak{G}_\alpha$.

For instance, the additive group of elements of R^n is the underlying group of a topological group which is the product of n groups identical with R and whose underlying space is the usual cartesian space R^n.

§II. LOCAL CHARACTERIZATION OF A TOPOLOGICAL GROUP

Let G be a group. To every element $\tau \varepsilon G$ there are associated two mappings of G onto itself, namely the *left translation* T_τ, defined by $T_\tau \sigma = \tau\sigma$ for every $\sigma \varepsilon G$, and the *right translation* T^*_τ, defined by $T^*_\tau \sigma = \sigma\tau$. We have

$$T_{\tau_1\tau_2} = T_{\tau_1} \circ T_{\tau_2}; \qquad T^*_{\tau_1\tau_2} = T^*_{\tau_2} \circ T^*_{\tau_1}$$

Moreover, $T_{\tau^{-1}}$ is the reciprocal mapping of T_τ, and $T^*_{\tau^{-1}}$ is the reciprocal mapping of T^*_τ.

If G is the underlying group of a topological group $\mathfrak{G}$, the mappings T_τ, T^*_τ are continuous mappings of $\mathfrak{G}$ onto itself and so are their reciprocal mappings. Hence T_τ and T^*_τ are homeomorphisms of $\mathfrak{G}$ with itself.

It follows that, if we know the complete family $\mathcal{V}$ of neighbourhoods of the neutral element ϵ, we can obtain the complete family of neighbourhoods of any point σ_0 by applying either of the operations T_{σ_0} or $T^*_{\sigma_0}$ to the sets of $\mathcal{V}$. This means that the topology of $\mathfrak{G}$ is entirely determined when the family $\mathcal{V}$ is given.

But, of course, this family $\mathcal{V}$ cannot be given arbitrarily; it has to satisfy certain conditions:

I. *The intersection of any two sets of* $\mathcal{V}$ *lies in* $\mathcal{V}$.

II. *The intersection of all sets of* $\mathcal{V}$ *is the set* $\{\epsilon\}$.

III. *Any set containing a set of* $\mathcal{V}$ *lies in* $\mathcal{V}$.

IV. *If* $V \varepsilon \mathcal{V}$, *there exists a set* $V_1 \varepsilon \mathcal{V}$ *such that* $V_1V_1 \subset V$. This follows at once from the continuity of the function $\sigma\tau$ at (ϵ, ϵ).

V. *The family of sets* V *such that* $V^{-1} \varepsilon \mathcal{V}$ *coincides with* $\mathcal{V}$. This follows at once from the continuity of the function σ^{-1} at ϵ.

VI. *If* $\sigma_0 \varepsilon G$, *the family of sets* $\sigma_0 V \sigma_0^{-1}$, *with* $V \varepsilon \mathcal{V}$, *coincides with* $\mathcal{V}$.

This last property can be proved in the following way: the set of all neighbourhoods of σ_0 coincides with the family of sets of the form $\sigma_0 V$ $(V \varepsilon \mathcal{V})$; therefore, every set of the form $V\sigma_0 (V \varepsilon \mathcal{V})$ can also be written in the form $\sigma_0 V'$, $V' \varepsilon \mathcal{V}$, and conversely.

We shall now prove that, if we have in a group G a family $\mathcal{V}$ of subsets which satisfies conditions I, II, III, IV, V, VI, we can introduce in G a topology which makes G a topological group and for which $\mathcal{V}$ is the family of neighbourhoods of ϵ.

Let $\mathfrak{U}$ be the family of subsets U of G which satisfy the following condition: if $\sigma \varepsilon U$, then U contains some set of the form $\sigma V (V \varepsilon \mathcal{V})$. It is clear that any union of sets of $\mathfrak{U}$ belongs to $\mathfrak{U}$ and that $G \varepsilon \mathfrak{U}$, $\phi \varepsilon \mathfrak{U}$. It follows immediately from I that the intersection of two sets of $\mathfrak{U}$

belongs to $\mathfrak{U}$. Let V be any set in $\mathfrak{V}$, and let U be the set of elements $\sigma\varepsilon G$ which have the following property: there exists a set $V_1\varepsilon\mathfrak{V}$ such that $\sigma V_1 \subset V$. We shall prove that $U\varepsilon\mathfrak{U}$. In fact, assume that $\sigma\varepsilon U$, $\sigma V_1 \subset V$, $V_1\varepsilon\mathfrak{V}$. Let V_2 be a set of $\mathfrak{V}$ such that $V_2V_2 \subset V_1$; if $\tau\varepsilon V_2$, we have $\tau V_2 \subset V$, whence $\tau\varepsilon U$, $\sigma V_2 \subset U$, which proves our assertion. If we observe that the sets of $\mathfrak{U}$ are permuted among themselves by any left translation T_σ, we see that any set of the form σV, $V\varepsilon\mathfrak{V}$, contains a set $U\varepsilon\mathfrak{U}$ such that $\sigma\varepsilon U$. Let σ_0, σ_1 be two distinct elements of G; we can find a set $V_1\varepsilon\mathfrak{V}$ such that $\sigma_0^{-1}\sigma_1$ does not belong to V_1. By IV and V, there exists a set V such that $VV^{-1} \subset V_1$; it follows easily that $V \cap \sigma_0^{-1}\sigma_1 V = \phi$, whence $\sigma_0 V \cap \sigma_1 V = \phi$. We can find sets U_0, U_1 in $\mathfrak{U}$ such that $\sigma_0\varepsilon U_0$, $\sigma_1\varepsilon U_1$, $U_0 \subset \sigma_0 V$, $U_1 \subset \sigma_1 V$, whence $U_0 \cap U_1 = \phi$. It follows that we can define a topology in G in which $\mathfrak{U}$ is the family of open sets. It is clear that the family of neighbourhoods cf a point $\sigma\varepsilon G$ in this topology is the family of sets σV, $V\varepsilon\mathfrak{V}$.

It remains to prove that the mapping $(\sigma, \tau) \to \sigma\tau^{-1}$ is continuous. Let (σ_0, τ_0) be a point of G $\times$ G, and let $\sigma_0\tau_0^{-1}V$ be a neighbourhood of $\sigma_0\tau_0^{-1}$ (i.e. $V\varepsilon\mathfrak{V}$). Let V_1 be a set in $\mathfrak{V}$ such that $V_1V_1 \subset V$; by VI, the set $\tau_0^{-1}V_1$ may be written in the form $V_2\tau_0^{-1}(V_2\varepsilon\mathfrak{V})$. We have $(\sigma_0V_2)(V_1^{-1}\tau_0)^{-1} = \sigma_0V_2\tau_0^{-1}V_1 \subset \sigma_0\tau_0^{-1}V_1V_1 \subset \sigma_0\tau_0^{-1}V$, which proves our assertion.

A subset $\mathfrak{V}'$ of the complete set of neighbourhoods $\mathfrak{V}$ of a point p in a topological space is said to be a *fundamental system of neighbourhoods* of p if every set of $\mathfrak{V}$ contains some set of $\mathfrak{V}'$. To the conditions I, II, III, IV, V, VI for the set of neighbourhoods of ϵ in a topological group, there correspond the following conditions for a fundamental system of neighbourhoods of ϵ:

I′. *If $V_1\varepsilon\mathfrak{V}'$, $V_2\varepsilon\mathfrak{V}'$, there exists a set $V_3\varepsilon\mathfrak{V}'$ such that $V_3 \subset V_1 \cap V_2$.*

II′. *The intersection of all sets of $\mathfrak{V}'$ is the set $\{\epsilon\}$.*

IV′. *If $V\varepsilon\mathfrak{V}'$, there exists a set $V_1\varepsilon\mathfrak{V}'$ such that $V_1V_1 \subset V$.*

V′. *If $V\varepsilon\mathfrak{V}'$, there exists a set $V_1\varepsilon\mathfrak{V}'$ such that $V_1^{-1} \subset V$.*

VI′. *If $V\varepsilon\mathfrak{V}'$, $\sigma_0\varepsilon G$, there exists a set $V_1\varepsilon\mathfrak{V}'$ such that $\sigma_0V_1\sigma_0^{-1} \subset V$.*

§III. Homogeneous Spaces. Factor Groups

Let $\mathfrak{G}$ be a topological group, and let $\mathfrak{H}$ be a closed subgroup of $\mathfrak{G}$. We say that two elements σ, τ are congruent modulo $\mathfrak{H}$ if the left cosets $\sigma\mathfrak{H}$, $\tau\mathfrak{H}$ coincide. This is obviously an equivalence relation among elements of $\mathfrak{G}$, and the corresponding equivalence classes are the left cosets modulo $\mathfrak{H}$.

To every $\rho \varepsilon \mathfrak{H}$ we associate the corresponding right translation T_ρ^* of $\mathfrak{G}$. A necessary and sufficient condition for σ and τ to belong to the same coset modulo $\mathfrak{H}$ is that there should exist a $\rho \varepsilon \mathfrak{H}$ such that $T_\rho^* \sigma = \tau$. The operations T_ρ^* form a group of homeomorphisms of $\mathfrak{G}$ onto itself. Moreover we observe that the set of all pairs (σ, τ) for which there exists a $\rho \varepsilon \mathfrak{H}$ such that $T_\rho^* \sigma = \tau$ is a closed subset of $\mathfrak{G} \times \mathfrak{G}$, since it is the reciprocal image of $\mathfrak{H}$ under the continuous mapping $(\sigma, \tau) \to \sigma^{-1}\tau$.

More generally, let $\mathfrak{B}$ be a topological space, and let H be a group of homeomorphisms of $\mathfrak{B}$ onto itself. The group H defines an equivalence relation in $\mathfrak{B}$, two points p and q being considered as equivalent if there exists an element $\eta \varepsilon H$ such that $\eta p = q$. Assume that the subset γ of $\mathfrak{B} \times \mathfrak{B}$ which is composed of the pairs (p, q) such that p is equivalent with q is closed in $\mathfrak{B} \times \mathfrak{B}$. Then, we shall define a topology in the set K of equivalence classes. Let $\mathcal{V}$ be the family of those subsets O of K for which the set $\bigcup_{X \varepsilon O} X$ is open in $\mathfrak{B}$. It is clear that any union or finite intersection of sets in $\mathcal{V}$ belongs to $\mathcal{V}$, and that $K \varepsilon \mathcal{V}$, $\phi \varepsilon \mathcal{V}$. Let U be any open subset of $\mathfrak{B}$; then the set of classes X which have an element in common with U belongs to $\mathcal{V}$. In fact, let O be this set, and let p be a point of $\bigcup_{X \varepsilon O} X$. The point p belongs to a class X which has a point q in common with U; there exists an element $\eta \epsilon H$ such that $p = \eta q$. The set $\eta(U)$ is open, contains p and is contained in $\bigcup_{X \varepsilon O} X$, which proves that the latter set is open. Let X_1, X_2 be two distinct equivalence classes, and let a_i be a point of X_i $(i = 1, 2)$. Then (a_1, a_2) does not belong to γ; since γ is closed, there exists open sets U_1, U_2 in $\mathfrak{B}$ such that $(a_1, a_2) \varepsilon U_1 \times U_2$, $\gamma \cap (U_1 \times U_2) = \phi$. Let O_i be the set of equivalence classes which have at least one point in common with U_i $(i = 1, 2)$; we have $O_i \varepsilon \mathcal{V}$, $X_i \varepsilon O_i$ $(i = 1, 2)$ and $O_1 \cap O_2 = \phi$. It follows that we can define a topology in K in which $\mathcal{V}$ is the family of open sets. Let $\bar{\omega}$ be the mapping which assigns to every $p \varepsilon \mathfrak{B}$ its equivalence class; it is clear that $\bar{\omega}$ is a continuous and interior mapping of $\mathfrak{B}$ onto K.

Returning to the consideration of the group $\mathfrak{G}$ and of its closed subgroup $\mathfrak{H}$, we see that we can define a topology in the set of all cosets modulo $\mathfrak{H}$. Let $\mathfrak{G}/\mathfrak{H}$ be the topological space obtained in this way. The space $\mathfrak{G}/\mathfrak{H}$ is called the factor space of $\mathfrak{G}$ by $\mathfrak{H}$. Any space which may be obtained in this way by means of a topological group $\mathfrak{G}$ and a closed subgroup of $\mathfrak{G}$ is called a *homogeneous space*.

The mapping which assigns to every $\sigma \varepsilon \mathfrak{G}$ its coset modulo $\mathfrak{H}$ is called the *natural mapping* of $\mathfrak{G}$ onto $\mathfrak{G}/\mathfrak{H}$. We shall denote this mapping by $\bar{\omega}$.

If $x = \sigma\mathfrak{H}\varepsilon\mathfrak{G}/\mathfrak{H}$ and $\tau\varepsilon\mathfrak{G}$, $\tau\sigma\mathfrak{H}$ is a coset modulo $\mathfrak{H}$ which we denote by τx. This shows that any element $\tau\varepsilon\mathfrak{G}$ defines a mapping of $\mathfrak{G}/\mathfrak{H}$ onto itself; in other words, the group $\mathfrak{G}$ *operates* on $\mathfrak{G}/\mathfrak{H}$.

If $y\varepsilon\mathfrak{G}/\mathfrak{H}$, there always exists a $\tau\varepsilon\mathfrak{G}$ such that $\tau x = y$; therefore, $\mathfrak{G}$ operates *transitively* on $\mathfrak{G}/\mathfrak{H}$. A necessary and sufficient condition for the equality $\tau x = x$ to hold is that $\tau\sigma\varepsilon\sigma\mathfrak{H}$, whence $\tau\varepsilon\sigma\mathfrak{H}\sigma^{-1}$. Therefore, if $x = \bar{\omega}(\sigma)$, the group of elements τ which leave x invariant is $\sigma\mathfrak{H}\sigma^{-1}$.

Now, we prove that the mapping $(\sigma, x) \rightarrow \sigma x$ of $\mathfrak{G} \times (\mathfrak{G}/\mathfrak{H})$ onto $\mathfrak{G}/\mathfrak{H}$ is continuous. Let U be an open subset of $\mathfrak{G}/\mathfrak{H}$, and let U_1 be the set of pairs (σ, x) such that $\sigma x\varepsilon U$. We have to prove that U_1 is open; let (σ_0, x_0) be a point of U_1, and let V be the open subset $\overset{-1}{\bar{\omega}}(U)$ of $\mathfrak{G}$. If τ_0 is any point such that $\bar{\omega}(\tau_0) = x_0$, V is a neighbourhood of $\sigma_0\tau_0$; therefore, there exist open sets V_1, V_2 such that $\sigma_0\varepsilon V_1$, $\tau_0\varepsilon V_2$, $V_1V_2 \subset V$. Let U'_2 be the set $\bar{\omega}(V_2)$; since $\bar{\omega}$ is an interior mapping, U'_2 is open in $\mathfrak{G}/\mathfrak{H}$. Therefore $V_1 \times U'_2$ is open in $\mathfrak{G} \times (\mathfrak{G}/\mathfrak{H})$; moreover, we have $(\sigma_0, x_0)\varepsilon V_1 \times U'_2$, $V_1 \times U'_2 \subset U_1$, which proves that U_1 is a neighbourhood of (σ_0, x_0). It follows that U_1 is open.

In particular, if we fix σ, the mapping $x \rightarrow \varphi_\sigma(x) = \sigma x$ of $\mathfrak{G}/\mathfrak{H}$ onto itself is continuous; $\varphi_{\sigma^{-1}}$ is the reciprocal mapping of φ_σ and is also continuous. It follows that, for every fixed σ, φ_σ is a homeomorphism of $\mathfrak{G}/\mathfrak{H}$ onto itself.

Let f be a mapping of $\mathfrak{G}$ into some set X. If the value of $f(\sigma)$ depends only upon the coset $\sigma\mathfrak{H}$ of σ, we can define a mapping f_1 of $\mathfrak{G}/\mathfrak{H}$ into X such that $f = f_1 \circ \bar{\omega}$: we set $f_1(\sigma\mathfrak{H}) = f(\sigma)$. Suppose that X is a topological space. Then, if f is continuous, f_1 is also continuous. In fact, let Y be any open subset of X; the set $\overset{-1}{f_1}(Y)$ coincides with $\bar{\omega}(\overset{-1}{f}(Y))$. Since f is continuous, $\overset{-1}{f}(Y)$ is open, and therefore also $\overset{-1}{f_1}(Y)$, which proves that f_1 is continuous. If f is an interior mapping, so is f_1; in fact, if U is an open subset of $\mathfrak{G}/\mathfrak{H}$, $f_1(U)$ coincides with $f(\overset{-1}{\bar{\omega}}(U))$ which is open because $\bar{\omega}$ is continuous and f is interior.

Proposition 1. *If $\mathfrak{H}$ and $\mathfrak{G}/\mathfrak{H}$ are compact, the group $\mathfrak{G}$ is compact.*

In fact, let Φ be a family of closed subsets of $\mathfrak{G}$ which has the finite intersection property (i.e. every finite sub family of Φ has a non-empty intersection). We have to prove that the intersection of all sets of Φ is not empty. We may assume without loss of generality that the intersection of a finite number of sets of Φ belongs to Φ. Let Ψ be the family of the sets $\bar{\omega}(F)$, for $F \varepsilon \Phi$; Ψ is a family of subsets of $\mathfrak{G}/\mathfrak{H}$ and has the finite intersection property. Since $\mathfrak{G}/\mathfrak{H}$ is compact, it contains a point x which is adherent to all sets of Ψ.

Let U be any neighbourhood of the neutral element ϵ in $\mathfrak{G}$, and let σ_0 be an element of G such that $\bar{\omega}(\sigma_0) = x$. Then $\bar{\omega}(U\sigma_0)$ is a neighbourhood of x, and therefore meets all the sets $\bar{\omega}(F)$, $F\varepsilon\Phi$; it follows that, if $F\varepsilon\Phi$, we have $F \cap U\sigma_0\mathfrak{H} \neq \phi$, or $U^{-1}F \cap \sigma_0\mathfrak{H} \neq \phi$. Let Φ_1 be the family of sets $U^{-1}F \cap \sigma_0\mathfrak{H}$, F running over all sets in Φ and U over all neighbourhoods of ϵ. Then Φ_1 has again the finite intersection property. In fact, if $F_1, \cdots, F_m\varepsilon\Phi$ and $U_1, \cdots, U_m$ are neighbourhoods of ϵ, the set $\bigcap_{i=1}^m (U_i^{-1}F_i \cap \sigma_0\mathfrak{H})$ contains $U^{-1}F \cap \sigma_0\mathfrak{H}$ (where $U = \bigcap_{i=1}^m U_i$ and $F = \bigcap_{i=1}^m F_i$), and the latter set is not empty because $F\varepsilon\Phi$ and U is a neighbourhood of ϵ. But $\sigma_0\mathfrak{H}$ is homeomorphic to $\mathfrak{H}$, and therefore compact. Consequently, there exists a point $\sigma_1\varepsilon\sigma_0\mathfrak{H}$ which is adherent to all sets of Φ_1, and, *a fortiori*, to all sets $U^{-1}F$. If we keep F fixed and let U run over all neighbourhoods of ϵ, U^{-1} also runs over all neighbourhoods of ϵ, and it follows that σ_1 is adherent to F. Since F is closed, we have $\sigma_1\varepsilon F$; Proposition 1 is thereby proved.

Spheres as homogeneous spaces

Let us consider the group $\mathfrak{G} = O(n)$, with $n \geqslant 2$. Those matrices $\sigma\varepsilon\mathfrak{G}$ which are of the form

$$\sigma = \begin{pmatrix} & & 0 \\ & & \cdot \\ & \bar{\sigma} & \cdot \\ & & \cdot \\ & & 0 \\ 0 \ldots\ldots & 0 & 1 \end{pmatrix} \tag{1}$$

obviously form a closed subgroup $\mathfrak{H}$ of $\mathfrak{G}$. The matrices $\bar{\sigma}$ which occur in the matrices $\sigma\varepsilon\mathfrak{H}$ are clearly the matrices of $O(n-1)$; it follows immediately that $\mathfrak{H}$ is isomorphic (as a topological group) with $O(n-1)$.

Let us now consider the real matrices of degree n as endomorphisms of the vector space R^n, in which we select the base $(\mathbf{e}_1, \cdots, \mathbf{e}_n)$ composed of those vectors whose coordinates consist of $n-1$ zeros and a 1. If σ is any element in $O(n)$, we set $\mathbf{a}(\sigma) = \sigma\mathbf{e}_n$. Since σ is orthogonal, $\mathbf{a}(\sigma)$ is a unit vector. If $\sigma\varepsilon\mathfrak{H}$, then $\sigma\mathbf{e}_n = \mathbf{e}_n$. Conversely, if $\sigma\mathbf{e}_n = \mathbf{e}_n$, and if $\sigma = (x_{ij})$, we have $x_{jn} = \delta_{jn} (1 \leqslant j \leqslant n)$. Since σ is orthogonal, we have $\Sigma_{i=1}^n x_{ni}^2 = 1$; since $x_{nn} = 1$, it follows that $x_{ni} = 0$ for $i < n$, whence $\sigma\epsilon\mathfrak{H}$. Therefore, a necessary and sufficient condition for the equality $\mathbf{a}(\sigma) = \mathbf{a}(\sigma')$ to hold is that $\sigma'\varepsilon\sigma\mathfrak{H}$. The element $\mathbf{a}(\sigma)$ depends only upon the coset x of σ modulo $\mathfrak{H}$; we may therefore set $\mathbf{a}(\sigma) = \mathbf{a}(x)$.

The set of unit vectors $\mathbf{a}\epsilon R^n$ is the unit sphere S^{n-1} of dimension $n - 1$. Hence, we have an obviously continuous mapping $x \to \mathbf{a}(x)$ of $\mathfrak{G}/\mathfrak{H}$ into S^{n-1}. Moreover, this mapping is univalent. Proposition 3a, §III, Chapt. I, p. 11 shows that every $\mathbf{a}\varepsilon S^{n-1}$ is the image of some element $x\varepsilon\mathfrak{G}/\mathfrak{H}$ under our mapping. Since $\mathfrak{G}/\mathfrak{H}$ is compact (being a continuous image of $O(n)$), the mapping $x \to \mathbf{a}(x)$ is topological.

We may identify $O(n - 1)$ with the subgroup $\mathfrak{H}$, and get the following result:

Proposition 2. *The factor space of $O(n)$ by $O(n - 1)$ is homeomorphic with S^{n-1} if $n \geqslant 2$.*

We observe next that, if $\mathbf{a}\varepsilon S^{n-1}$, there always exists an element $\sigma\varepsilon SO(n)$ such that $\sigma\mathbf{e}_n = \mathbf{a}$. In fact, let us select an element $\sigma_1\varepsilon O(n)$ with the required property; if $\boxed{\sigma_1} = -1$, we replace σ_1 by $\sigma = \sigma_1\sigma_0$, with

$$\sigma_0 = \begin{pmatrix} -1 & 0\ldots\ldots 0 \\ 0 & 1\ldots\ldots 0 \\ \ldots & \ldots\ldots\ldots \\ 0 & 0\ldots\ldots 1 \end{pmatrix}$$

We have $\sigma\varepsilon SO(n)$, $\sigma\mathbf{e}_n = \mathbf{a}$. Therefore, we have

Proposition 2a. *If $n \geqslant 2$, the factor space of $SO(n)$ by $SO(n - 1)$ is homeomorphic with S^{n-1}.*

If we observe that the set of unit vectors of C^n is homeomorphic with S^{2n-1}, we obtain, by entirely similar arguments,

Proposition 3. *The factor spaces of $U(n)$ by $U(n - 1)$ and of $SU(n)$ by $SU(n - 1)$ are homeomorphic with S^{2n-1} for $n \geqslant 2$.*

Finally, let us consider the group $Sp(n)$, with $n > 1$. The elements of $Sp(n)$ are matrices of degree n with coefficients in the division algebra Q of quaternions. Let us introduce the vector space Q^n over Q, as we did in Chapter I, §VII, p. 18. Those elements $\sigma\varepsilon Sp(n)$ which leave the basic element $\mathbf{e}_n = (0, \cdots, 0, 1)$ invariant are the matrices of the form (1) with $\tilde{\sigma}\epsilon Sp(n - 1)$. They form a subgroup of $Sp(n)$ which we may identify with $Sp(n - 1)$.

We have defined (Chapt. I, §VIII, p. 21) a one-to-one correspondence between Q^n and C^{2n}; this correspondence allows us to introduce in Q^n a topology, and the operations of $Sp(n)$ are continuous mappings of Q^n (with this topology) onto itself. The images of $\mathbf{e}_n$

under these operations are all the unit vectors of Q^n, as follows from Proposition 2, §VII, Chapt. I, p. 21. On the other hand, Formula (2), §VIII, Chapt. I, p. 22 shows that the unit vectors of Q^n are those which correspond to the unit vectors of C^{2n}; they are the elements of a set which is homeomorphic with S^{4n-1}. Hence:

Proposition 4. *The factor space of $Sp(n)$ by $Sp(n-1)$ is homeomorphic with the sphere S^{4n-1}, if $n \geqslant 2$.*

Factor groups

Let us now consider the case where $\mathfrak{H}$ is a closed distinguished subgroup of $\mathfrak{G}$. Let G and H be the underlying groups of $\mathfrak{G}$ and $\mathfrak{H}$; then the factor group G/H is a group whose set of elements is the set of points of the space $\mathfrak{G}/\mathfrak{H}$; the natural mapping $\bar{\omega}$ of $\mathfrak{G}$ onto $\mathfrak{G}/\mathfrak{H}$ is a homomorphism of G onto G/H. We shall see that the group G/H, together with the space $\mathfrak{G}/\mathfrak{H}$, gives a topological group. Let x_0 and y_0 be elements of G/H; we set $z_0 = x_0 y_0^{-1}$ and we denote by U an open subset of $\mathfrak{G}/\mathfrak{H}$ containing z_0. Let σ_0, τ_0 be elements of G such that $\bar{\omega}(\sigma_0) = x_0$, $\bar{\omega}(\tau_0) = y_0$; then $\overset{-1}{\bar{\omega}}(U)$ is an open set in $\mathfrak{G}$ and contains $\sigma_0\tau_0^{-1}$. It follows that there exist open sets V_1, V_2 in $\mathfrak{G}$, such that $\sigma_0 \varepsilon V_1$, $\tau_0 \varepsilon V_2$, $V_1 V_2^{-1} \subset \overset{-1}{\bar{\omega}}(U)$. The sets $U_1 = \bar{\omega}(V_1)$, $U_2 = \bar{\omega}(V_2)$ are open in $\mathfrak{G}/\mathfrak{H}$, and the conditions $x \varepsilon U_1$, $y \varepsilon U_2$ imply $xy^{-1} \varepsilon U$, which proves that the mapping $(x, y) \to xy^{-1}$ of $(\mathfrak{G}/\mathfrak{H}) \times (\mathfrak{G}/\mathfrak{H})$ onto $\mathfrak{G}/\mathfrak{H}$ is continuous.

The topological group which we have defined is called the *factor group* of $\mathfrak{G}$ by $\mathfrak{H}$; it is denoted by $\mathfrak{G}/\mathfrak{H}$.

For instance, let H be the subgroup of R^n composed of the points with integral coordinates. The group H is obviously a closed discrete subgroup of R^n. The factor group R^n/H is called the n-dimensional *torus* and is denoted by T^n. The group T^1 is also denoted by T; it is homeomorphic to the circumference of a circle in R^2. It is easy to see that T^n is isomorphic (as a topological group) with the product of n times the group T.

Let φ be a continuous homomorphism of a topological group $\mathfrak{G}$ into some other topological group $\mathfrak{G}_1$. The kernel of this homomorphism (i.e. the set $\mathfrak{H}$ of elements of $\mathfrak{G}$ which are mapped upon the neutral element by φ) is a distinguished subgroup of $\mathfrak{G}$, and is closed because φ is continuous. The element $\varphi(\sigma)$ depends only upon the coset $\sigma\mathfrak{H}$ of σ modulo $\mathfrak{H}$, and hence φ defines a continuous homomorphism φ_1 of $\mathfrak{G}/\mathfrak{H}$ into $\mathfrak{G}_1$. The homomorphism φ_1 is univalent, but it should be observed that φ_1 is not necessarily a homeomorphism.

Proposition 5. *Let φ be a homomorphism of a topological group $\mathfrak{G}$ into a topological group $\mathfrak{G}_1$. If φ is continuous at the neutral element ϵ of $\mathfrak{G}$, it is continuous everywhere.*

Let σ_0 be any element of $\mathfrak{G}$, and let $\varphi(\sigma_0)V_1$ be a neighbourhood of $\varphi(\sigma_0)$ in $\mathfrak{G}_1$ (V_1 being a neighbourhood of the neutral element in $\mathfrak{G}_1$). By assumption, there exists a neighbourhood V of ϵ in $\mathfrak{G}$ such that $\varphi(V) \subset V_1$, whence $\varphi(\sigma_0 V) \subset \varphi(\sigma_0)V_1$, which proves that φ is continuous at σ_0.

§IV. Components of a Topological Group

Proposition 1. *The component of the neutral element ϵ in a topological group $\mathfrak{G}$ is a closed distinguished subgroup of $\mathfrak{G}$.*

In fact, let K be this component, and let τ be an element in K. The right translation associated with τ^{-1} being a homeomorphism of $\mathfrak{G}$ onto itself, the set $K\tau^{-1}$ is connected and contains $\tau\tau^{-1} = \epsilon$; it follows that $K\tau^{-1} \subset K$, whence $KK^{-1} \subset K$, which shows that K is a subgroup of $\mathfrak{G}$. Now, let ρ be any element in $\mathfrak{G}$; the mapping $\sigma \rightarrow \rho\sigma\rho^{-1}$ is a homeomorphism of $\mathfrak{G}$ onto itself. It follows that $\rho K\rho^{-1}$ is connected; since $\epsilon \in \rho K \rho^{-1}$, we have $\rho K\rho^{-1} \subset K$, which proves that K is distinguished.

The factor group $\mathfrak{G}/K$ is called the *group of components* of $\mathfrak{G}$; its elements are the cosets of K, which are also the components of $\mathfrak{G}$. If the group $\mathfrak{G}$ is locally connected (i.e., if there exists a connected neighbourhood V of ϵ), the group $\mathfrak{G}/K$ is discrete. In fact, the image of V under the natural mapping of $\mathfrak{G}$ onto $\mathfrak{G}/K$ is a point (because $V \subset K$) and is a neighbourhood of the neutral element in $\mathfrak{G}/K$ (because the natural mapping is interior).

Theorem 1. *In a connected topological group, any neighborhood of the neutral element is a set of generators of the group.*

In fact, let V be a neighbourhood of the neutral element in a connected group $\mathfrak{G}$, and let H be the subgroup generated by the elements of V. If $\sigma \in H$, we have also $V\sigma \in H$; therefore H is an open set. But, *any open subgroup H of a topological group $\mathfrak{G}$ is also closed,* for every coset modulo H is an open set, and H is the complement in $\mathfrak{G}$ of the union of the cosets different from H. It follows that the set H is open and closed in $\mathfrak{G}$; $\mathfrak{G}$ being connected, and H being not empty, we have $H = \mathfrak{G}$.

Remark. More generally, we have the following result: Let $\mathfrak{G}$ be a topological group, and assume that V is a connected neighbourhood of the neutral element ϵ in $\mathfrak{G}$. Let W be any neighbourhood of ϵ; then, any element $\sigma \in V$ may be written in the form $\sigma_1\sigma_2 \cdots \sigma_m$, with $\sigma_i \in W$ $(1 \leqslant i \leqslant m)$ and $\sigma_1\sigma_2 \cdots \sigma_i \in V$ $(1 \leqslant i \leqslant m)$. To prove this result, we may assume without loss of generality that $W = W^{-1}$ and that $W \subset V$ (if these conditions were not satisfied, we would replace W by a smaller neighbourhood). Let E be the set of elements of V which may be written in the form indicated above. It is clear that, if $\sigma \varepsilon E$, we have also $\sigma W \cap V \subset E$; therefore E is relatively open in V. Let σ_0 be a point of V which is adherent to E; then $\sigma_0 W$ has a point σ in common with E. We have $\sigma_0 \varepsilon \sigma W^{-1} = \sigma W$, whence $\sigma_0 \epsilon E$, which proves that E is also relatively closed in V. Since $\epsilon \varepsilon E$ and V is connected, we have $E = V$, which proves our assertion.

Proposition 2. *Let $\mathfrak{G}$ be a topological group, and let $\mathfrak{H}$ be a closed subgroup of $\mathfrak{G}$. If the group $\mathfrak{H}$ and the factor space $\mathfrak{G}/\mathfrak{H}$ are connected, then $\mathfrak{G}$ is connected.*

In fact, assume that $\mathfrak{G} = U \cup V$, where U and V are non empty open sets. The natural mapping of $\mathfrak{G}$ onto $\mathfrak{G}/\mathfrak{H}$ maps U and V onto open subsets U_1 and V_1 of $\mathfrak{G}/\mathfrak{H}$, and we have $\mathfrak{G}/\mathfrak{H} = U_1 \cup V_1$. It follows that $U_1 \cap V_1$ contains at least one element $\sigma_1\mathfrak{H}$ of $\mathfrak{G}/\mathfrak{H}$. This means that $\sigma_1\mathfrak{H}$ meets both U and V. We have $\sigma_1\mathfrak{H} = (\sigma_1\mathfrak{H} \cap U) \cup (\sigma_1\mathfrak{H} \cap V)$. On the other hand, $\sigma_1\mathfrak{H}$ is homeomorphic with $\mathfrak{H}$, and hence connected. It follows that U and V have at least one point in common in $\sigma_1\mathfrak{H}$. Proposition 2 is thereby proved.

Lemma 1. *The sphere $S^n (n \geqslant 1)$ is connected.*

S^n is the subset of R^{n+1} defined by the equation

$$x_1^2 + \cdots + x_{n+1}^2 = 1$$

Let E be the set composed of the points of S^n for which $x_{n+1} \geqslant 0$. If we map the point $(x_1, \cdots, x_{n+1}) \epsilon E$ upon the point $(x_1, \cdots, x_n) \epsilon R^n$, we clearly obtain a homeomorphism of E with the set B^n composed of the points $(x_1, \cdots, x_n)$ such that $\Sigma_{i=1}^n x_i^2 \leqslant 1$. The set B^n is obviously connected; hence E is connected. Similarly, the lower hemisphere E' of S^n, defined by the condition $x_{n+1} \leqslant 0$ is connected. Since $n \geqslant 1$, the set $E \cap E'$ is not empty. It follows that S^n is connected.

Lemma 2. *The groups $SO(1)$, $U(1)$, $SU(1)$, $Sp(1)$ are connected.*

The groups $SO(1)$, $SU(1)$ contain only their neutral elements. The group $U(1)$ is the multiplicative group of complex numbers of absolute value 1; it is homeomorphic to S^1, and hence connected. The group $Sp(1)$ is the multiplicative group of quaternions of norm 1. A quaternion of norm 1 may be written in the form $a_0e_0 + a_1e_1 + a_2e_2 + a_3e_3$, with $\Sigma_{i=0}^3 a_i^2 = 1$. It follows easily that $Sp(1)$ is homeomorphic with S^3, and hence connected.

Proposition 3. *The groups $SO(n)$, $U(n)$, $SU(n)$, $Sp(n)$ are all connected for $n \geqslant 1$.*

This is proved by induction on n, making use of Proposition 2, Lemmas 1 and 2 and of Propositions 2a, 3, 4, §III, p. 33.

On the other hand, the group $O(n)$ has exactly two connected components. In fact, it contains a matrix of determinant -1, namely

$$\sigma_0 = \begin{pmatrix} -1 & 0 \ldots\ldots 0 \\ 0 & 1 \ldots\ldots 0 \\ \ldots & \ldots\ldots\ldots \\ 0 & 0 \ldots\ldots 1 \end{pmatrix}$$

Since the determinant is a continuous function which does not vanish on $O(n)$, $O(n)$ cannot be connected. On the other hand, $O(n)$ contains $SO(n)$ as a distinguished subgroup of index 2. Hence it has exactly two connected components, one of which is $SO(n)$, and the other the set of orthogonal matrices of determinant -1.

§V. Local Isomorphism. Examples

Let $\mathfrak{G}$ be a connected topological group, and let ϵ be the neutral element of $\mathfrak{G}$. We know that any neighbourhood V of ϵ is a set of generators of $\mathfrak{G}$ (Theorem 1, §IV, p. 35). We shall now inquire about the relations between these generators.

Let us assume that we have an analytic apparatus which permits the computation of products $\sigma\tau$ only in the case where σ, τ and $\sigma\tau$ lie in V (we shall see later that this situation arises in many cases). Then, to each pair (σ, τ) of elements of V such that $\sigma\tau \in V$, there corresponds a relation $\sigma\tau = \rho$ between our generators of $\mathfrak{G}$. The question is: are all the relations which hold between elements of V consequences of relations of the type we have just described?

We shall see in a moment that the answer to this question is negative in general. Before doing that, we shall first formulate our problem in a different, but equivalent, way.

Definition 1. *Let $\mathfrak{G}$, $\mathfrak{G}_1$ be two topological groups. A local isomorphism of $\mathfrak{G}$ into $\mathfrak{G}_1$ is a homeomorphism f of some neighbourhood V of the neutral element ϵ of $\mathfrak{G}$ onto a neighbourhood V_1 of the neutral element ϵ_1 of V_1 which has the following properties:*

1) *the conditions $\sigma \in V$, $\tau \in V$, $\sigma\tau \in V$ imply $f(\sigma\tau) = f(\sigma)f(\tau)$;*

2) *the conditions $\sigma \in V$, $\tau \in V$, $f(\sigma\tau) \in V_1$ imply $\sigma\tau \in V$.*

In terms of this notion, our original question may be formulated as follows: f being a local isomorphism of $\mathfrak{G}$ with $\mathfrak{G}_1$, is it always possible to extend the domain of definition of f to all of $\mathfrak{G}$ in such a way that f becomes an isomorphism of the underlying group of $\mathfrak{G}$ with the underlying group of $\mathfrak{G}_1$?

We shall give two examples which show that this extension is not always possible:

1) Let φ be the mapping which assigns to every real number x its residue class $\varphi(x)$ modulo 1; let f be the contraction of φ to the interval $]-\frac{1}{4}, +\frac{1}{4}[$. It is clear that f is a local isomorphism of R into T. But f cannot be extended to an isomorphism of R with T, because these groups are not isomorphic.

2) Let us consider the group $Sp(1)$, i.e. the multiplicative group of the quaternions $q = ae_0 + be_1 + ce_2 + de_3$ such that $a^2 + b^2 + c^2 + d^2 = 1$. To every such quaternion q there is associated a linear mapping T_q of Q (the algebra of quaternions) onto itself, defined by the formula $T_q(r) = qrq^{-1} (r \varepsilon Q)$. We have

$$T_q(r^\iota) = qr^\iota q^{-1} = (q^{-1})^\iota r^\iota q^\iota = (T_q(r))^\iota$$

because $q^\iota = q^{-1}$. It follows that T_q maps onto itself the set P of *pure quaternions* (a pure quaternion is a quaternion of the form $x_1e_1 + x_2e_2 + x_3e_3$); in fact, the pure quaternions p are characterized by the condition $p\iota = -p$. If

$$T_q(x_1e_1 + x_2e_2 + x_3e_3) = x_1'e_1 + x_2'e_2 + x_3'e_3$$

we have

(1) $$x_i' = \Sigma_{j=1}^3 a_{ij}(q)x_j$$

Let $\theta(q)$ denote the matrix $(a_{ij}(q))$. We have $T_{qq'} = T_q \circ T_{q'}$ whence $\theta(qq') = \theta(q)\theta(q')$. Hence the mapping $q \to \theta(q)$ is a representation of $Sp(1)$ by matrices of degree 3.

If r, r' are any two quaternions, we have $T_q(rr') = T_q(r)T_q(r')$. Since the coefficient of e_0 in the expression of $(\Sigma_1^3 x_ie_i)(\Sigma_1^3 y_ie_i)$ is $-\Sigma_1^3 x_iy_i$, we see that the linear substitution (1), performed on the variables x and y, leaves invariant the expression $\Sigma_1^3 x_iy_i$. In other words, the matrix $\theta(q)$ is orthogonal.

We know that $Sp(1)$ is connected (Lemma 2, §IV, p. 36). The mapping $q \to \theta(q)$ being clearly continuous, it follows that $\theta(q)$ belongs to the connected component of the neutral element in $O(3)$, *i.e.* to $SO(3)$. We shall now prove that $\theta(Sp(1))$ is the whole group $SO(3)$. An easy computation shows that

$$\theta(\text{Cos}\,\lambda e_0 + \text{Sin}\,\lambda e_1) = \begin{pmatrix} 1 & 0 & 0 \\ 0 & \text{Cos}\,2\lambda & -\text{Sin}\,2\lambda \\ 0 & \text{Sin}\,2\lambda & \text{Cos}\,2\lambda \end{pmatrix}$$

It follows that $\theta(Sp(1))$ contains the group g_1 of all rotations around the x_1-axis; similarly, we would see that $\theta(Sp(1))$ contains the group g_2 of rotations around the x_2-axis. Our assertion will therefore be proved if we show that g_1 and g_2 generate $SO(3)$. Let r be any rotation, and let M_1 be the extremity of the unit vector of origin O carried by the x_1-axis. We can find an operation s_1 of g_1 such that $s_1(rM_1)$ is a point of the x_1x_3-plane; this point being at unit distance from the origin, we can find an operation $s_2 \varepsilon g_2$ such that $s_2s_1rM_1 = M_1$; it follows that $s_2s_1r\varepsilon g_1$, which shows that r is in the group generated by g_1 and g_2.

Let us now determine the kernel of the representation $q \to \theta(q)$. If $\theta(q)$ is the unit matrix, q must commute with every pure quaternion, in particular with e_1, e_2, e_3. We see at once that this condition implies $b = c = d = 0$, whence $a = \pm 1$. It follows that the kernel of our representation consists of the two quaternions e_0, $-e_0$.

Let V be a compact neighbourhood of e_0 in $Sp(1)$ such that $-e_0$ does not belong to VVV^{-1}; then θ maps V in a continuous univalent way. Since V is compact, the contraction of θ to V is a homeomorphism of V with some subset of $SO(3)$. We shall prove that $\theta(V)$ is a neighbourhood of the neutral element in $SO(3)$. Let V_1 be an open neighbourhood of e_0 in $Sp(1)$ such that $V_1 \subset V$. The complement A of $V_1 \cup (-e_0)V_1$ in $Sp(1)$ is compact. Let U run over all compact neighbourhoods of the unit matrix in $SO(3)$; if the family of sets $\overset{-1}{\theta}(U) \cap A$ had the finite intersection property, there would exist a point $q\varepsilon A$ such that $\theta(q)\varepsilon U$ for every U, whence $\theta(q) = E$ (the unit matrix), and this is impossible. Since any finite intersection of sets of the form $\overset{-1}{\theta}(U) \cap A$ is again of the same form, it follows that there exists a neighbourhood U of E in $SO(3)$ such that $\overset{-1}{\theta}(U) \cap A = \phi$, whence $\overset{-1}{\theta}(U) \subset V \cup (-e_0)V$. Since $U = \theta(\overset{-1}{\theta}(U))$ and $\theta(V) = \theta((-e_0)V)$, it follows that $\theta(V) \supset U$, which proves our assertion.

Let f be the contraction of θ to V. If $q\varepsilon V$, $q'\varepsilon V$, $f(qq')\varepsilon f(V)$, there exists an element r in V such that $f(qq') = f(r)$, i.e. $qq'r^{-1} = \pm e_0$. Since $-e_0$ does not belong to VVV^{-1}, we have $qq'r^{-1} \neq -e_0$, whence $qq' = r\varepsilon V$. It follows that f is a local isomorphism of $Sp(1)$ into $SO(3)$.

Suppose now that it would be possible to extend f to an isomorphism θ' of $Sp(1)$ with $SO(3)$; since θ and θ' are both homomorphisms and coincide on V which is a set of generators of $Sp(1)$, we would have $\theta = \theta'$, which is impossible, since $\theta(-e_0)$ is the unit matrix.

Let H be the subgroup of $Sp(1)$ which is composed of the elements

e_0, $-e_0$. There corresponds to θ a continuous univalent homomorphism θ_1 of $Sp(1)/H$ onto $SO(3)$. Since $Sp(1)/H$ is compact, θ_1 is also a homeomorphism. Hence, we have proved

Proposition 1. *The group $SO(3)$ is isomorphic (as a topological group) with the factor group of $Sp(1)$ by the subgroup composed of the elements e_0 and $-e_0$.*

A necessary and sufficient condition for the equality $\theta(q) = \theta(q')$ to hold is that $q = \pm q'$. If we represent $q = ae_0 + be_1 + ce_2 + de_3$ by the point $(a, b, c, d) \varepsilon S^3$, we see that $SO(3)$ is homeomorphic to the space obtained by identifying the pairs of antipodal points on S^3, i.e. to the three dimensional projective space.

§VI. NOTION OF COVERING SPACE

Definition 1. *A topological space is said to be locally connected if any neighbourhood of any point p of the space contains a connected neighbourhood of the point.*

Proposition 1. *In a locally connected space, every component of an open set is an open set.*

In fact, let K be a component of an open set U. If $p \varepsilon K$, U is a neighbourhood of p, and therefore U contains a connected neighbourhood V of p. Since K is a component and $V \cap K \neq \phi$, we have $V \subset K$, and p is interior to K. Proposition 1 is thereby proved.

Remark. It follows immediately that any neighbourhood of a point p in a locally connected space contains an open connected neighbourhood of the point.

Definition 2. *Let f be a continuous mapping of a space $\tilde{\mathfrak{B}}$ into a space $\mathfrak{B}$. A subset E of $\mathfrak{B}$ is said to be evenly covered by $\tilde{\mathfrak{B}}$ (with respect to f) if $\overset{-1}{f}(E)$ is not empty and every component of $\overset{-1}{f}(E)$ is mapped topologically onto E by the mapping f.*

It is clear that any set which is evenly covered is *ipso facto* connected.

Definition 3. *Let $\mathfrak{B}$ be a topological space. A covering space $(\tilde{\mathfrak{B}}, f)$ of $\mathfrak{B}$ is a pair formed by a connected and locally connected space $\tilde{\mathfrak{B}}$ and a continuous mapping f of $\tilde{\mathfrak{B}}$ onto $\mathfrak{B}$ which has the following property: each point of $\mathfrak{B}$ has a neighbourhood which is evenly covered by $\tilde{\mathfrak{B}}$ (with respect to f).*

It is clear that a space cannot have a covering space unless it is connected and locally connected. Conversely, if $\mathfrak{B}$ is a connected and locally connected space, $\mathfrak{B}$ admits at least one covering space, viz. $(\mathfrak{B}, e)$, where e is the identity mapping.

If $(\tilde{\mathfrak{B}}, f)$ is a covering space of $\mathfrak{B}$, f is an interior mapping. In fact, let $\tilde{U}$ be an open set of $\tilde{\mathfrak{B}}$, and let $p = f(\tilde{p})$ (with $\tilde{p} \varepsilon \tilde{U}$) be any point of

$f(\tilde{U})$. The point p has a neighbourhood V which is evenly covered by $\tilde{\mathfrak{V}}$. Let $\tilde{V}$ be the component of $\tilde{p}$ in $\overset{-1}{f}(V)$; then $\tilde{V} \cap \tilde{U}$ is relatively open in $\tilde{V}$. Since f maps $\tilde{V}$ topologically onto V, the set $f(\tilde{V} \cap \tilde{U})$ is relatively open in V, and is therefore a neighbourhood of p. Since $f(\tilde{V} \cap \tilde{U}) \subset f(\tilde{U})$, $f(\tilde{U})$ is a neighbourhood of p, which proves our assertion.

Lemma 1. *Let f be a continuous mapping of a locally connected space $\tilde{\mathfrak{V}}$ into a space $\mathfrak{V}$. Let $\tilde{p}$ be a point of $\tilde{\mathfrak{V}}$, and let V be a neighbourhood of $p = f(\tilde{p})$ in $\mathfrak{V}$. The component $\tilde{V}$ of $\tilde{p}$ in $\overset{-1}{f}(V)$ is a neighbourhood of $\tilde{p}$ in $\tilde{\mathfrak{V}}$.*

The set V contains an open set U such that $p \varepsilon U$. We have $\tilde{p} \varepsilon \overset{-1}{f}(U)$; let $\tilde{U}$ be the component of $\tilde{p}$ in $\overset{-1}{f}(U)$. The set $\overset{-1}{f}(U)$ is open because f is continuous; $\tilde{U}$ is open by Proposition 1. Since $\tilde{U} \subset \tilde{V}$, Lemma 1 is proved.

Lemma 1 shows that, if $(\tilde{\mathfrak{V}}, f)$ is a covering space of $\mathfrak{V}$, f is a *local homeomorphism*, i.e. every point of $\tilde{\mathfrak{V}}$ has a neighbourhood which is mapped topologically by f. This condition, however, is not sufficient to make $(\tilde{\mathfrak{V}}, f)$ a covering space of $\mathfrak{V}$, as will be shown by the following example. Let f_1 be the mapping of R into T which assigns to every $x \varepsilon R$ its residue class modulo 1; it is easy to verify that (R, f_1) is a covering space of T. Let f_2 be the mapping of R^2 into T^2 which is defined by $f_2(x, y) = (f_1(x), f_1(y))$; here again, it is easy to see that (R^2, f_2) is a covering space of T^2. Let now $\tilde{\mathfrak{V}}$ be the space obtained by removing one point from R^2; $\tilde{\mathfrak{V}}$ is connected and locally connected. If f is the contraction of f_2 to $\tilde{\mathfrak{V}}$, f is a local homeomorphism and maps $\tilde{\mathfrak{V}}$ onto T^2; nevertheless, $(\tilde{\mathfrak{V}}, f)$ is *not* a covering space of T^2.

Lemma 2. *Let f be a continuous mapping of a space $\tilde{\mathfrak{V}}$ into a space $\mathfrak{V}$. If E is a subset of $\mathfrak{V}$ which is evenly covered by $\tilde{\mathfrak{V}}$ (with respect to f), then every connected subset F of E is also evenly covered. The components of $\overset{-1}{f}(F)$ are the intersections of the components of $\overset{-1}{f}(E)$ with $\overset{-1}{f}(F)$.*

Let $\tilde{E}_\nu$ be the components of $\overset{-1}{f}(E)$, ν running over some set of indices. Set $\tilde{F}_\nu = \tilde{E}_\nu \cap \overset{-1}{f}(F)$. Then f maps $\tilde{F}_\nu$ topologically onto F, which proves that $\tilde{F}_\nu$ is connected. Conversely, any connected subset of $\overset{-1}{f}(F)$ is contained in some component $\tilde{E}_\nu$ of $\overset{-1}{f}(E)$: Lemma 2 is proved.

Lemma 3. *Assume that* $(\tilde{\mathfrak{V}}, f)$ *is a covering space of the space* $\mathfrak{V}$. *If* p *is any point of* $\mathfrak{V}$, *every neighbourhood of* p *contains an open connected neighbourhood which is evenly covered by* $\tilde{\mathfrak{V}}$.

Lemma 3 follows immediately from Lemma 2 and Proposition 1.

Lemma 4. *Let* f_1 *be a continuous mapping of a locally connected space* $\tilde{\mathfrak{V}}_1$ *into a connected space* $\mathfrak{V}$. *Assume that every point* $p \varepsilon \mathfrak{V}$ *has a neighbourhood which is evenly covered by* $\tilde{\mathfrak{V}}_1$ *(with respect to* f_1*). Let* $\tilde{\mathfrak{V}}$ *be a component of* $\tilde{\mathfrak{V}}_1$, *and let* f *be the contraction of* f_1 *to* $\tilde{\mathfrak{V}}$. *Then* $(\tilde{\mathfrak{V}}, f)$ *is a covering space of* $\mathfrak{V}$, *and* $\tilde{\mathfrak{V}}$ *is open in* $\tilde{\mathfrak{V}}_1$.

We first prove that $f(\tilde{\mathfrak{V}}) = \mathfrak{V}$. Let p be a point of $\mathfrak{V}$, and let V be a neighbourhood of p which is evenly covered by $\tilde{\mathfrak{V}}_1$. Let $\tilde{V}_\nu$ be the components of $\overset{-1}{f_1}(V)$, ν running over a set of indices N. If a set $\tilde{V}_\nu$ meets $\tilde{\mathfrak{V}}$, it is entirely contained in $\tilde{\mathfrak{V}}$, whence

$$\tilde{\mathfrak{V}} \cap \overset{-1}{f_1}(V) = \bigcup_{\nu \varepsilon N'} \tilde{V}_\nu$$

where N' is the set of indices ν such that $\tilde{V}_\nu \cap \tilde{\mathfrak{V}} \neq \phi$. It follows that, if V meets $f(\tilde{\mathfrak{V}})$, we have $V \subset f(\tilde{\mathfrak{V}})$; in particular, if p is adherent to $f(\tilde{\mathfrak{V}})$, then p is interior to $f(\tilde{\mathfrak{V}})$. It follows immediately that $f(\tilde{\mathfrak{V}})$ is open and closed in $\mathfrak{V}$, whence $f(\tilde{\mathfrak{V}}) = \mathfrak{V}$, since $\mathfrak{V}$ is connected.

The components of $\overset{-1}{f}(V)$ are the sets $\tilde{V}_\nu$ for $\nu \varepsilon N'$, because each $\tilde{V}_\nu$ is a maximal connected subset of $\overset{-1}{f_1}(V)$, and, *a fortiori*, of $\overset{-1}{f}(V)$. It follows that $(\tilde{\mathfrak{V}}, f)$ is a covering space of $\mathfrak{V}$. The fact that $\tilde{\mathfrak{V}}$ is open follows immediately from Lemma 1.

Lemma 5. *Assume that* $(\tilde{\mathfrak{V}}, f)$ *is a covering space of the space* $\mathfrak{V}$. *Let* $\mathfrak{X}$ *be a connected and locally connected subspace of* $\mathfrak{V}$. *If* $\tilde{\mathfrak{X}}$ *is any component of* $\overset{-1}{f}(\mathfrak{X})$, $\tilde{\mathfrak{X}}$ *is relatively open in* $\overset{-1}{f}(\mathfrak{X})$; *if* g *is the contraction of* f *to* $\tilde{\mathfrak{X}}$, $(\tilde{\mathfrak{X}}, g)$ *is a covering space of* $\mathfrak{X}$.

Let p be a point of $\mathfrak{X}$, and let V be an open neighbourhood of p which is evenly covered by $\tilde{\mathfrak{V}}$ (Cf. Lemma 3). Since $\mathfrak{X}$ is locally connected, we can find a connected neighbourhood X of p with respect to $\mathfrak{X}$ such that $X \subset V$. The components $\tilde{X}_\nu$ of $\overset{-1}{f}(X)$ are the intersections of $\overset{-1}{f}(X)$ with the components $\tilde{V}_\nu$ of $\overset{-1}{f}(V)$ (Cf. Lemma 2). If $\tilde{p}_\nu$ is the point of $\tilde{X}_\nu$ which is mapped on p by f, $\tilde{p}_\nu$ is interior to $\tilde{V}_\nu$ and $\tilde{X}_\nu$ is therefore a neighbourhood of $\tilde{p}_\nu$ with respect to $\overset{-1}{f}(\mathfrak{X})$. It follows easily that $\overset{-1}{f}(\mathfrak{X})$ is locally connected, and that every point of $\mathfrak{X}$ has a neighbourhood which is evenly covered by $\overset{-1}{f}(\mathfrak{X})$ with respect to the contraction of f to $\overset{-1}{f}(\mathfrak{X})$. Lemma 5 then follows from Lemma 4.

In Definition 3, we have required the space $\tilde{\mathfrak{V}}$ to be locally connected. The following lemma describes the situation which we obtain if we omit this condition.

Lemma 6. *Let $\mathfrak{V}$ be a connected and locally connected space. Assume that a space $\tilde{\mathfrak{V}}$ has a continuous mapping f onto $\mathfrak{V}$ such that every point of $\mathfrak{V}$ has a neighbourhood which is evenly covered by $\tilde{\mathfrak{V}}$ with respect to f. Let $\mathcal{K}$ be the family of subsets of $\tilde{\mathfrak{V}}$ which can be represented as unions of components of open sets of $\tilde{\mathfrak{V}}$. Then $\mathcal{K}$ can be taken to be the family of open sets of a space $\tilde{\mathfrak{V}}'$ which has the same points as $\tilde{\mathfrak{V}}$. The space $\tilde{\mathfrak{V}}'$ is locally connected. Every point of $\mathfrak{V}$ has a neighbourhood which is evenly covered by $\tilde{\mathfrak{V}}'$ with respect to f.*

It is clear that any union of sets of $\mathcal{K}$ is in $\mathcal{K}$ and that every open set of $\tilde{\mathfrak{V}}$ is in $\mathcal{K}$. It follows that, $\tilde{p}_1$ and $\tilde{p}_2$ being any two points of $\tilde{\mathfrak{V}}$ distinct from each other, there exist sets $\tilde{K}_1$ and $\tilde{K}_2$ in $\mathcal{K}$ such that $\tilde{p}_1 \varepsilon \tilde{K}_1$, $\tilde{p}_2 \varepsilon \tilde{K}_2$, $\tilde{K}_1 \cap \tilde{K}_2 = \phi$.

Now, let $\tilde{K}_1$ and $\tilde{K}_2$ be any sets in $\mathcal{K}$ which have a point $\tilde{p}$ in common. It follows immediately from Lemma 2 that $p = f(\tilde{p})$ has an open connected neighbourhood U_3 in $\mathfrak{V}$ which is evenly covered by $\tilde{\mathfrak{V}}$ with respect to f. On the other hand, we can find open subsets $\tilde{U}_i$ of $\tilde{\mathfrak{V}}$ such that the component of $\tilde{p}$ in $\tilde{U}_i$ is contained in $\tilde{K}_i$ $(i = 1, 2)$. Let $\tilde{K}$ be the component of $\tilde{p}$ in $\tilde{U}_1 \cap \tilde{U}_2 \cap \overset{-1}{f}(U_3)$. Then $\tilde{K}$ belongs to $\mathcal{K}$, contains $\tilde{p}$ and is contained in $\tilde{K}_1 \cap \tilde{K}_2$. This proves that $\tilde{K}_1 \cap \tilde{K}_2 \varepsilon \mathcal{K}$; the first assertion of Lemma 6 follows immediately from this. Let $\tilde{K}_3$ be the component of $\tilde{p}$ in $\overset{-1}{f}(U_3)$; then $\tilde{K}$ is also the component of $\tilde{p}$ in $\tilde{U}_1 \cap \tilde{U}_2 \cap \tilde{K}_3$. By assumption, f maps $\tilde{K}_3$ (considered as a subspace of $\tilde{\mathfrak{V}}$) topologically onto U_3. Hence $\tilde{K}_3$ is locally connected and therefore $\tilde{K}$ is relatively open in $\tilde{K}_3$. Taking $\tilde{K}_1 = \tilde{K}_2$, we see that the topologies induced by $\tilde{\mathfrak{V}}$ and $\tilde{\mathfrak{V}}'$ on the set $\tilde{K}_3$ coincide with each other, and this proves that $\tilde{\mathfrak{V}}'$ is locally connected. Furthermore, it is clear that U_3 is evenly covered by $\tilde{\mathfrak{V}}'$ with respect to f. Lemma 6 is now completely proved.

§VII. Simply Connected Spaces. The Principle of Monodromy

In conformity with the general notion of isomorphism, we shall say that two covering spaces $(\tilde{\mathfrak{V}}_1, f_1)$ and $(\tilde{\mathfrak{V}}_2, f_2)$ of the same space $\mathfrak{V}$ are *isomorphic* if there exists a homeomorphism φ of $\tilde{\mathfrak{V}}_1$ with $\tilde{\mathfrak{V}}_2$ such that $f_1 = f_2 \circ \varphi$.

Definition 1. *A space $\mathfrak{V}$ is said to be simply connected if it is connected and locally connected and if every covering space of $\mathfrak{V}$ is isomorphic to the trivial covering space $(\mathfrak{V}, e)$, where e is the identity mapping.*

The word "every," used in connection with the class of all covering spaces of a space $\mathfrak{V}$, might arouse the suspicions of a trained logician, because the notion of "all" covering spaces involves in particular the notion of all sets with a given cardinal number. We shall avoid this difficulty by showing that, $\mathfrak{V}$ being given, we can construct by legitimate logical processes a set $\mathfrak{C}$ of covering spaces of $\mathfrak{V}$ such that any given covering space can be proved to be isomorphic to one of these. We may then restrict the meaning of the "every" of Definition 1 to only those covering spaces which are elements of $\mathfrak{C}$.

Let p_0 be a point of $\mathfrak{V}$. We construct the set Σ composed of all finite sequences $S = (p_0, U_0, p_1, U_1, \cdots, p_n, U_n, p_{n+1})$ composed alternatively of points p_i and of open sets U_i and which satisfy the following conditions: 1) the first term of S is p_0; 2) both p_i and p_{i+1} belong to U_i $(0 \leqslant i \leqslant n)$; 3) the last term of S is a point. Let $(\tilde{\mathfrak{V}}, f)$ be a covering space of $\mathfrak{V}$; we shall prove that the cardinal number of $\tilde{\mathfrak{V}}$ is at most equal to the cardinal number of Σ. Let Σ' be the subset of Σ composed of the sequences $S = (p_0, U_0, p_1, U_1, \cdots, p_n, U_n, p_{n+1})$ which have the property that each U_i is evenly covered by $\tilde{\mathfrak{V}}$ with respect to f. Let $\tilde{p}_0$ be a fixed point of $\tilde{\mathfrak{V}}$ such that $f(\tilde{p}_0) = p_0$. If $S \varepsilon \Sigma'$, we define $\tilde{p}_k$ $(1 \leqslant k \leqslant n + 1)$ by induction on k in the following way: $\tilde{p}_k$ being already defined, we denote by $\tilde{U}_k$ the component of $\tilde{p}_k$ in $\overset{-1}{f}(U_k)$, and we denote by $\tilde{p}_{k+1}$ the point of $\tilde{U}_k$ which is mapped upon p_{k+1} by f ($\tilde{U}_k$ contains one and only one point with this property because U_k is evenly covered). We denote the point $\tilde{p}_{n+1}$ by $\varphi(S)$; φ is a mapping of Σ' into $\tilde{\mathfrak{V}}$. We shall prove that this mapping is *onto* $\tilde{\mathfrak{V}}$. Let $\tilde{p}$ be a point of $\tilde{\mathfrak{V}}$; the point $p = f(\tilde{p})$ belongs to at least one open set U which is evenly covered by $\tilde{\mathfrak{V}}$ (Lemma 3, §VI, p. 42), and the component $\tilde{U}$ of $\tilde{p}$ in $\overset{-1}{f}(U)$ is open (Proposition 1, §VI, p. 40). If $\tilde{U}$ meets $\varphi(\Sigma')$, it is entirely contained in $\varphi(\Sigma')$. In fact, assume that $\varphi(S) = \tilde{q}$ is a point of $\tilde{U}$, with $S = (p_0, U_0, p_1, U_1, \cdots, p_n, U_n, p_{n+1}) \varepsilon \Sigma'$. Then $p_{n+1} = f(\tilde{q})$ is in U and, r being any point of U, the sequence $S_r = (p_0, U_0, p_1, U_1, \cdots, p_n, U_n, p_{n+1}, U, r)$ belongs to Σ'; the point $\tilde{r} = \varphi(S_r)$ belongs to $\tilde{U}$, and $f(\tilde{r}) = r$, which proves our assertion. It follows that every point of $\tilde{\mathfrak{V}}$ which is adherent to $\varphi(\Sigma')$ is interior to $\varphi(\Sigma')$; therefore, $\varphi(\Sigma')$ is open and closed in $\tilde{\mathfrak{V}}$, whence $\varphi(\Sigma') = \tilde{\mathfrak{V}}$. We conclude that the cardinal number of $\tilde{\mathfrak{V}}$ is at most equal to the cardinal number of Σ', and, *a fortiori*, of Σ.

We can legitimately speak of the set of all topological spaces whose points are elements of Σ (such a space is determined by giving a subset A of Σ and a family of subsets of A). We may therefore construct the set of all pairs composed of a space $\tilde{\mathfrak{V}}$ whose points belong to Σ and of a mapping of $\tilde{\mathfrak{V}}$ into $\mathfrak{V}$. Finally, we may single out among these pairs those which are covering spaces of $\mathfrak{V}$; we obtain in this way a constructible set $\mathfrak{C}$ of covering spaces of $\mathfrak{V}$. It follows from what we have said that any given covering space of $\mathfrak{V}$ is isomorphic to a member of the set $\mathfrak{C}$.

The following lemma is sometimes useful to prove that a space is simply connected:

Lemma 1. *Let* $(\tilde{\mathfrak{V}}, f)$ *be a covering space of a space* $\mathfrak{V}$. *If* $\tilde{\mathfrak{V}}$ *contains an open set* A *which is mapped in a univalent way onto* $\mathfrak{V}$ *by* f, *then* f *is a homeomorphism of* $\tilde{\mathfrak{V}}$ *with* $\mathfrak{V}$.

Since f is continuous and interior, the contraction of f to A is a homeomorphism of A with $\mathfrak{V}$. Therefore, we have only to prove that $A = \tilde{\mathfrak{V}}$; this in turn will be established if we show that A is closed in $\tilde{\mathfrak{V}}$. Let $\tilde{p}$ be a point of $\tilde{\mathfrak{V}}$ which is adherent to A, and let V be a neighbourhood of $f(\tilde{p})$ which is evenly covered by $\tilde{\mathfrak{V}}$. We denote by $\tilde{V}$ the component of $\tilde{p}$ in $\overset{-1}{f}(V)$; both $\tilde{V}$ and the set $\tilde{V}' = A \cap \overset{-1}{f}(V)$ are mapped topologically onto V by f. On the other hand, $\tilde{V}$ is a neighbourhood of $\tilde{p}$ (Lemma 1, §VI, p. 41) and therefore meets A; it follows that $\tilde{V} \cap \tilde{V}' \neq \phi$. Since $\tilde{V}'$ is connected, we have $\tilde{V}' \subset \tilde{V}$; since f maps $\tilde{V}$ in a univalent way, we have $\tilde{V}' = \tilde{V}$, $\tilde{p} \varepsilon A$; Lemma 1 is thereby proved.

Proposition 1. *If* $\mathfrak{V}_1$ *and* $\mathfrak{V}_2$ *are simply connected spaces, the space* $\mathfrak{V}_1 \times \mathfrak{V}_2$ *is likewise simply connected.*

It is clear that $\mathfrak{V}_1 \times \mathfrak{V}_2$ is connected and locally connected. Let $(\mathfrak{W}, f)$ be a covering space of $\mathfrak{V}_1 \times \mathfrak{V}_2$; by a *horizontal fiber* we shall mean a component of a set of the form $\overset{-1}{f}(\mathfrak{V}_1 \times \{v_2\})$, with some $v_2 \varepsilon \mathfrak{V}_2$, by *a vertical fiber*, a component of a set of the form $\overset{-1}{f}(\{v_1\} \times \mathfrak{V}_2)$, with some $v_1 \varepsilon \mathfrak{V}_1$. It follows immediately from Lemma 5, §VI, p. 42 that f maps every horizontal fiber topologically onto $\mathfrak{V}_1$ (and every vertical fiber topologically onto $\mathfrak{V}_2$). Let $\tilde{\mathfrak{V}}_2^0$ be a fixed vertical fiber. We denote by A the union of the horizontal fibers which meet $\tilde{\mathfrak{V}}_2^0$. It is clear that f maps A in a univalent way onto $\mathfrak{V}_1 \times \mathfrak{V}_2$. Proposition 1 will follow from Lemma 1 above if we can prove that A is open in $\mathfrak{W}$.

Let $\tilde{\mathfrak{V}}_1$ be a horizontal fiber which is contained in A, and let E be the set of points of $\tilde{\mathfrak{V}}_1$ which are interior to A. The set E is relatively open in $\tilde{\mathfrak{V}}_1$. If we can prove that E is closed and not empty, it will follow that $E = \tilde{\mathfrak{V}}_1$ and that A is open.

Let w be any point of $\tilde{\mathfrak{V}}_1$; if $f(w) = (v_1, v_2)$, we can find connected neighbourhoods V_1, V_2 of v_1, v_2 with respect to $\mathfrak{V}_1$, $\mathfrak{V}_2$ respectively which are such that $V_1 \times V_2$ is evenly covered by $\mathfrak{W}$. Let W be the component of w in $\overset{-1}{f}(V_1 \times V_2)$. If $(v_1', v_2') \varepsilon V_1 \times V_2$, we set $\tilde{V}_1(v_2') = W \cap \overset{-1}{f}(V_1 \times \{v_2'\})$, $\tilde{V}_2(v_1') = W \cap \overset{-1}{f}(\{v_1'\} \times V_2)$; f maps $\tilde{V}_1(v_2')$ topologically onto $V_1 \times \{v_2'\}$ and $\tilde{V}_2(v_1')$ onto $\{v_1'\} \times V_2$. It follows that $\tilde{V}_1(v_2')$ is contained in a horizontal fiber $\tilde{\mathfrak{V}}_1(v_2')$ and that $\tilde{V}_2(v_1')$ is contained in a vertical fiber $\tilde{\mathfrak{V}}_2(v_1')$. The fibers $\tilde{\mathfrak{V}}_1(v_2')$ and $\tilde{\mathfrak{V}}_2(v_1')$ have in common the point of W which is mapped upon (v_1', v_2') by f. On the other hand, we have

$$W = \textstyle\bigcup_{v_2' \varepsilon V_2} \tilde{V}_1(v_2') = \bigcup_{v_1' \varepsilon V_1} \tilde{V}_2(v_1') \tag{1}$$

Assume first that w is the point where $\tilde{\mathfrak{V}}_1$ meets $\tilde{\mathfrak{V}}_2^0$; then $\tilde{\mathfrak{V}}_2(v_1)$ coincides with $\tilde{\mathfrak{V}}_2^0$, whence $\tilde{\mathfrak{V}}_1(v_2') \subset A$ for all $v_2' \varepsilon V_2$. It follows from Formula (1) that $W \subset A$, whence $w \varepsilon E$: E is not empty.

Assume next that w is adherent to E; then $\tilde{V}_1(v_2)$ has a point w' in common with E. We set $f(w') = (v_1^*, v_2)$. We can find a neighbourhood V_2^* of v_2 with respect to $\mathfrak{V}_2$ such that $V_2^* \subset V_2$ and that $W \cap \overset{-1}{f}(\{v_1^*\} \times V_2^*)$ is contained in A. If $v_2^* \varepsilon V_2^*$, the horizontal fiber $\tilde{\mathfrak{V}}_1(v_2^*)$ meets A, whence $\tilde{\mathfrak{V}}_1(v_2^*) \subset A$. The set $W \cap \overset{-1}{f}(V_1 \times V_2^*) = W'$ is the union of the sets $\tilde{V}_1(v_2^*)$ for all $v_2^* \varepsilon V_2^*$, whence $W' \subset A$. Since W' is a neighbourhood of w, we have $w \epsilon E$, which proves that E is closed. Proposition 1 is thereby proved.

We shall now prove the fundamental property of simply connected spaces:

Theorem 2 (Principle of monodromy). *Let $\mathfrak{V}$ be a simply connected space. Assume that we have assigned to every $p \varepsilon \mathfrak{V}$ a non empty set E_p (E_p is an abstract set, not related to $\mathfrak{V}$). Assume furthermore that we have assigned to every point (p, q) of a certain subset D of $\mathfrak{V} \times \mathfrak{V}$ a mapping φ_{pq} of E_p into E_q, in such a way that the following conditions are satisfied:*

1) *The set D is a connected neighbourhood of the diagonal in $\mathfrak{V} \times \mathfrak{V}$ (the diagonal is the set of all pairs (p, p), $p \varepsilon \mathfrak{V}$);*

2) *Each* φ_{pq} *is a one-to-one mapping of* E_p *onto* E_q; φ_{pp} *is the identity mapping;*

3) *If* φ_{pq}, φ_{qr}, φ_{pr} *are all defined, we have* $\varphi_{pr} = \varphi_{qr} \circ \varphi_{pq}$. *Then there exists a mapping* ψ *which assigns to every* $p \varepsilon \mathfrak{V}$ *an element* $\psi(p) \varepsilon E_p$ *in such a way that* $\psi(q) = \varphi_{pq}(\psi(p))$ *whenever* φ_{pq} *is defined. Moreover, if* p_0 *is a given point of* $\mathfrak{V}$, ψ *may be selected in such a way that* $\psi(p_0)$ *is any preassigned element* $e^0_{p_0}$ *of* E_{p_0}, *and* ψ *is then uniquely determined.*

Let $\tilde{V}$ be the union of all sets $\{p\} \times E_p$, for $p \varepsilon \mathfrak{V}$. We shall define a topology in $\tilde{V}$. Let $\mathfrak{U}$ be the family of those subsets $\tilde{U}$ of $\tilde{V}$ which satisfy the following condition: if $(p, e_p) \varepsilon \tilde{U}$, there exists a neighbourhood N of p in $\mathfrak{V}$ such that $N \times N \subset D$ and $(q, \varphi_{pq}e_p) \varepsilon \tilde{U}$ for all $q \varepsilon N$. It is clear that $\tilde{V}$ and the empty set belong to $\mathfrak{U}$, that any union of sets of $\mathfrak{U}$ belongs to $\mathfrak{U}$ and that the intersection of two sets of $\mathfrak{U}$ belongs to $\mathfrak{U}$.

Let $\bar{\omega}$ be the mapping of $\tilde{V}$ into $\mathfrak{V}$ which is defined by $\bar{\omega}(p, e_p) = p$. The following statements follow immediately from the definition: if U is an open subset of $\mathfrak{V}$, $\overset{-1}{\bar{\omega}}(U)$ belongs to $\mathfrak{U}$; if $\tilde{U} \varepsilon \mathfrak{U}$, $\bar{\omega}(\tilde{U})$ is open.

Let U be an open subset of $\mathfrak{V}$ such that $U \times U \subset D$. Let p be a point of U and let e_p be an element of E_p; we denote by $\tilde{U}(p, U, e_p)$ the set composed of the $(q, \varphi_{pq}(e_p))$ for $q \varepsilon U$. This set belongs to $\mathfrak{U}$. In fact, let $(q, \varphi_{pq}(e_p))$ be a point of $\tilde{U}(p, U, e_p)$; if r is any point of U, φ_{pq}, φ_{qr} and φ_{pr} are defined, and we have

$$(r, \varphi_{qr}(\varphi_{pq}(e_p))) = (r, \varphi_{pr}(e_p)) \varepsilon \tilde{U}(p, U, e_p),$$

which proves our assertion.

Let (p, e_p) and $(p', e'_{p'})$ be distinct points of $\tilde{V}$. If $p \neq p'$, we can find open sets U', U'' in $\mathfrak{V}$ such that $p \varepsilon U'$, $p' \varepsilon U''$, $U' \cap U'' = \phi$. We have $(p, e_p) \varepsilon \overset{-1}{\bar{\omega}}(U')$, $(p', e'_{p'}) \varepsilon \overset{-1}{\bar{\omega}}(U'')$, $\overset{-1}{\bar{\omega}}(U') \cap \overset{-1}{\bar{\omega}}(U'') = \phi$. Assume now that $p = p'$, whence $e_p \neq e'_p$. If U is an open set containing p such that $U \times U \subset D$, the sets $\tilde{U}(p, U, e_p)$ and $\tilde{U}(p, U, e'_p)$ belong to $\mathfrak{U}$ and are disjoint because the mappings φ_{pq} are univalent. It follows that $\mathfrak{U}$ can be taken as the family of open sets in a topology on $\tilde{V}$. Let $\tilde{\mathfrak{V}}$ be the topological space obtained in this way. It is clear that $\bar{\omega}$ is a continuous interior mapping of $\tilde{\mathfrak{V}}$ onto $\mathfrak{V}$.

Every point $p \varepsilon \mathfrak{V}$ has a connected open neighbourhood U such that $U \times U \subset D$. The set $\overset{-1}{\bar{\omega}}(U)$ is the union of the sets $\tilde{U}(p, U, e_p)$ for all $e_p \varepsilon E_p$ (because each φ_{pq} maps E_p onto E_q). These sets are open in $\tilde{\mathfrak{V}}$ and $\bar{\omega}$ maps each of them in a univalent way onto U. It follows that $\bar{\omega}$ maps each $\tilde{U}(p, U, e_p)$ topologically onto U. Since U is connected, the sets $\tilde{U}(p, U, e_q)$ are the components of $\overset{-1}{\bar{\omega}}(U)$, and U is evenly covered by $\mathfrak{B}$ with respect to $\bar{\omega}$.

Let $\tilde{\mathfrak{B}}_0$ be the component of $(p_0, e^0_{p_0})$ in $\tilde{\mathfrak{B}}$, and let $\bar{\omega}_0$ be the contraction of $\bar{\omega}$ to $\tilde{\mathfrak{B}}_0$. By Lemma 5, §VI, p. 42, $(\tilde{\mathfrak{B}}_0, \bar{\omega}_0)$ is a covering space of $\mathfrak{B}$. Since $\mathfrak{B}$ is simply connected, $\bar{\omega}_0$ is a homeomorphism. We define the mapping ψ by

$$\overset{-1}{\bar{\omega}_0}(p) = (p, \psi(p))$$

Let D^* be the set of all pairs $(p, q)\epsilon D$ such that $\psi(q) = \varphi_{pq}(\psi(p))$, and let (p_1, q_1) be any element of D. We can find open connected neighbourhoods U_1 and V_1 of p_1 and q_1 such that $U_1 \times U_1 \subset D$, $V_1 \times V_1 \subset D$, $U_1 \times V_1 \subset D$. Assume that $U_1 \times V_1$ has an element (p_2, q_2) in common with D^*, i.e. $\psi(q_2) = \varphi_{p_2q_2}(\psi(p_2))$. The set $\tilde{U}(p_1, U_1, \psi(p_1))$ is connected, and therefore contained in $\tilde{\mathfrak{B}}_0$, whence

$$\psi(p_2) = \varphi_{p_1p_2}(\psi(p_1)) \tag{1}$$

The mappings $\varphi_{p_1p_2}, \varphi_{p_2q_2}, \varphi_{p_1q_2}, \varphi_{q_1q_2}, \varphi_{p_1q_1}$ being defined, we have

$$\psi(q_2) = \varphi_{p_2q_2}(\varphi_{p_1p_2}(\psi(p_1))) = \varphi_{p_1q_2}(\psi(p_1)) = \varphi_{q_1q_2}(\varphi_{p_1q_1}(\psi(p_1))) \tag{2}$$

On the other hand, we have (by the same argument which was used to prove (1)):

$$\psi(q_2) = \varphi_{q_1q_2}(\psi(q_1)) \tag{3}$$

Because $\varphi_{p_1q_1}$ is a one-to-one mapping of E_{p_1} onto E_{q_1}, it follows from (2) and (3) that $\varphi_{p_1q_1}(\psi(p_1)) = \psi(q_1)$, i.e. $(p_1, q_1)\varepsilon D^*$. It follows immediately that D^* is relatively open and closed in D, whence $D^* = D$.

It remains only to prove the uniqueness of the mapping ψ. Let ψ' be any mapping which satisfies the same conditions as ψ (including $\psi'(p_0) = e^0_{p_0}$). Let A be the set of points p such that $\psi'(p) = \psi(p)$; we know already that A is not empty. Let p be any point of $\mathfrak{B}$ and let N be a neighbourhood of p such that $N \times N \subset D$. Assume that N has a point p_1 in common with A; then $\varphi_{p_1p}(\psi'(p)) = \psi'(p_1) = \psi(p_1) = \varphi_{p_1p}(\psi(p))$, whence $\psi(p) = \psi'(p)$. It follows immediately that A is open and closed in $\mathfrak{B}$, whence $A = \mathfrak{B}$. Theorem 2 is proved.

Definition 2. *Let $\mathfrak{G}$ be a topological group. A local homomorphism of $\mathfrak{G}$ into a group H is a mapping η of a neighbourhood V of the neutral element of $\mathfrak{G}$ into H which satisfies the condition that $\eta(\sigma\tau) = \eta(\sigma)\eta(\tau)$ whenever σ, τ and $\sigma\tau$ belong to V.*

Theorem 3. *Let $\mathfrak{G}$ be a simply connected topological group. Let η be a local homomorphism of $\mathfrak{G}$ into a group H. If the set on which η is defined is connected, it is possible to extend η to a homomorphism of the whole of $\mathfrak{G}$ into H.*

Let V be the set on which η is defined. We denote by D the subset of $\mathfrak{G} \times \mathfrak{G}$ composed of the pairs (σ, τ) such that $\tau\sigma^{-1} \varepsilon V$. It is clear that D is a neighbourhood of the diagonal in $\mathfrak{G} \times \mathfrak{G}$. Moreover, D may be represented as the union of the sets $\{\sigma\} \times V\sigma$, $\sigma \varepsilon \mathfrak{G}$. Since each of these sets is connected (and meets the diagonal of $\mathfrak{G} \times \mathfrak{G}$, which is connected), it follows that D is connected.

If $(\sigma, \tau) \varepsilon D$, we denote by $\varphi_{\sigma\tau}$ the mapping $\alpha \rightarrow \eta(\tau\sigma^{-1})\alpha$ of H into itself. If (σ, τ), (τ, ρ) and (σ, ρ) are all three in D, we have

$$\varphi_{\sigma\rho}(\alpha) = \eta(\rho\sigma^{-1})\alpha = \eta(\rho\tau^{-1})\eta(\tau\sigma^{-1})\alpha = \varphi_{\tau\rho}(\varphi_{\sigma\rho}(\alpha))$$

We can therefore apply Theorem 2 and obtain a mapping ψ of $\mathfrak{G}$ into H such that $\psi(\epsilon)$ is the neutral element of H and such that

$$\psi(\tau) = \eta(\tau\sigma^{-1})\psi(\sigma)$$

whenever $\tau\sigma^{-1}\varepsilon V$. Putting $\sigma = \epsilon$, we see that ψ coincides with η on V. The formula $\psi(\zeta\sigma) = \psi(\zeta)\psi(\sigma)$ holds when $\zeta \varepsilon V$. Since $\mathfrak{G}$ is connected, every $\rho \varepsilon \mathfrak{G}$ may be written in the form $\zeta_1 \cdots \zeta_h$ with $\zeta_i \varepsilon V (1 \leqslant i \leqslant h)$ (this follows easily from Theorem 1, §IV, p. 35 if we observe that $V \cap V^{-1}$ is a neighbourhood of ϵ). An easy induction shows that

$$\psi(\zeta_1 \cdots \zeta_h\sigma) = \psi(\zeta_1) \cdots \psi(\zeta_h)\psi(\sigma)$$

Putting $\sigma = \varepsilon$, we get $\psi(\rho) = \psi(\zeta_1) \cdots \psi(\zeta_h)$, whence $\psi(\rho\sigma) = \psi(\rho)\psi(\sigma)$: ψ is a homomorphism. Theorem 3 is thereby proved.

Scholium. *Let $\mathfrak{G}$ be a simply connected topological group. If a connected topological group $\mathfrak{G}_1$ is locally isomorphic to $\mathfrak{G}$, $\mathfrak{G}_1$ is isomorphic to the factor group of $\mathfrak{G}$ by a discrete subgroup of the center of $\mathfrak{G}$.*

Let η be a local isomorphism of a connected neighbourhood of the neutral element ϵ of $\mathfrak{G}$ into $\mathfrak{G}_1$. Then η can be extended to a homomorphism ψ of $\mathfrak{G}$ into $\mathfrak{G}_1$. Because ψ is a homomorphism and is continuous at the neutral element, ψ is continuous everywhere. (Cf. Proposition 5, §III, p. 35.) The set $\psi(\mathfrak{G})$ is a subgroup of $\mathfrak{G}_1$ and contains a neighbourhood of the neutral element ϵ_1 of $\mathfrak{G}_1$; $\mathfrak{G}_1$ being connected, it follows that $\psi(\mathfrak{G}) = \mathfrak{G}_1$ (Cf. Theorem 1, §IV, p. 35). Because ψ maps a neighbourhood of ϵ in $\mathfrak{G}$ onto a neighbourhood of ϵ_1 in $\mathfrak{G}_1$, ψ maps any open set onto an open set. It follows easily that $\mathfrak{G}_1$ is isomorphic to $\mathfrak{G}/K$, where K is the kernel of ψ.

There exists a neighbourhood of ϵ in $\mathfrak{G}$ whose intersection with K contains only ϵ; therefore K is discrete. The fact that K belongs to the center of $\mathfrak{G}$ follows from

Proposition 2. *A discrete distinguished subgroup K of a connected topological group $\mathfrak{G}$ belongs to the center of $\mathfrak{G}$.*

Let κ be any element of K. Let N be a neighbourhood of κ in $\mathfrak{G}$ such that $N \cap K = \{\kappa\}$, and let V be a neighbourhood of ϵ in $\mathfrak{G}$ such that $V\kappa V^{-1} \subset N$. Because K is distinguished, we have $\sigma\kappa\sigma^{-1} = \kappa$ for all $\sigma \varepsilon V$. The elements of $\mathfrak{G}$ which commute with κ form a subgroup G' of $\mathfrak{G}$ which contains V. Since $\mathfrak{G}$ is connected, we have $G' = \mathfrak{G}$. Proposition 2 is thereby proved.

§VIII. The Poincaré Group, Covering Groups

Let $\mathfrak{V}$ be a space which admits a simply connected covering space (i.e. a covering space $(\tilde{\mathfrak{V}}, f)$ such that $\tilde{\mathfrak{V}}$ is simply connected). We shall then prove that this covering space is unique, except for isomorphisms.

We must first prove

Proposition 1. *Let $\mathfrak{W}$ be a simply connected space. Let $\mathfrak{V}$ be a space, and assume that $(\tilde{\mathfrak{V}}, f)$ is a covering space of $\mathfrak{V}$. Let φ be a continuous mapping of $\mathfrak{W}$ into $\mathfrak{V}$. Then there exists a continuous mapping $\tilde{\varphi}$ of $\mathfrak{W}$ into $\tilde{\mathfrak{V}}$ such that $\varphi = f \circ \tilde{\varphi}$. If w_0 is a point of $\mathfrak{W}$ and if $\tilde{p}_0$ is any point of $\tilde{\mathfrak{V}}$ such that $f(\tilde{p}_0) = p_0 = \varphi(w_0)$, $\tilde{\varphi}$ may be constructed so as to map w_0 upon $\tilde{p}_0$, and is then uniquely determined.*

Let $\mathfrak{X}_1$ be the set of pairs $(w, \tilde{p}) \varepsilon \mathfrak{W} \times \tilde{\mathfrak{V}}$ such that $\varphi(w) = f(\tilde{p})$. The contraction to $\mathfrak{X}_1$ of the projection of $\mathfrak{W} \times \tilde{\mathfrak{V}}$ onto $\mathfrak{W}$ is a continuous mapping ψ_1 of $\mathfrak{X}_1$ onto $\mathfrak{W}$. If $w \varepsilon \mathfrak{W}$, we can find a connected neighbourhood V of $p = \varphi(w)$ in $\mathfrak{V}$ which is evenly covered by $\tilde{\mathfrak{V}}$; let $\tilde{V}_\nu$ be the components of $\overset{-1}{f}(V)$, ν running over some set of indices. Let W be a connected neighbourhood of w in $\mathfrak{W}$ such that $\varphi(W) \subset V$; if $w' \varepsilon W$, we denote by $\tilde{w}'_\nu$ the point $(w', \tilde{p}'_\nu)$, where $\tilde{p}'_\nu$ is the point of $\tilde{V}_\nu$ defined by $f(\tilde{p}'_\nu) = \varphi(w')$. The mapping $w' \to \tilde{w}'_\nu$ maps W continuously upon a subset $\tilde{W}_\nu$ of $\mathfrak{X}_1$ and $\psi_1(\tilde{w}'_\nu)$ is the point w'. It follows that ψ_1 maps $\tilde{W}_\nu$ topologically onto W. The set $\overset{-1}{\psi_1}(W)$ is the union of the sets $\tilde{W}_\nu$; if $\tilde{W}'$ is any connected subset of $\overset{-1}{\psi_1}(W)$, the mapping $(w, \tilde{p}) \to \tilde{p}$ maps $\tilde{W}'$ onto a connected subset of $\overset{-1}{f}(V)$, i.e. upon a subset of some $\tilde{V}_\nu$. It follows that the sets $\tilde{W}_\nu$ are the components of $\overset{-1}{\psi_1}(W)$ and that W is evenly covered by $\mathfrak{X}_1$ with respect to ψ_1.

Let $\mathfrak{X}$ be the component of $(w_0, \tilde{p}_0)$ in $\mathfrak{X}_1$, and let ψ be the contraction of ψ_1 to $\mathfrak{X}$. By Lemma 5, §VI, p. 42, $(\mathfrak{X}, \psi)$ is a covering space of $\mathfrak{W}$. Since $\mathfrak{W}$ is simply connected, ψ is a homeomorphism. We can now define $\tilde{\varphi}$ by the formula $\overset{-1}{\psi}(w) = (w, \tilde{\varphi}(w))$. It is clear that $\tilde{\varphi}$ has the required properties. The uniqueness of $\tilde{\varphi}$ will follow from

Lemma 1. *Assume that $(\tilde{\mathfrak{V}}, f)$ is a covering space of a space $\mathfrak{V}$. Let φ, φ' be continuous mappings of a connected space $\mathfrak{W}$ into $\tilde{\mathfrak{V}}$ such that $f \circ \varphi = f \circ \varphi'$. If $\varphi(w_0) = \varphi'(w_0)$ for at least one point w_0, then $\varphi = \varphi'$.*

Let A be the set of points such that $\varphi(w) = \varphi'(w)$; A is clearly closed and not empty. Lemma 1 will be proved if we can show that A is open. If $w \varepsilon A$, then $v = f(\varphi(w))$ has a neighbourhood V which is evenly covered by $\tilde{\mathfrak{V}}$. The component $\tilde{V}$ of $\varphi(w) = \varphi'(w)$ in $\overset{-1}{f}(V)$ is a neighbourhood of $\varphi(w)$ in $\tilde{\mathfrak{V}}$ (Cf. Lemma 1, §VI, p. 41). It follows that there exists a neighbourhood W of w in $\mathfrak{W}$ such that $\varphi(W) \subset \tilde{V}$, $\varphi'(W) \subset \tilde{V}$. Because f maps $\tilde{V}$ topologically, $w' \varepsilon W$ implies $\varphi(w') = \varphi'(w')$, whence $W \subset A$: Lemma 1 is proved.

Remark. The statement of Proposition 1 is to a certain degree similar to the principle of monodromy. In fact, Proposition 1 can be deduced from the principle of monodromy in the case where we assume that $\mathfrak{W}$ is normal.

Let now $(\tilde{\mathfrak{V}}, f)$ and $(\tilde{\mathfrak{V}}', f')$ be simply connected covering spaces of a space $\mathfrak{V}$. Let p be a point of $\mathfrak{V}$, and let $\tilde{p}$, $\tilde{p}'$ be points of $\tilde{\mathfrak{V}}$, $\tilde{\mathfrak{V}}'$ respectively such that $f(\tilde{p}) = p$, $f'(\tilde{p}') = p$. By Proposition 1, there exist continuous mappings; φ of $\tilde{\mathfrak{V}}$ into $\tilde{\mathfrak{V}}'$, and φ' of $\tilde{\mathfrak{V}}'$ into $\tilde{\mathfrak{V}}$ such that

$$f' \circ \varphi = f; \qquad f \circ \varphi' = f'; \qquad \varphi(\tilde{p}) = \tilde{p}'; \qquad \varphi'(\tilde{p}') = \tilde{p}$$

Then, $\varphi' \circ \varphi$ is a continuous mapping θ of $\tilde{\mathfrak{V}}$ into itself such that $f \circ \theta = f$ and $\theta(\tilde{p}) = \tilde{p}$. By Proposition 1, θ is the identity mapping of $\tilde{\mathfrak{V}}$. In the same way, we see that $\varphi \circ \varphi'$ is the identity mapping of $\tilde{\mathfrak{V}}'$. It follows that φ is a homeomorphism and that $\varphi' = \overset{-1}{\varphi}$. We have proved

Proposition 2. *If a space $\mathfrak{V}$ admits a simply connected covering space, it admits only one (except for isomorphisms).*

Let us furthermore apply our previous considerations to the case where $\tilde{\mathfrak{V}} = \tilde{\mathfrak{V}}'$, $f = f'$. We see that, if $\tilde{p}$, $\tilde{p}'$ are points of $\tilde{\mathfrak{V}}$ such that $f(\tilde{p}) = f(\tilde{p}')$, there exists a uniquely defined homeomorphism φ of $\tilde{\mathfrak{V}}$ with itself such that $f \circ \varphi = f$, $\varphi(\tilde{p}) = \tilde{p}'$.

Definition 1. *Let* $\mathfrak{V}$ *be a space which admits a simply connected covering space* $(\tilde{\mathfrak{V}}, f)$. *The group of those homeomorphisms* φ *of* $\tilde{\mathfrak{V}}$ *with itself such that* $f \circ \varphi = f$ *is called the Poincaré group (or fundamental group) of* $\mathfrak{V}$.

The Poincaré groups of any two simply connected covering spaces of the same space $\mathfrak{V}$ are isomorphic (as follows immediately from Proposition 2). If we think of the abstract group which is isomorphic to the Poincaré group of any simply connected covering space, we call this group the Poincaré group (or fundamental group) of $\mathfrak{V}$. We have proved

Proposition 3. *Assume that the space* $\mathfrak{V}$ *admits a simply connected covering space* $(\tilde{\mathfrak{V}}, f)$. *If* $\tilde{p}$, $\tilde{p}'$ *are points of* $\tilde{\mathfrak{V}}$ *such that* $f(\tilde{p}') = f(\tilde{p})$, *then there exists a uniquely determined operation of the Poincaré group of* $\mathfrak{V}$ *which maps* $\tilde{p}$ *upon* $\tilde{p}'$.

Lemma 2. *Let* $\mathfrak{V}_1$, $\mathfrak{V}_2$ *be two spaces; assume that* $(\tilde{\mathfrak{V}}_i, f_i)$ *is a covering space of* $\mathfrak{V}_i$ $(i = 1, 2)$. *If we set* $f(\tilde{v}_1, \tilde{v}_2) = (f_1(\tilde{v}_1), f_2(\tilde{v}_2))$, *then* $(\tilde{\mathfrak{V}}_1 \times \tilde{\mathfrak{V}}_2, f)$ *is a covering space of* $\tilde{\mathfrak{V}}_1 \times \tilde{\mathfrak{V}}_2$.

Let $v = (v_1, v_2)$ be any point of $\mathfrak{V}_1 \times \mathfrak{V}_2$. We can find a neighbourhood V_i of v_i with respect to $\mathfrak{V}_i$ which is evenly covered by $\tilde{\mathfrak{V}}_i$ $(i = 1, 2)$. If $\tilde{V}_i$ is any component of $\overset{-1}{f_i}(V_i)$ $(i = 1, 2)$, f maps $\tilde{V}_1 \times \tilde{V}_2$ topologically onto $V_1 \times V_2$. The set $\tilde{V}_1 \times \tilde{V}_2$, being connected, is contained in a component $\tilde{V}$ of $\overset{-1}{f}(V_1 \times V_2)$; the projection of $\tilde{\mathfrak{V}}_1 \times \tilde{\mathfrak{V}}_2$ onto $\tilde{\mathfrak{V}}_i$ maps $\tilde{V}$ into a component of $\overset{-1}{f_i}(V_i)$, whence $\tilde{V} = \tilde{V}_1 \times \tilde{V}_2$. Any point of $\overset{-1}{f}(V_1 \times V_2)$ belongs to a set of the form $\tilde{V}_1 \times \tilde{V}_2$; it follows that the sets $\tilde{V}_1 \times \tilde{V}_2$ are the components of $\overset{-1}{f}(V_1 \times V_2)$ and that $V_1 \times V_2$ is evenly covered by $\tilde{\mathfrak{V}}_1 \times \tilde{\mathfrak{V}}_2$ with respect to f. Since $\tilde{\mathfrak{V}}_1 \times \tilde{\mathfrak{V}}_2$ is connected and locally connected, Lemma 2 is proved.

Proposition 4. *Assume that the spaces* $\mathfrak{V}_1$, $\mathfrak{V}_2$ *both admit simply connected covering spaces. Then* $\mathfrak{V}_1 \times \mathfrak{V}_2$ *admits a simply connected covering space and its Poincaré group is isomorphic to the product of the Poincaré groups of* $\mathfrak{V}_1$ *and* $\mathfrak{V}_2$.

Let $(\tilde{\mathfrak{V}}_i, f_i)$ be a simply connected covering space of $\mathfrak{V}_i$ $(i = 1, 2)$. If we define the mapping f as in Lemma 2, $(\tilde{\mathfrak{V}}_1 \times \tilde{\mathfrak{V}}_2, f)$ is a covering space of $\mathfrak{V}_1 \times \mathfrak{V}_2$, and is simply connected by Proposition 1, §VII, p. 45. Let F_i be the fundamental group of $(\tilde{\mathfrak{V}}_i, f_i)$ $(i = 1, 2)$; if $\varphi_i \varepsilon F_i$, the mapping φ of $\tilde{\mathfrak{V}}_1 \times \tilde{\mathfrak{V}}_2$ into itself defined by $\varphi(\tilde{v}_1, \tilde{v}_2) = (\varphi_1(\tilde{v}_1), \varphi_2(\tilde{v}_2))$ clearly belongs to the Poincaré group F of $(\tilde{\mathfrak{V}}_1 \times \tilde{\mathfrak{V}}_2, f)$. The mapping $(\varphi_1, \varphi_2) \to \varphi$ is easily seen to be an isomorphism of $F_1 \times F_2$ with a subgroup of F. If $(\tilde{v}_1, \tilde{v}_2)$ and $(\tilde{v}'_1, \tilde{v}'_2)$ are any points of $\tilde{\mathfrak{B}}_1 \times \tilde{\mathfrak{B}}_2$ such that $f(\tilde{v}_1, \tilde{v}_2) = f(\tilde{v}'_1, \tilde{v}'_2)$, we have $f_i(\tilde{v}_i) = f_i(\tilde{v}'_i)$ $(i = 1, 2)$, and there exists an element $\varphi_i \varepsilon F_i$ such that $\varphi_i(\tilde{v}_i) = \tilde{v}'_i$. The operation φ of F which corresponds to (φ_1, φ_2) maps $(\tilde{v}_1, \tilde{v}_2)$ upon $(\tilde{v}'_1, \tilde{v}'_2)$. Taking Proposition 3 into account, it follows that the mapping $(\varphi_1, \varphi_2) \to \varphi$ maps $F_1 \times F_2$ *onto* F. Proposition 4 is proved.

Let us now consider the notion of a covering space in the case where the space which is covered is a topological group.

Definition 2. *Let $\mathfrak{G}$ be a topological group. By a covering group of $\mathfrak{G}$, we mean a pair $(\tilde{\mathfrak{G}}, f)$ composed of a topological group $\tilde{\mathfrak{G}}$ and of a homomorphism f of $\tilde{\mathfrak{G}}$ into $\mathfrak{G}$ such that $(\tilde{\mathfrak{G}}, f)$ is a covering space of $\mathfrak{G}$.*

Proposition 5. *Assume that a topological group $\mathfrak{G}$ has a simply connected covering space $(\tilde{\mathfrak{G}}, f)$. It is then possible to define a multiplication in $\tilde{\mathfrak{G}}$ which turns the space $\tilde{\mathfrak{G}}$ into a topological group and the covering space $(\tilde{\mathfrak{G}}, f)$ into a covering group.*

Let ϵ be the neutral element of $\mathfrak{G}$, and let $\tilde{\epsilon}$ be any element of $\tilde{\mathfrak{G}}$ such that $f(\tilde{\epsilon}) = \epsilon$. The space $\tilde{\mathfrak{G}} \times \tilde{\mathfrak{G}}$ is simply connected (Proposition 1, §VII, p. 45); by Proposition 1 above, there exists a continuous mapping φ of $\tilde{\mathfrak{G}} \times \tilde{\mathfrak{G}}$ into $\tilde{\mathfrak{G}}$ such that $f(\varphi(\tilde{\sigma}, \tilde{\tau})) = f(\tilde{\sigma})(f(\tilde{\tau}))^{-1}$, $\varphi(\tilde{\epsilon}, \tilde{\epsilon}) = \tilde{\epsilon}$. We have $f(\varphi(\tilde{\sigma}, \tilde{\epsilon})) = f(\tilde{\sigma})$ and $\varphi(\tilde{\epsilon}, \tilde{\epsilon}) = \tilde{\epsilon}$. Making use of the uniqueness statement in Proposition 1 (applied this time to the mapping $\tilde{\sigma} \to f(\varphi(\tilde{\sigma}, \tilde{\epsilon}))$), we conclude that $\varphi(\tilde{\sigma}, \tilde{\epsilon}) = \tilde{\sigma}$. We set

$$\tilde{\tau}^{-1} = \varphi(\tilde{\epsilon}, \tilde{\tau}) \qquad \tilde{\sigma}\tilde{\tau} = \varphi(\tilde{\sigma}, \tilde{\tau}^{-1})$$

whence $f(\tilde{\tau}^{-1}) = (f(\tilde{\tau}))^{-1}$, $f(\tilde{\sigma}\tilde{\tau}) = f(\tilde{\sigma})f(\tilde{\tau})$. Making use again of the uniqueness statement in Proposition 1, we derive easily the formulas $(\tilde{\sigma}\tilde{\tau})\tilde{\rho} = \tilde{\sigma}(\tilde{\tau}\tilde{\rho})$, $\tilde{\sigma}\tilde{\epsilon} = \tilde{\epsilon}\tilde{\sigma} = \tilde{\sigma}$. The continuous mapping $\tilde{\sigma} \to \tilde{\sigma}\tilde{\sigma}^{-1}$ maps the connected space $\tilde{\mathfrak{G}}$ into the discrete space $\overset{-1}{f}(\epsilon)$ and maps $\tilde{\epsilon}$ upon itself, whence $\tilde{\sigma}\tilde{\sigma}^{-1} = \tilde{\epsilon}$; in the same way, we would prove that $\tilde{\sigma}^{-1}\tilde{\sigma} = \tilde{\epsilon}$. It follows that our law of composition $(\tilde{\sigma}, \tilde{\tau}) \to \tilde{\sigma}\tilde{\tau}$ turns $\tilde{\mathfrak{G}}$ into a topological group: Proposition 5 is proved.

The law of composition which we have defined in $\tilde{\mathfrak{G}}$ depends upon the choice of $\tilde{\epsilon}$. Neverless, we shall prove

Proposition 6. *If a topological group admits a simply connected covering group $(\tilde{\mathfrak{G}}, f)$, this covering group is unique except for isomorphisms;* i.e. *if $(\tilde{\mathfrak{G}}', f')$ is an other simply connected covering group of $\mathfrak{G}$, there exists an isomorphism θ of the topological group $\tilde{\mathfrak{G}}$ with $\tilde{\mathfrak{G}}'$ such that $f = f' \circ \theta$.*

Let $\tilde{\epsilon}$, $\tilde{\epsilon}'$, ϵ be the neutral elements of $\tilde{\mathfrak{G}}$, $\tilde{\mathfrak{G}}'$, $\mathfrak{G}$ respectively. We can find neighbourhoods $\tilde{V}$, $\tilde{V}'$, V of $\tilde{\epsilon}$, $\tilde{\epsilon}'$, ϵ such that the contractions of f, f' to $\tilde{V}$, $\tilde{V}'$ are local isomorphisms of these sets with V. It follows that we can find a local isomorphism θ of $\tilde{V}$ with $\tilde{V}'$ and a local isomorphism θ' of $\tilde{V}'$ with $\tilde{V}$ such that $\theta' \circ \theta$ and $\theta \circ \theta'$ are the identity mappings of $\tilde{V}$, $\tilde{V}'$ respectively and furthermore such that $f' \circ \theta$ coincides with f on $\tilde{V}$. By Theorem 3, §VII, p. 49, θ and θ' may be extended to homomorphisms of $\tilde{\mathfrak{G}}$ into $\tilde{\mathfrak{G}}'$ and of $\tilde{\mathfrak{G}}'$ into $\tilde{\mathfrak{G}}$ respectively; we shall also denote by θ, θ' these extended homomorphisms. Because $\tilde{V}$, $\tilde{V}'$ are sets of generators of $\tilde{\mathfrak{G}}$, $\tilde{\mathfrak{G}}'$ respectively (Theorem 1, §IV, p. 35), $\theta' \circ \theta$ and $\theta \circ \theta'$ are the identity mappings of $\tilde{\mathfrak{G}}$ and $\tilde{\mathfrak{G}}'$ respectively. It follows that θ is an isomorphism of $\tilde{\mathfrak{G}}$ (considered as a topological group) with $\tilde{\mathfrak{G}}'$. Moreover, since f and $f' \circ \theta$ are both homomorphisms of $\tilde{\mathfrak{G}}$ into $\mathfrak{G}$, they coincide everywhere. Proposition 6 is thereby proved.

Proposition 7. *Assume that the topological group $\mathfrak{G}$ admits a simply connected covering group $(\tilde{\mathfrak{G}}, f)$. Then the Poincaré group of $\mathfrak{G}$ is isomorphic to the kernel of the homomorphism f; in particular, this group is abelian.*

Let D be the kernel of f; if $\delta \varepsilon D$, the left translation φ_δ associated with δ in $\tilde{\mathfrak{G}}$ is a homeomorphism of $\tilde{\mathfrak{G}}$ with itself such that $f \circ \varphi_\delta = f$. It follows that φ_δ belongs to the Poincaré group of $(\tilde{\mathfrak{G}}, f)$. Making use of Proposition 3 above, we see easily that the mapping $\delta \rightarrow \varphi_\delta$ is an isomorphism of D with the Poincaré group of $(\tilde{\mathfrak{G}}, f)$. Proposition 7 then follows from Proposition 2, §VII, p. 50.

§IX. Existence of Simply Connected Covering Spaces

Definition 1. *A space $\mathfrak{V}$ is said to be locally simply connected if every point of $\mathfrak{V}$ has at least one simply connected neighbourhood.*

Observe that we do not require that every neighbourhood of the point should contain a simply connected neighbourhood.

A locally simply connected space is of course *ipso facto* locally connected.

Theorem 4. *Let $\mathfrak{V}$ be a connected and locally simply connected space. Then $\mathfrak{V}$ has a simply connected covering space.*

We select arbitrarily a point $v_0 \varepsilon \mathfrak{V}$. By a *specified covering space* of $\mathfrak{V}$ we shall mean a triple $(\tilde{\mathfrak{V}}, \tilde{v}^0, f)$ such that $(\tilde{\mathfrak{V}}, f)$ is a covering space of $\mathfrak{V}$ and $\tilde{v}^0$ is a point of $\tilde{\mathfrak{V}}$ with $f(\tilde{v}^0) = v_0$. Two specified covering spaces $(\tilde{\mathfrak{V}}, \tilde{v}^0, f)$ and $(\tilde{\mathfrak{V}}_1, \tilde{v}_1^0, f_1)$ are said to be of the same type if there exists a homeomorphism η of $\tilde{\mathfrak{V}}$ with $\tilde{\mathfrak{V}}_1$ such that $f_1 \circ \eta = f$ and $\eta(\tilde{v}^0) = \tilde{v}_1^0$. We know that we can construct a set of covering spaces of $\mathfrak{V}$ such that every covering space is isomorphic to one of them; it follows easily that we can construct a complete system of representatives $(\tilde{\mathfrak{V}}_\alpha, \tilde{v}^0_\alpha, f_\alpha)$ for all types of specified covering spaces (α running over some set of indices). Let $\mathfrak{X}_1$ be the subset of the product space $\Pi_\alpha \tilde{\mathfrak{V}}_\alpha$ composed of those elements $x = (\cdot\cdot\cdot\ \tilde{v}_\alpha\ \cdot\cdot\cdot)$ for which the $f_\alpha(\tilde{v}_\alpha)$ are all equal; if $x \varepsilon \mathfrak{X}_1$, we denote by $f^*(x)$ the common value of the elements $f_\alpha(\tilde{v}_\alpha)$. It is clear that f^* is a continuous mapping of $\mathfrak{X}_1$ onto $\mathfrak{V}$ and that the point $x^0 = (\cdot\cdot\cdot\ \tilde{v}^0_\alpha\ \cdot\cdot\cdot)$ belongs to $\mathfrak{X}_1$.

Let v be a point of $\mathfrak{V}$, and let V be a simply connected neighbourhood of v. We denote by $\tilde{V}_{\alpha,\nu}$ the components of $\overset{-1}{f_\alpha}(V)$, ν running over a set of indices N_α which depends on α. It follows immediately from Lemma 5, §VI, p. 42 tht fa_α maps each $\tilde{V}_{\alpha,\nu}$ topologically onto V.

Now set $Z = \Pi_\alpha N_\alpha$; if $\zeta \varepsilon Z$, we denote by $\zeta(\alpha)$ the α-coordinate of ζ and set

$$\tilde{V}_\zeta = \mathfrak{X}_1 \cap \Pi_\alpha \tilde{V}_{\alpha,\zeta(\alpha)}$$

Let f_ζ be the contraction of f^* to $\tilde{V}_\zeta$. It is clear that f_ζ is univalent and continuous. On the other hand, if $v \varepsilon V$, the α-coordinate of $\overset{-1}{f_\zeta}(v)$ is the point of $\tilde{V}_{\alpha,\zeta(\alpha)}$ which is mapped on v under f_α; this point is a continuous function of v, and it follows that $\overset{-1}{f_\zeta}$ is continuous. Therefore f_ζ is a topological mapping of $\tilde{V}_\zeta$ onto V. In particular, we see that $\tilde{V}_\zeta$ is connected. Clearly,

$$\overset{-1}{f^*}(V) = \bigcup_{\zeta \varepsilon Z} \tilde{V}_\zeta$$

If we assign to every $x = (\cdot\cdot\cdot\ \tilde{v}_\alpha\ \cdot\cdot\cdot) \varepsilon \mathfrak{X}_1$ the element $g^*_\alpha(x) = \tilde{v}_\alpha$, we obtain a continuous mapping g^*_α of $\mathfrak{X}_1$ into $\tilde{\mathfrak{V}}_\alpha$. The set $\overset{-1}{g^*_\alpha}(\tilde{V}_{\alpha,\nu})$ is the union of the sets $\tilde{V}_\zeta$ for which $\zeta(\alpha) = \nu$. If $\tilde{K}$ is any component of $\overset{-1}{f^*}(V)$, $g^*_\alpha(\tilde{K})$ is a connected subset of $\overset{-1}{f_\alpha}(V)$ and is therefore contained in some $\tilde{V}_{\alpha,\nu}$. It follows that each $\tilde{V}_\zeta$ is a component of $\overset{-1}{f^*}(V)$ and that the components of $\overset{-1}{g^*_\alpha}(\tilde{V}_{\alpha,\nu})$ are the sets $\tilde{V}_\zeta$ such that $\zeta(\alpha) = \nu$. From the facts that f_ζ is a homeomorphism and that f_α and g^*_α are continuous, we deduce easily that g^*_α maps $\tilde{V}_\zeta$ topologically onto $\tilde{V}_{\alpha,\zeta(\alpha)}$.

Let $\mathfrak{X}'_1$ be the space which has the same points as $\mathfrak{X}_1$, and, for open sets, the unions of components of open sets of $\mathfrak{X}_1$ (cf. Lemma 6, §VI, p. 43). Then $\mathfrak{X}'_1$ is locally connected. Every point of $\mathfrak{V}$ has a neighbourhood which is evenly covered by $\mathfrak{X}'_1$ with respect to f^*;

every point of $\tilde{\mathfrak{V}}_\alpha$ has a neighbourhood which is evenly covered by $\mathfrak{X}'_1$ with respect to $\overset{*}{g}_\alpha$. Let $\mathfrak{X}$ be the component of x^0 in $\mathfrak{X}'_1$, and let f and g_α be the contractions of f^* and $\overset{*}{g}_\alpha$ to $\mathfrak{X}$. Then $(\mathfrak{X}, f)$ is a covering space of $\mathfrak{V}$ and $(\mathfrak{X}, g_\alpha)$ is a covering space of $\tilde{\mathfrak{V}}_\alpha$ (cf. Lemma 5, §VI, p. 40). Theorem 4 will be proved if we can show that $\mathfrak{X}$ is simply connected.

Let $(\tilde{\mathfrak{X}}, \varphi)$ be a covering space of $\mathfrak{X}$. We shall see that $(\tilde{\mathfrak{X}}, f \circ \varphi)$ is a covering space of $\mathfrak{V}$. Let x be any point of $\mathfrak{X}$, and let V be a simply connected neighbourhood of $f(x)$ in $\mathfrak{V}$. Denote by $\tilde{V}$ the component of x in $\overset{-1}{f}(V)$; then it follows from Lemma 1 and 5, §VI, p. 40 that $\tilde{V}$ is a simply connected neighbourhood of x in $\mathfrak{X}$ and is therefore evenly covered by $\tilde{\mathfrak{X}}$ with respect to φ. This proves our assertion.

Let $\tilde{x}^0$ be a point of $\tilde{\mathfrak{X}}$ such that $\varphi(\tilde{x}^0) = x^0$. Then $(\tilde{\mathfrak{X}}, \tilde{x}_0, f \circ \varphi)$ is a specified covering of $\mathfrak{V}$, and, as such, it is of the same type as $(\tilde{\mathfrak{V}}_\alpha, \tilde{v}^0_\alpha, f_\alpha)$ for some α. There exists a homeomorphism h of $\tilde{\mathfrak{V}}_\alpha$ with $\tilde{\mathfrak{X}}$ such that $(f \circ \varphi) \circ h = f_\alpha$ and $h(\tilde{v}^0_\alpha) = \tilde{x}^0$. We set $\psi = \varphi \circ h$; then $(\tilde{\mathfrak{V}}_\alpha, \psi)$ is a covering space of $\mathfrak{X}$. We have $\psi(\tilde{v}^0_\alpha) = x_0$ and $f \circ \psi = f_\alpha$. In order to prove that φ is a homeomorphism, it will be sufficient to prove that ψ is a homeomorphism. The mapping $g_\alpha \circ \psi$ maps $\tilde{\mathfrak{V}}_\alpha$ continuously into itself, and we have $f_\alpha \circ (g_\alpha \circ \psi) = f \circ \psi = f_\alpha$, $(g_\alpha \circ \psi)(\tilde{v}^0_\alpha) = \tilde{v}^0_\alpha$. By Lemma 1, §VIII, p. 50, it follows that $g^0_\alpha \circ \psi$ is the identity mapping of $\tilde{\mathfrak{V}}_\alpha$, which proves that ψ is univalent. Since $(\tilde{\mathfrak{V}}_\alpha, \psi)$ is a covering space of $\mathfrak{X}$, it follows that ψ is a homeomorphism. Theorem 4 is thereby proved.

§X. THE POINCARÉ GROUPS OF SOME SPECIAL SPACES

Proposition 1. *The additive group R of real numbers is simply connected.*

Let $(\tilde{R}, f)$ be a covering group of R. We can find a neighbourhood $\tilde{V}$ of the neutral element in $\tilde{R}$ which is mapped topologically by f upon an interval $]-a, +a[$ in R (with $a > 0$). Because f is a homomorphism, the elements of $\tilde{V} \cap \overset{-1}{f}(]-a/2, +a/2[)$ commute with each other. It follows that $\tilde{R}$ is abelian; we shall write the law of composition in $\tilde{R}$ additively. Let $\tilde{d}$ be an element of $\tilde{R}$ such that $f(\tilde{d}) = 0$; we may write $\tilde{d}$ in the form $\tilde{d}_1 + \cdots + \tilde{d}_h$ with $\tilde{d}_i \varepsilon \tilde{V}$ $(1 \leqslant i \leqslant h)$. Set $f(\tilde{d}_i) = d_i$; we have $d_1 + \cdots + d_h = 0$. Let k be an integer such that $|k^{-1}(d_1 + \cdots + d_i)| < a$ $(1 \leqslant i \leqslant h)$, and let $\tilde{x}_i$ be the element of $\tilde{V}$ which is mapped upon $k^{-1}d_i$ by f. Because f is a local isomorphism, we have $f(\tilde{x}_1 + \cdots + \tilde{x}_h) = f(\tilde{x}_1) + \cdots$

$+ f(\tilde{x}_h) = 0$, whence $\tilde{x}_1 + \cdots + \tilde{x}_h = \tilde{0}$ (where $\tilde{0}$ is the neutral element of $\tilde{R}$). Again because f is a local isomorphism, we have $\tilde{d}_i = k\tilde{x}_i$, whence $\tilde{d} = \Sigma\tilde{d}_i = k\Sigma\tilde{x}_i = \tilde{0}$. This proves that f is univalent. Proposition 1 is thereby proved.

Lemma 1. *Assume that $(\tilde{\mathfrak{V}}, f)$ is a covering space of the space $\mathfrak{V}$. Let A and B be two closed, connected and locally connected subsets of $\mathfrak{V}$ which are both evenly covered by $\tilde{\mathfrak{V}}$ with respect to f. If $A \cap B$ is connected and not empty, the set $A \cup B$ is evenly covered by $\tilde{\mathfrak{V}}$.*

Let $\tilde{A}_\nu$ be the components of $\overset{-1}{f}(A)$, ν running over some set of indices N (we assume that $\tilde{A}_\nu \neq \tilde{A}_{\nu'}$ for $\nu \neq \nu'$). Set $C = A \cap B$, $\tilde{A}_\nu \cap \overset{-1}{f}(C) = \tilde{C}_\nu$; we know that f maps $\tilde{A}_\nu$ topologically onto A; it follows that f maps $\tilde{C}_\nu$ topologically onto C. In particular, $\tilde{C}_\nu$ is connected; as such, it belongs to a uniquely determined component $\tilde{B}_\nu$ of $\overset{-1}{f}(B)$. It is clear that $\tilde{B}_\nu \neq \tilde{B}_{\nu'}$, for $\nu \neq \nu'$ and that every component of $\overset{-1}{f}(B)$ occurs among the sets $\tilde{B}_\nu$. We set $\tilde{K}_\nu = \tilde{A}_\nu \cup \tilde{B}_\nu$, $\tilde{L}_\nu = \bigcup_{\nu' \neq \nu} \tilde{K}_{\nu'}$; $\tilde{K}_\nu$ is clearly a closed set. We have $\tilde{L}_\nu = (\bigcup_{\nu' \neq \nu} \tilde{A}_{\nu'}) \cup (\bigcup_{\nu' \neq \nu} \tilde{B}_{\nu'})$; we know that $\tilde{A}_\nu$ is relatively open in $\overset{-1}{f}(A)$ and that $\tilde{B}_\nu$ is relatively open in $\overset{-1}{f}(B)$ (Cf. Lemma 5, §VI, p. 40). It follows that the sets $\bigcup_{\nu' \neq \nu'} \tilde{A}_\nu$, $\bigcup_{\nu' \neq \nu} \tilde{B}_{\nu'}$, and therefore also $\tilde{L}_\nu$, are closed.

Because $\tilde{K}_\nu \cup \tilde{L}_\nu = \overset{-1}{f}(A \cup B)$, $\tilde{K}_\nu$ is relatively open and closed in $\overset{-1}{f}(A \cup B)$. It follows that the sets $\tilde{K}_\nu$ are the components of $\overset{-1}{f}(A \cup B)$. If f_ν is the contraction of f to $\tilde{K}_\nu$, $(\tilde{K}_\nu, f_\nu)$ is a covering space of $A \cup B$ (Lemma 5, §VI, p. 40). On the other hand, f_ν maps $\tilde{K}_\nu$ in a univalent way. In fact, assume that $f(\tilde{p}) = f(\tilde{p}')$, with $\tilde{p}$, $\tilde{p}' \varepsilon \tilde{K}_\nu$. If p, p' belong to the same one of the sets $\tilde{A}_\nu$, $\tilde{B}_\nu$, we have clearly $\tilde{p} = \tilde{p}'$; if not, we have $f(\tilde{p}) = f(\tilde{p}') \varepsilon C$ and again $\tilde{p} = \tilde{p}'$ because f_ν maps both $\tilde{A}_\nu$ and $\tilde{B}_\nu$ in a univalent way. It follows that f maps $\tilde{K}_\nu$ topologically onto $A \cup B$: Lemma 1 is proved.

Proposition 2. *Any interval in R is simply connected.*

Let us consider first the case of the half-open interval, $\mathfrak{V} =]a, b]$ (where $a < b$). Let $(\tilde{\mathfrak{V}}, f)$ be a covering space of $\mathfrak{V}$. There is then some closed neighbourhood $[c, b]$ of b evenly covered by $\tilde{\mathfrak{V}}$. Now $]a, b[$ is homeomorphic to R. It is easy to see that $]a, b[$ is evenly covered by $\tilde{\mathfrak{V}}$. Select a c' such that $c < c' < b$. Then $]a, c']$ is evenly covered by $\tilde{\mathfrak{V}}$. The set $]a, c'] \cap [c, b] = [c, c']$ is connected and not empty. Therefore, using Lemma 1, we see that $\mathfrak{V}$ is evenly covered by $\tilde{\mathfrak{B}}$. It is then easy to see that $\mathfrak{B}$ is simply connected. A similar argument now applied to a instead of b shows that $[a, b]$ is simply connected. Proposition 2 is proved.

Corollary. *The product of a finite number of intervals in R is simply connected.*

It follows that an open or closed ball in R^n is simply connected (the open ball of center p and of radius r is defined to be the set of points whose distances from p are $<r$; the closed ball is the adherence of the open ball).

Proposition 3. *The Poincaré group of S^1 is isomorphic to the additive group of integers; if $n > 1$, S^n is simply connected.*

S^1 is homeomorphic to T^1, which is the factor group of R by the group of integers. Since R is simply connected, the Poincaré group of T^1 is isomorphic to the additive group of integers (Cf. Proposition 7, §VIII, p. 50). If $n > 1$, we denote by A and B the subsets of R^{n+1} defined by the conditions

$$A: \qquad x_{n+1} \geqslant 0, \qquad \Sigma_1^{n+1} x_i^2 = 1$$
$$B: \qquad x_{n+1} \leqslant 0, \qquad \Sigma_1^{n+1} x_i^2 = 1$$

The mapping $(x_1, \cdots, x_n, x_{n+1}) \to (x_1, \cdots, x_n)$ maps A and B topologically onto a closed ball in R^n; it follows that A and B are simply connected. The set $A \cap B$ is homeomorphic to S^{n-1}, which is connected for $n > 1$. Proposition 3 follows therefore from Lemma 1.

Proposition 4. *Let $\mathfrak{G}$ be a connected and locally connected topological group, and let $\mathfrak{H}$ be a closed locally connected subgroup of $\mathfrak{G}$. Let $\mathfrak{H}_0$ be the component of the neutral element in $\mathfrak{H}$. Then there exists a mapping f of $\mathfrak{G}/\mathfrak{H}_0$ into $\mathfrak{G}/\mathfrak{H}$ such that $(\mathfrak{G}/\mathfrak{H}_0, f)$ is a covering space of $\mathfrak{G}/\mathfrak{H}$. If $\mathfrak{H}$ is distinguished, $(\mathfrak{G}/\mathfrak{H}_0, f)$ is a covering group of $\mathfrak{G}/\mathfrak{H}$.*

Since $\mathfrak{H}$ is locally connected, we know that $\mathfrak{H}_0$ is relatively open in $\mathfrak{H}$ (Proposition 1, §VI, p. 40). It follows that there exists a neighbourhood V of the neutral element in $\mathfrak{G}$ such that $V^{-1}V \cap \mathfrak{H} \subset \mathfrak{H}_0$. We may assume without loss of generality that V is open and connected.

Let $\bar{\omega}$ and $\bar{\omega}_0$ be the natural mappings of $\mathfrak{G}$ onto $\mathfrak{G}/\mathfrak{H}$ and $\mathfrak{G}/\mathfrak{H}_0$ respectively. If $u \varepsilon \mathfrak{G}/\mathfrak{H}_0$, u is a coset modulo $\mathfrak{H}_0$, say $u = \sigma\mathfrak{H}_0$; this coset is entirely contained in the coset $\sigma\mathfrak{H}$ modulo $\mathfrak{H}$. If $w = \sigma\mathfrak{H}$, we set $w = f(u)$. Then f is a mapping of $\mathfrak{G}/\mathfrak{H}_0$ onto $\mathfrak{G}/\mathfrak{H}$. Since $\bar{\omega}$ and $\bar{\omega}_0$ are both continuous and interior, it is easy to see that f is continuous and interior.

If $\sigma \varepsilon \mathfrak{G}$, we set $W(\sigma) = \bar{\omega}(\sigma V)$. We select a complete set of representatives Δ for all cosets of $\mathfrak{H}$ modulo $\mathfrak{H}_0$; we have $\mathfrak{H} = \bigcup_{\delta \varepsilon \Delta} \delta \mathfrak{H}_0$.

The set $\overset{-1}{f}(W(\sigma))$ is the union of the sets $\tilde{W}_\delta(\sigma) = \bar{\omega}_0(\sigma V \delta)$, $\delta \varepsilon \Delta$. Each of these sets is mapped onto $W(\sigma)$ by f. The sets $\tilde{W}_\delta(\sigma)$ are mutually disjoint and each of them is mapped in a univalent way by f. In fact, assume that $\bar{\omega}(\sigma\tau_1\delta) = \bar{\omega}(\sigma\tau_2\delta')$, $\tau_1,\tau_2 \varepsilon V$, $\delta,\delta' \varepsilon \Delta$. We have $\sigma\tau_1\delta = \sigma\tau_2\delta'\eta$, $\eta\varepsilon\mathfrak{H}$, whence $\tau_2^{-1}\tau_1 = \delta'\eta\delta^{-1}$. On the other hand, $\delta'\eta\delta^{-1}$ belongs to $\mathfrak{H}$, whence $\tau_2^{-1}\tau_1 \varepsilon V^{-1}V \cap \mathfrak{H} \subset \mathfrak{H}_0$. If we assume $\delta = \delta'$, we see that $\eta\varepsilon\mathfrak{H}_0$ (we know that $\mathfrak{H}_0$ is a closed distinguished subgroup of $\mathfrak{H}$; cf. Proposition 1, §IV, p. 35), whence $\bar{\omega}_0(\sigma\tau_1\delta) = \bar{\omega}_0(\sigma\tau_2\delta)$, which proves that each $\tilde{W}_\delta(\eta)$ is mapped in a univalent way. If, on the other hand, we assume that $\bar{\omega}_0(\sigma\tau_1\delta) = \bar{\omega}_0(\sigma\tau_2\delta')$, we have $\eta\varepsilon\mathfrak{H}_0$, whence $\delta'\eta\delta^{-1}\varepsilon\delta'\mathfrak{H}_0\delta^{-1} = (\delta'\mathfrak{H}_0\delta'^{-1})(\delta'\delta^{-1}) = \mathfrak{H}_0\delta'\delta^{-1}$; we therefore have $\mathfrak{H}_0\delta'\delta^{-1} \cap \mathfrak{H}_0 \neq \phi$, whence $\delta' = \delta$. This proves that the sets $\tilde{W}_\delta(\sigma)$ are mutually disjoint.

Each $\tilde{W}_\delta(\sigma)$ is open in $\mathfrak{G}/\mathfrak{H}_0$; therefore f maps $\tilde{W}_\delta(\sigma)$ onto $W(\sigma)$ in a continuous, interior and univalent way, i.e. topologically. Since each $\tilde{W}_\delta(\sigma)$ is connected (it is a continuous image of $\sigma V\delta$), we see that the components of $\overset{-1}{f}(W(\sigma))$ are the sets $\tilde{W}_\delta(\sigma)$. If follows immediately that $(\mathfrak{G}/\mathfrak{H}_0, f)$ is a covering space of $\mathfrak{G}/\mathfrak{H}$.

If $\mathfrak{H}$ is a distinguished subgroup of $\mathfrak{G}$, $\mathfrak{H}_0$ is also distinguished. In fact, if $\sigma\epsilon\mathfrak{G}$, $\sigma\mathfrak{H}_0\sigma^{-1}$ is a connected subset of $\mathfrak{H}$ and contains the neutral element, whence $\sigma\mathfrak{H}_0\sigma^{-1} \subset \mathfrak{H}_0$. Furthermore, the mapping f is clearly a homomorphism of $\mathfrak{G}/\mathfrak{H}_0$ onto $\mathfrak{G}/\mathfrak{H}$. Proposition 4 is thereby proved.

Corollary 1. *The notation being as in Proposition* 4, *if* $\mathfrak{G}/\mathfrak{H}$ *is simply connected, then* $\mathfrak{H}$ *is connected.*

In fact, if $\mathfrak{G}/\mathfrak{H}$ is simply connected, f must be univalent, whence $\mathfrak{H} = \mathfrak{H}_0$.

Corollary 2. *Let* $\mathfrak{G}$ *be a connected and locally connected topological group. If,* $\mathfrak{H}$ *is a discrete distinguished subgroup of* $\mathfrak{G}$, *and if* f *is the natural mapping of* $\mathfrak{G}$ *onto* $\mathfrak{G}/\mathfrak{H}$, *then* $(\mathfrak{G}, f)$ *is a covering group of* $\mathfrak{G}/\mathfrak{H}$.

Using the notation of Proposition 4, it is clear that $\mathfrak{G}/\mathfrak{H}_0 = \mathfrak{G}$ and that the mapping f which has been constructed in the proof is the natural mapping of $\mathfrak{G}$ onto $\mathfrak{G}/\mathfrak{H}$.

Proposition 5. *Let* $\mathfrak{G}$ *be a connected and locally connected topological group, and let* $\mathfrak{H}$ *be a closed locally connected subgroup of* $\mathfrak{G}$. *Assume that* $\mathfrak{G}/\mathfrak{H}$ *is simply connected and that* $\mathfrak{G}$ *and* $\mathfrak{H}$ *are locally simply connected. Then the Poincaré group of* $\mathfrak{G}$ *is isomorphic to a factor group of the Poincaré group of* $\mathfrak{H}$.

Let $(\tilde{\mathfrak{G}}, g)$ be a simply connected covering group of $\mathfrak{G}$. We set $\tilde{\mathfrak{H}} = \overset{-1}{g}(\mathfrak{H})$; it is clear that g maps every coset of $\tilde{\mathfrak{G}}$ modulo $\tilde{\mathfrak{H}}$ onto a coset of $\mathfrak{G}$ modulo $\mathfrak{H}$ and maps two distinct cosets onto two distinct cosets. Let $\bar{\omega}$ and $\tilde{\omega}$ be the natural mappings of $\mathfrak{G}$ and $\tilde{\mathfrak{G}}$ onto $\mathfrak{G}/\mathfrak{H}$ and $\tilde{\mathfrak{G}}/\tilde{\mathfrak{H}}$ respectively; we see that there exists a univalent mapping g^* of $\tilde{\mathfrak{G}}/\tilde{\mathfrak{H}}$ onto $\mathfrak{G}/\mathfrak{H}$ such that

$$g^*(\tilde{\omega}(\tilde{\sigma})) = \bar{\omega}(g(\tilde{\sigma}))(\tilde{\sigma}\epsilon\tilde{\mathfrak{G}})$$

From the fact $\bar{\omega}$ and $\tilde{\omega}$ are continuous interior mappings, it follows immediately that g^* is also continuous and interior. Therefore g^* is a homeomorphism, from which we conclude that $\tilde{\mathfrak{G}}/\tilde{\mathfrak{H}}$ is simply connected.

The group $\tilde{\mathfrak{H}}$ is locally connected. In fact, let $\tilde{V}$ be a neighbourhood of the neutral element $\tilde{\epsilon}$ of $\tilde{\mathfrak{G}}$ which is mapped topologically by g. The set $g(\tilde{V}) \cap \mathfrak{H}$ contains a locally connected neighbourhood W of $\epsilon = g(\tilde{\epsilon})$ with respect to $\mathfrak{H}$. The set $\tilde{V} \cap \overset{-1}{g}(W)$ is homeomorphic to W, and therefore locally connected; this set being a neighbourhood of $\tilde{\epsilon}$ with respect to $\tilde{\mathfrak{H}}$, our assertion is proved.

It follows from Corollary 1 to Proposition 4 that $\tilde{\mathfrak{H}}$ is connected. Let g_0 be the contraction of g to $\tilde{\mathfrak{H}}$; it follows from Lemma 5, §VI, p. 40 that $(\tilde{\mathfrak{H}}, g_0)$ is a covering group of $\mathfrak{H}$.

The Poincaré group of $\mathfrak{G}$ is isomorphic to the kernel F of the homomorphism g, and we have $F \subset \tilde{\mathfrak{H}}$. Let now $(\tilde{\mathfrak{H}}_1, g_1)$ be a simply connected covering group of $\mathfrak{H}$. It is easy to see that we can find a local isomorphism η of a connected neighbourhood W_1 of the neutral element of $\tilde{\mathfrak{H}}_1$ into a neighbourhood of the neutral element of $\tilde{\mathfrak{H}}$ such that $g_0(\eta(\rho)) = g_1(\rho)$ for all $\rho\varepsilon W_1$. By Theorem 3, §VII, p. 50, we can extend η to a homomorphism h of $\tilde{\mathfrak{H}}_1$ into $\tilde{\mathfrak{H}}$. The set of elements $\rho\varepsilon\tilde{\mathfrak{H}}_1$ for which $g_0(h(\rho)) = g_1(\rho)$ is a subgroup of $\tilde{\mathfrak{H}}_1$ and contains W_1; by Theorem 1, §IV, p. 35 it follows that this set coincides with $\tilde{\mathfrak{H}}_1$, whence $g_0 \circ h = g_1$. Let F_1 and H be the kernels of the homomorphisms g_1 and h respectively; the kernel of g_0 being F, it is clear that F is isomorphic to F_1/H. Since F_1 is isomorphic to the Poincaré group of $\mathfrak{H}$, Proposition 5 is proved.

It follows immediately from Propositions 3 and 5 that the Poincaré groups of $SO(n)$ (for $n \geqslant 3$), $SU(n)$ and $Sp(n)$ (for $n > 1$) are isomorphic to factor groups of the Poincaré groups of $SO(n - 1)$, $SU(n - 1)$ and $Sp(n - 1)$ respectively (Cf. Propositions 2*a*, 3, 4, §III, p. 29). The group $SU(1)$ contains only one element and is therefore simply connected. The group $Sp(1)$ is isomorphic with S^3 and therefore simply connected. Thus:

Proposition 6. *The groups $SU(n)$ and $Sp(n)$ are simply connected for every $n \geqslant 1$.*

The Poincaré group of SO(3) is of order 2 by Proposition 1, §V, p. 35. Therefore, for every $n > 3$, the Poincaré group of $SO(n)$ is of order either 1 or 2; we shall prove in the next section that this group is actually of order 2.

We pass now to the consideration of $U(n)$. We represent by $\rho(\varphi)$ the matrix

$$\begin{pmatrix} \exp 2\pi\sqrt{-1}\,\varphi & 0 & \cdots & 0 \\ 0 & 1 & & \vdots \\ \vdots & & \ddots & \vdots \\ 0 & 0 & \cdots & 1 \end{pmatrix}$$

The matrices of the form $\rho(\varphi)$ form a subgroup $\mathfrak{g}$ of $U(n)$ which is isomorphic with T^1. Let σ be any matrix of $U(n)$; then $|\sigma|$ is of absolute value 1, and therefore there exists a number φ such that $\sigma = \rho(\varphi)\tau$, $\tau \varepsilon SU(n)$. Since $\mathfrak{g} \cap SU(n)$ contains only the unit matrix, every $\sigma \varepsilon U(n)$ may be written in one and only one way in the form $\sigma = \rho\tau$, $\rho \varepsilon \mathfrak{g}$, $\tau \varepsilon SU(n)$. The mapping $(\rho, \tau) \to \rho\tau$ of $\mathfrak{g} \times SU(n)$ onto $U(n)$ is continuous and univalent; since $\mathfrak{g} \times SU(n)$ is compact, our mapping is a homeomorphism, and we have proved

Proposition 7. *The underlying space of $U(n)$ is homeomorphic to $T^1 \times SU(n)$. The Poincaré group of $U(n)$ is isomorphic to the additive group of integers.*

§XI. The Clifford Numbers.

Let K be a field of characteristic $\neq 2$. We shall construct an algebra $\mathfrak{o}$ over K, which will contain a unit element e_0 and which will be generated by e_0 and by n other elements $e_1, \cdots, e_n$ (where n is any integer >0) such that the identity

$$(\Sigma_{i=1}^{n} x_i e_i)^2 = -e_0 \Sigma_{i=1}^{n} x_i^2$$

holds for any $x_1, \cdots, x_n \varepsilon K$; i.e. we shall have

$$(1) \quad e_0 e_0 = e_0; \qquad e_0 e_i = e_i e_0 = e_i; \qquad e_i e_j + e_j e_i = 0 \qquad (i \neq j)$$
$$e_i^2 = -e_0 \qquad (1 \leqslant i, j \leqslant n)$$

It follows easily that every element of $\mathfrak{o}$ will be a linear combination of e_0 and of the products $e_{i_1} \cdots e_{i_m}$, with $1 \leqslant i_1 < \cdots < i_m \leqslant n$.

We now proceed to the actual construction of $\mathfrak{o}$. To every subset A of the set $N = \{1, 2, \cdots, n\}$ we associate a symbol e_A, and we consider these symbols e_A as forming a base of a vector space over K.

This vector space has therefore the dimension 2^n. If A and B are two subsets of N, we denote by $A + B$ the set of those elements which occur either in A or in B, but not in both. If $j \in N$, we denote by $p(A, j)$ the number of elements $i \epsilon A$ such that $i \geqslant j$, and we set

$$p(A, B) = \Sigma_{j \in B} p(A, j) \qquad \zeta(A, B) = (-1)^{p(A,B)}$$

We define the multiplication of the basic elements e_A by the formula

$$e_A e_B = \zeta(A, B) e_{A+B}$$

We shall prove that this multiplication is associative. Let $\mathfrak{K}$ be the field of characteristic 2 with 2 elements 0 and 1; if $A \subset N$, we denote by f_A the mapping of N into $\mathfrak{K}$ defined by $f_A(i) = 1$ if $i \in A$, $f_A(i) = 0$ if i does not belong to A. Since $1 + 1 = 0$, we have $f_{A+B} = f_A + f_B$, whence $f_{(A+B)+C} = f_A + f_B + f_C = f_{A+(B+C)}$ and $(A + B) + C = A + (B + C)$.

On the other hand, we have

$$p(A, B + C) = \Sigma_{j \in B+C} p(A, j) \equiv p(A, B) + p(A, C) \pmod{2}$$
$$p(A + B, C) = \Sigma_{j \in C} p(A + B, j) \equiv p(A, C) + p(B, C) \pmod{2}$$

and it follows that $(e_A e_B)e_C$ and $e_A(e_B e_C)$ are both equal to

$$\zeta(A, B)\zeta(B, C)\zeta(A, C)e_{(A+B)+C}.$$

We have therefore defined an associative algebra $\mathfrak{o}$ over K. If we set $e_0 = e_\phi$, $e_i = e_{\{i\}}$ $(1 \leqslant i \leqslant n)$, we have

$$e_A = e_{i_1} \cdots e_{i_m} \quad \text{if} \quad A = \{i_1, \cdots, i_m\}, \quad i_1 < \cdots < i_m.$$

and the formulas (2) hold.

The elements of the algebra $\mathfrak{o}$ are called the *Clifford numbers*.

Now we shall determine the center of $\mathfrak{o}$ and the ideals in $\mathfrak{o}$. To every $h (1 \leqslant h \leqslant n)$ we associate the linear mapping Q_h of $\mathfrak{o}$ into itself defined by $Q_h(x) = \frac{1}{2}(x - e_h x e_h)$. We shall compute $Q_h(e_A)$; denoting by $s(A)$ the number of elements in A, we find easily that:

if $s(A) \equiv 0 \pmod{2}$, then $Q_h(e_A) = 0$ if $h \in A$, and $Q_h(e_A) = e_A$ if h does not belong to A;

if $s(A) \equiv 1 \pmod{2}$, then $Q_h(e_A) = e_A$ if $h \epsilon A$ and $Q_h(e_A) = 0$ if h does not belong to A.

Let Q be the linear mapping $Q_1 \circ \cdots \circ Q_n$. If $n \equiv 0 \pmod{2}$, we have $Q(e_0) = e_0$, $Q(e_A) = 0$ for $A \neq \phi$. If $n \equiv 1 \pmod{2}$, we have $Q(e_0) = e_0$, $Q(e_N) = e_N$, $Q(e_A) = 0$ for $A \neq \phi, N$. It follows in particular from this that e_N belongs to the center of $\mathfrak{o}$ if $n \equiv 1 \pmod{2}$.

Let $\mathfrak{c}$ be the center of $\mathfrak{o}$. If $x \in \mathfrak{c}$, we have $Q_h(x) = x$ for every h, whence $Q(x) = x$. It follows immediately that $\mathfrak{c} = Ke_0$ if n is even, and $\mathfrak{c} = Ke_0 + Ke_N$ if n is odd.

Let now $\mathfrak{a}$ be any ideal $\neq \{0\}$ in $\mathfrak{o}$, and let $x = \Sigma_A c_A e_A$ be any element $\neq 0$ in $\mathfrak{a}$. Assume that $c_{A_0} \neq 0$; then $e_{A_0}^{-1} x = \Sigma_A c'_A e_A$ belongs to $\mathfrak{a}$, and the same holds for $Q(e_{A_0}^{-1}x)$. We have $c'_\phi \neq 0$. If n is even, we have $Q(e_{A_0}^{-1}x) = c'_\phi e_0$, whence $e_0 \varepsilon \mathfrak{a}$ and therefore $\mathfrak{a} = \mathfrak{o}$. We have proved

Proposition 1. *If n is even, the center of the algebra of Clifford numbers is Ke_0 (where K is the basic field). The only ideals in the algebra are $\{0\}$ and the whole algebra.*

If n is odd, then $Q(e_{A_0}^{-1}x) = c'_\phi e_0 + c'_N e_N$. It is easy to see that $e_N^2 = (-1)^{n(n+1)/2}$. If $(-1)^{n(n+1)/2}$ is not a square in K, the center $\mathfrak{c} = Ke_0 + Ke_N$ of $\mathfrak{o}$ is a field; since $\mathfrak{a} \cap \mathfrak{c}$ is an ideal $\neq \{0\}$ in $\mathfrak{c}$, we have $\mathfrak{a} \cap \mathfrak{c} = \mathfrak{c}$, whence $\mathfrak{a} = \mathfrak{o}$. Assume now that $(-1)^{n(n+1)/2} = j^2$, $j \varepsilon K$. Then the elements $u = \frac{1}{2}(e_0 + je_N)$ and $v = \frac{1}{2}(e_0 - je_N)$ are orthogonal idempotents in $\mathfrak{c}$ (i.e. $u^2 = u$, $v^2 = v$, $uv = 0$), and we have $\mathfrak{c} = Ku + Kv$. The ideals $\neq \{0\}$ in $\mathfrak{c}$ are Ku, Kv and $\mathfrak{c}$. It follows that $\mathfrak{a}$ contains one of the elements u and v. Assume that $\mathfrak{a}$ contains u; if there exists an element $y \varepsilon \mathfrak{a}$ such that $x = yv \neq 0$, then $\mathfrak{a} \cap \mathfrak{c}$ contains $Q(e_{A_0}^{-1}x) = Q(e_{A_0}^{-1}y)v \varepsilon Kv$ (observe that v belongs to the center, whence $Q(zv) = Q(z)v$ for every z). It follows that, if $u \varepsilon \mathfrak{a}$, we have either $\mathfrak{a} = \{\mathfrak{o}\}$ or $\mathfrak{a}v = \{0\}$. In the latter case, we have clearly $\mathfrak{a} = \mathfrak{o}u$. We have proved

Proposition 2. *If n is odd, the center of the algebra of Clifford numbers is $Ke_0 + Ke_N$. If $(-1)^{n(n+1)/2}$ is not a square in K, the only ideals in $\mathfrak{o}$ are $\{0\}$ and $\mathfrak{o}$. If $(-1)^{n(n+1)/2} = j^2$, $j \varepsilon K$, the ideals in $\mathfrak{o}$ are $\{0\}$, $\mathfrak{o}$, $\mathfrak{o}u$ and $\mathfrak{o}v$, where $u = \frac{1}{2}(e_0 + je_N)$, $v = \frac{1}{2}(e_0 - je_N)$.*

If $x \varepsilon \mathfrak{o}$, the mapping $y \to \theta(x)y = xy$ is a linear endomorphism of the vector space $\mathfrak{o}$ over K. After having arranged the basic elements e_A in a certain order (in an arbitrary way), we can represent this endomorphism by a matrix of degree 2^n; we shall also denote this matrix by $\theta(x)$. We obtain in this way a representation of the algebra $\mathfrak{o}$ by matrices; this representation is called the *regular representation.* We shall denote by $\Delta(x)$ the determinant of $\theta(x)$. If n is odd, we denote by K' the field obtained by adjunction of $\sqrt{-1}$ to K; the linear combinations of the elements e_A with coefficients in K' form an algebra, which is the algebra $\mathfrak{o}'$ of Clifford numbers over K'. The matrix $\theta(x)$ may be considered as defining a linear endomorphism of $\mathfrak{o}'$. If we set $u' = \frac{1}{2}(e_0 + je_N)$, $v' = \frac{1}{2}(e_0 - je_N)$ (where j is an element of K' such that $j^2 = (-1)^{n(n+1)/2}$), it is clear that $\theta(x)$ maps into themselves the subspaces $\mathfrak{o}'u$ and $\mathfrak{o}'v$ of $\mathfrak{o}'$; we shall denote by $\theta'(x)$ and $\theta''(x)$ the contractions of $\theta(x)$ to $\mathfrak{o}'u$ and $\mathfrak{o}'v$, and by $\Delta'(x)$, $\Delta''(x)$ the determinants of the endomorphisms $\theta'(x)$, $\theta''(x)$. If D is any one of the functions Δ, Δ', Δ'', we have $D(xy) = D(x)D(y)$.

An element $x \varepsilon \mathfrak{o}$ is said to be *regular* if it has an inverse, i.e. if there exists an element $x^{-1} \varepsilon \mathfrak{o}$ such that $xx^{-1} = x^{-1}x = e_0$. If x is regular, we have $\Delta(x)\Delta(x^{-1}) = 1$, whence $\Delta(x) \neq 0$. Conversely, if $\Delta(x) \neq 0$, $\theta(x)$ is a regular matrix, and therefore $\theta(x)$ maps $\mathfrak{o}$ *onto* itself in a univalent way. It follows that there exists an element x^{-1} such that $xx^{-1} = e_0$; we have $x(x^{-1}x) = x = xe_0$, whence also $x^{-1}x = e_0$: x is regular.

We shall now assume that the basic field K is the field R of real numbers. The regular elements of $\mathfrak{o}$ form a multiplicative group which we shall denote by $\mathfrak{o}^*$. The contraction of θ to $\mathfrak{o}^*$ is a faithful representation of $\mathfrak{o}^*$.

If $x \varepsilon \mathfrak{o}^*$, the mapping $y \to xyx^{-1}$ is an endomorphism of $\mathfrak{o}$, which we may also represent by a matrix $\psi(x)$. The mapping $x \to \psi(x)$ is a representation of $\mathfrak{o}^*$ whose kernel is the intersection of $\mathfrak{o}^*$ with the center $\mathfrak{c}$ of $\mathfrak{o}$.

If we assign to an element $x = \Sigma_A c_A e_A$ of $\mathfrak{o}$ the point of R^{2^n} whose coordinates are the coefficients c_A (after having arranged the sets A in some order), we obtain a one-to-one correspondence between $\mathfrak{o}$ and R^{2^n}. We may define a topology in $\mathfrak{o}$ by the requirement that this correspondence shall be a homeomorphism. The operations in $\mathfrak{o}$ (addition, multiplication between elements of $\mathfrak{o}$, multiplication by real numbers) are obviously continuous with respect to this topology. Furthermore, if $x \varepsilon \mathfrak{o}^*$, x^{-1} is a continuous function of x. In fact, we have seen that x^{-1} is the unique solution of the equation $\theta(x)y = e_0$; since the coefficients of the matrix $\theta(x)$ are linear functions of the coefficients c_A of x, the coefficients of x^{-1}, expressed as a linear combination of the basic elements e_A, are rational functions of the quantities c_A and the denominators of these functions are equal to $\Delta(x)$. Since $\Delta(x) \neq 0$ on $\mathfrak{o}^*$, x^{-1} is a continuous function of x on $\mathfrak{o}^*$. It follows that $\mathfrak{o}^*$, considered as a subspace of $\mathfrak{o}$, becomes a topological group, and that θ and ψ are continuous representations of $\mathfrak{o}^*$.

We observe furthermore that the mapping $x \to \theta(x)$ not only is continuous, but is a homeomorphism of $\mathfrak{o}$ with some subspace of the space of all matrices of degree 2^n with coefficients in R. In fact, we have $\theta(x)e_0 = x$, which shows that the coefficients of x are also the coefficients of a certain column of the matrix $\theta(x)$.

We have

$$\theta\left(e_0 + x + \frac{x^2}{2!} + \cdots + \frac{x^m}{m!}\right) = \theta(e_0) + \theta(x) + \frac{1}{2}(\theta(x))^2 + \cdots + \frac{1}{m!}(\theta(x))^m$$

If m tends to infinity, the right side tends to $\exp \theta(x)$. Therefore $e_0 + x + x^2/2! + \cdots + x^m/m!$ tends to a limit, $\exp x$, such that $\theta(\exp x) = \exp \theta(x)$. We have $\exp(x+y) = (\exp x)(\exp y)$ *if* $xy = yx$; in particular $(\exp(-x)) = (\exp x)^{-1}$, which shows that $\exp x \varepsilon \mathfrak{o}^*$. We have (by Corollary 1 to Proposition 2, §II, Chap. I, p. 5) $\Delta(\exp x) = \exp Sp\theta(x) \neq 0$.

We shall now compute $\psi(\exp x)$. The mapping $y \rightarrow xy - yx$ is a linear endomorphism of $\mathfrak{o}$, which we denote by $X(x)$. Let y_0 be any element of $\mathfrak{o}$, and set $y(t) = (\exp tx)y_0(\exp(-tx))$, where t is any real number. We have

$$\begin{aligned} y(t+h) &= (\exp hx)y(t)(\exp(-hx)) \\ &= (e_0 + hx + \cdots)y(t)(e_0 - hx + \cdots) \end{aligned}$$

whence

$$\frac{dy}{dt} = \lim_{h\rightarrow 0} \frac{y(t+h) - y(t)}{h} = xy(t) - y(t)x = X(x)y(t)$$

We know that the solution of this differential equation (which is equivalent to a system of 2^n linear homogeneous differential equations for the coefficients of $y(t)$) is given by the formula $y(t) = (\exp tX(x))y_0$. Therefore we have

$$\psi(\exp tx) = \exp(tX(x)).$$

Let $\mathfrak{M}$ be the vector subspace of $\mathfrak{o}$ which is spanned by $e_1, \cdots, e_n$. We shall consider the set of those elements $x \varepsilon \mathfrak{o}^*$ which are such that $\psi(x)$ maps $\mathfrak{M}$ into itself. This set is obviously a subgroup of $\mathfrak{o}^*$.

Definition 1. *Let G be the group of elements $x \varepsilon \mathfrak{o}^*$ such that $\psi(x)(\mathfrak{M}) \subset \mathfrak{M}$, $\Delta(x) = 1$ and (if n is odd), $\Delta'(x) = \Delta''(x) = 1$. The component of e_0 in G (considered as a topological subgroup of $\mathfrak{o}^*$) is called the spinor group. This group will be denoted by Spin (n).*

If x is any element of $\mathfrak{o}^*$ such that $\psi(x)\mathfrak{M} \subset \mathfrak{M}$, the contraction of $\psi(x)$ to $\mathfrak{M}$ is an endomorphism of $\mathfrak{M}$ which we shall denote by $\varphi(x)$. If $\varphi(x)e_i = \Sigma_{j=1}^n a_{ji}e_j$, we have $\varphi(x)(\Sigma_{i=1}^n x_ie_i) = \Sigma_{j=1}^n e_j(\Sigma_{i=1}^n a_{ji}x_i)$. Since $\psi(x)$ is an automorphism of the algebra $\mathfrak{o}$, we have $\psi(x)(\Sigma_1^n x_ie_i)^2 = (\psi(x)(\Sigma_{i=1}^n x_ie_i))^2 = -\Sigma_{j=1}^n(\Sigma_{i=1}^n a_{ji}x_i)^2 \cdot e_0$, whence $\Sigma_{j=1}^n(\Sigma_{i=1}^n a_{ji}x_i)^2 = \Sigma_{i=1}^n x_i^2$. It follows immediately that $\varphi(x)$ is represented by an orthogonal matrix in terms of the base $\{e_1, \cdots, e_n\}$ of $\mathfrak{M}$. Therefore, we have $\varphi(G) \subset O(n)$; the mapping $x \rightarrow \varphi(x)$ being clearly continuous, we have $\varphi(\text{Spin}(n)) \subset SO(n)$. We shall prove that $\varphi(\text{Spin}(n)) = SO(n)$.

Denote by $\mathfrak{M}_2$ the vector space spanned by the elements e_ie_j with $i \neq j$ $(1 \leqslant i, j \leqslant n)$. The dimension of this space is $n(n-1)/2$.

We have

$$X(e_ie_j)e_k = \begin{cases} 0 & \text{if} \quad k \neq i, j \\ 2e_j & \text{if} \quad k = i \\ -2e_i & \text{if} \quad k = j \end{cases} \qquad (1 \leqslant i, j, k \leqslant n, i \neq j)$$

It follows that, if $x \varepsilon \mathfrak{M}_2$, we have $X(x)\mathfrak{M} \subset \mathfrak{M}$. We have proved that $\psi(\exp tx) = \exp tX(x)$; it follows that $x \varepsilon \mathfrak{M}_2$ implies $\psi(\exp tx)\mathfrak{M} \subset \mathfrak{M}$ for every t. We have $Sp\ \theta(e_ie_j) = Sp\ \theta(e_i)\theta(e_j) = Sp\ \theta(e_j)\theta(e_i) = Sp\ \theta(e_je_i)$; since $e_ie_j + e_je_i = 0$, we have $Sp\ \theta(e_ie_j) = 0$ $(1 \leqslant i, j \leqslant n; i \neq j)$. By the same argument, we see that, if n is odd, $Sp\ \theta'(e_ie_j) = Sp\ \theta(e_ie_j) = 0$. Making use of the formula $\boxed{\exp U} = \exp SpU$, we see that $x \varepsilon \mathfrak{M}_2$ implies $\Delta(\exp x) = \Delta'(\exp x) = \Delta''(\exp x) = 1$. It follows that $x \varepsilon \mathfrak{M}_2$ implies $\exp tx \varepsilon G$ for all real t, and therefore also $\exp x \varepsilon$ Spin (n). If $x \varepsilon \mathfrak{M}_2$, we denote by $X_1(x)$ the matrix which represents (in terms of the base $\{e_1, \cdots, e_n\}$ in $\mathfrak{M}$) the contraction to $\mathfrak{M}$ of the endomorphism $X(x)$. Since $\exp tX_1(x) \varepsilon SO(n)$ for every t, $X_1(x)$ is skew symmetric; the equality $X_1(x) = 0$ implies that x belongs to the center of $\mathfrak{o}$, whence $x = 0$. It follows that $x \to X_1(x)$ is a univalent linear mapping of $\mathfrak{M}_2$ into the space of skew symmetric matrices of degree n. On the other hand, $\mathfrak{M}_2$ and the space of skew symmetric matrices of degree n have the same dimension $n(n-1)/2$; so the latter is covered by the mapping. Since $\exp X_1(x) \varepsilon \varphi(\text{Spin } (n))$, it follows by Proposition 4, §II, Chapt. I, p. 5 that $\varphi(\text{Spin } (n))$ contains a neighbourhood of the neutral element in $SO(n)$. But $\varphi(\text{Spin } (n))$ is a subgroup of $SO(n)$, and $SO(n)$ is connected; by Theorem 1, §IV, p. 35, we see that $\varphi(\text{Spin } (n)) = SO(n)$.

Let φ_1 be the contraction of φ to Spin (n). The mapping φ_1 of Spin (n) onto $SO(n)$ is obviously continuous. This mapping is also interior. In fact, let V be a neighbourhood of e_0 in $\mathfrak{o}$. Since the function $\exp x$ is continuous, there exists a neighbourhood U of 0 in $\mathfrak{M}_2$ such that $\exp x \varepsilon V$ for all $x \varepsilon U$. We know that, if X runs over all elements of a neighbourhood of 0 in the space of skew symmetric matrices, the set of the corresponding elements $\exp X$ is a neighbourhood of the neutral element in $SO(n)$. Therefore $\varphi_1(V \cap \text{Spin } (n))$ contains a neighbourhood of the neutral element in $SO(n)$, which proves our assertion. It follows immediately that $SO(n)$ is isomorphic (as a topological group) with the factor group of Spin (n) by the kernel F of the homomorphism φ_1.

The elements of G which are mapped upon the unit matrix by φ are the elements of $G \cap \mathfrak{c}$ (where $\mathfrak{c}$ is the center of $\mathfrak{o}$). If a is a real number, we have $\Delta(ae_0) = a^{2^n}$; therefore, the only elements of the form ae_0 which belong to G are $\pm e_0$. If $n \equiv 1$ (mod. 4), we can write

$$ae_0 + be_N = (a - \sqrt{-1}\, b)u + (a + \sqrt{-1}\, b)v$$

where $u = \frac{1}{2}(e_0 + \sqrt{-1}\, e_N)$, $v = \frac{1}{2}(e_0 - \sqrt{-1}\, e_N)$; since u is the unit element of $\mathfrak{o}u$, we have $\Delta'(u) = 1$, from which it follows easily that $\Delta'(ae_0 + be_N) = (a - \sqrt{-1}\, b)^{2^{n-1}}$, and similarly $\Delta''(ae_0 + be_N) = (a + \sqrt{-1}\, b)^{2^{n-1}}$. It follows that $\mathfrak{c} \cap G$ is the group of elements $ae_0 + be_N$ such that $(a + \sqrt{-1}\, b)^{2^{n-1}} = 1$; it is a cyclic group of order 2^{n-1}.

If $n \equiv 3$ (mod 4), we find in the same way that $\Delta'(ae_0 + be_N) = (a + b)^{2^{n-1}}$, $\Delta''(ae_0 + be_N) = (a - b)^{2^{n-1}}$. It follows that $\mathfrak{c} \cap G$ is composed of the elements $\pm e_0$, $\pm f$.

In any case, the group F, which is a subgroup of $G \cap \mathfrak{c}$, is a finite group. It follows that F is discrete; therefore (Spin (n), φ_1) is a covering group of $SO(n)$. An easy computation gives $\exp te_1e_2 = (\cos t)e_0 + (\sin t)e_1e_2$; it follows that $-e_0 = \exp \pi e_1e_2 \varepsilon$ Spin (n), whence $-e_0 \varepsilon F$. Since F contains an element $\neq e_0$, the group $SO(n)$ cannot be simply connected. But we know that the Poincaré group of $SO(n)$ is of order at most 2 for $n \geqslant 3$. Therefore we have proved

Proposition 3. *The Poincaré group of $SO(n)$ is of order* 2 *for* $n \geqslant 3$. *The group* Spin (n) *is simply connected if* $n \geqslant 3$.

CHAPTER III

Manifolds

Summary. The manifolds to be considered are exclusively "analytic manifolds." They are defined in §I; our method of definition seems slightly preferable to the method of Whitney in that it is "intrinsic"; i.e. it does not require *a posteriori* identifications.

We define in §IV the notion of tangent space to an abstractly given manifold; to every analytic mapping Φ of a manifold $\mathcal{V}$ into another manifold $\mathcal{W}$ is associated a differential mapping $d\Phi$ which maps the tangent space to $\mathcal{V}$ into the tangent space to $\mathcal{W}$. The differentials of functions are considered as a special case of these differential mappings.

In §V, we introduce the notion of an infinitesimal transformation, which is defined as a law which assigns to every point of the manifold a tangent vector at this point; we define the "bracket operation" for infinitesimal transformations, and we discuss the effect of a mapping on this operation.

In §§VI, VII, VIII we study the notion of a distribution on a manifold $\mathcal{V}$. A distribution is defined as a law which assigns to every point P of $\mathcal{V}$ a sub-space $\mathfrak{M}_P$ of the tangent space at P. An integral manifold of this distribution is a sub-manifold of $\mathcal{V}$ which admits $\mathfrak{M}_P$ as tangent space at any one of its points P. The existence of such integral manifolds depends upon certain integrability conditions, which we express by saying that the distribution must be "involutive" (Definition 5, §VI, p. 85). We prove in §VII that the condition of being "involutive" is actually sufficient for a distribution to have integral manifolds. The integral manifolds are first obtained locally; then by a topological process of "piecing together," we construct in §VIII the "complete" integral manifolds in the large.

In §IX we consider those manifolds for which the second axiom of denumerability of Hausdorff holds true. We use this axiom only to prove Proposition 1, §IX, p. 94; but do not know whether this axiom is necessary even there.

§I. Axiomatic Definition of a Manifold

Let $\mathfrak{V}$ be a topological space. We denote by p a point of $\mathfrak{V}$, and consider $k+1$ real valued functions, $f_0, f_1, \cdots, f_k$, which are all defined in some neighbourhood of p. We shall say that f_0 is *analytically dependent on $f_1, \cdots, f_k$ in the neighbourhood of p*, or *around p*, if there exists a neighbourhood V of p and a function $F(u_1, \cdots, u_k)$ of k real arguments, such that the following conditions are satisfied:

1) *The functions $f_0, f_1, \cdots, f_k$ are defined on V.*

2) *The domain of definition of F includes all systems of values of the form $u_1 = f_1(q), \cdots, u_k = f_k(q)$, for $q \in V$.*

3) *If* $q \varepsilon V$ *we have*

$$f_0(q) = F(f_1(q), \cdots, f_k(q)).$$

4) *The function* F *is analytic at the point* $u_1 = f_1(p), \cdots, u_k = f_k(p)$.[1] Let us now assume that $\mathfrak{V}$ is *connected* and that we have assigned to each point $p \varepsilon \mathfrak{V}$ a class $\mathfrak{A}(p)$ of real valued functions, satisfying the following conditions:

I. *Each function in* $\mathfrak{A}(p)$ *is defined in some neighbourhood of* p (this neighbourhood may depend on the function).

II. *Any function which depends analytically around* p *on a finite number of functions in* $\mathfrak{A}(p)$ *is itself in* $\mathfrak{A}(p)$.

III. *It is possible to find an ordered system* $(f_1, \cdots, f_n)$ *of functions in* $\mathfrak{A}(p)$, *a neighbourhood* V *of* p, *and a number* $a > 0$ *with the following properties:*

1) *The functions* $f_1, \cdots, f_n$ *are defined on* V.

2) *If we assign to each point* $q \varepsilon V$ *the point* $\Phi(q) \varepsilon R^n$ *whose coordinates are* $x_1 = f_1(q), \cdots, x_n = f_n(q)$, *the mapping* Φ *is a homeomorphism of* V *with the subset of* R^n *composed of the points* $(x_1, \cdots, x_n)$ *such that*

$$|x_1 - f_1(p)| < a, \qquad \cdots, \qquad |x_n - f_n(p)| < a.$$

3) *If* $q \varepsilon V$ *the functions* $f_1, \cdots, f_n$ *belong to* $\mathfrak{A}(q)$, *and every function in* $\mathfrak{A}(q)$ *depends analytically on* $f_1, \cdots, f_n$ *around* q.

Under these conditions we shall say that we have defined a *manifold* $\mathcal{V}$.[2] Therefore, to define a manifold we must first give a topological space $\mathfrak{V}$ and then select for every point $p \varepsilon \mathfrak{V}$ a certain class $\mathfrak{A}(p)$ of real valued functions.

The space $\mathfrak{V}$ is called the *underlying topological space* of the manifold. The class $\mathfrak{A}(p)$ is called the *class of analytic functions on* $\mathcal{V}$ *at the point* p.

The underlying space $\mathfrak{V}$ of a manifold cannot be an arbitrary topological space for we have required that it be connected and it follows from III that every point of $\mathfrak{V}$ has a neighbourhood which is homeomorphic to a cube in some cartesian space.

[1] This means that F may be represented in a neighbourhood of this system of values by a convergent power series

[2] This definition is equivalent to the classical one given by Whitney in "Differentiable Manifolds" (Annals of Math., vol. 37, 1936). It should be observed that we limit ourselves to the consideration of analytic manifolds.

We observe that, if an ordered system $(f_1, \cdots, f_n)$, a neighbourhood V, and a number $a > 0$ have the properties 1), 2), and 3) of condition III, these properties also hold if, without changing a or V, we perform an arbitrary permutation on the functions $f_1, \cdots, f_n$. On the other hand, property 2) implies that the functions $f_1, \cdots, f_n$ are distinct. Therefore properties 1), 2), 3), are properties of the finite set $\{f_1, \cdots, f_n\}$ (and, of course, of V and a).

Definition 1. *If the properties* 1), 2), 3), *of condition III hold for the system* $(f_1, \cdots, f_n)$, *the neighbourhood* V, *and the number* a, *we shall say that the finite set* $\{f_1, \cdots, f_n\}$ *is a system of coordinates on* $\mathcal{V}$ *at the point* p, *and that* V *is a cubic neighbourhood of* p *with respect to this system of coordinates. The number* a *is called the breadth of the neighbourhood* V *with respect to the system of coordinates* $\{f_1, \cdots, f_n\}$.

Remarks. 1) If $\{f_1, \cdots, f_n\}$ is a system of coordinates at p, and if V is a cubic neighbourhood of p with respect to this system, the set $\{f_1, \cdots, f_n\}$ is also a system of coordinates at every point of V.

2) If $\{f_1, \cdots, f_n\}$ is a system of coordinates at p, any neighbourhood of p contains a cubic neighbourhood with respect to this system.

3) If f is a function which is analytic at p on $\mathcal{V}$, there exists a neighbourhood V of p such that f is also analytic at every point $q \varepsilon V$. In fact, let $\{f_1, \cdots, f_n\}$ be a system of coordinates at p; there exists a neighbourhood V_1 of p such that $f, f_1, \cdots, f_n$ are defined on V_1 and

$$(1) \qquad f(q) = f^*(f_1(q), \cdots, f_n(q)), \qquad \text{for} \qquad q \varepsilon V_1,$$

where $f^*(u_1, \cdots, u_n)$ is a function of n arguments, which is analytic at the point $u_1 = f_1(p), \cdots, u_n = f_n(p)$. This function is also defined and analytic at all points of a neighbourhood U of this point in R^n; we can find a cubic neighbourhood V of p with respect to the system $\{f_1, \cdots, f_n\}$ such that $V \subset V_1$ and such that $q \varepsilon V$ implies $(f_1(q), \cdots, f_n(q)) \varepsilon U$. Then f is analytic at every point $q \varepsilon V$.

We shall say that (1) is the *expression* of f in terms of the coordinates $f_1, \cdots, f_n$. It should be observed that the function f^* actually depends on the way in which the functions of the system of coordinates are ordered.

Proposition 1. *Let* $\{x_1, \cdots, x_n\}$ *be a system of coordinates at the point* p *on the manifold* $\mathcal{V}$. *Let* $f_1, \cdots, f_m$ *be a finite number of functions, belonging to* $\mathfrak{A}(p)$. *In order that* $\{f_1, \cdots, f_m\}$ *should be a system of coordinates at* p, *the following conditions are necessary and sufficient:*

1) $m = n$, *and,*

2) *if* $f_i = f_i^*(x_1, \cdots, x_n)$ *is the expression of* f_i *in terms of the*

coordinates $x_1, \cdots, x_n$, *the functional determinant*,

$$\frac{D(f_i^*, \cdots, f_n^*)}{D(x_1, \cdots, x_n)},$$

is $\neq 0$ *for* $x_1 = x_1(p), \cdots, x_n = x_n(p)$.

1) The conditions are necessary. In fact, if $\{f_1, \cdots, f_m\}$ is a system of coordinates at p, the function x_i may be expressed in the neighbourhood of p as a function $x_i = g_i(f_1, \cdots, f_m)$, where $g_i(u_1, \cdots, u_m)$ is a function of m real arguments, defined and analytic in a neighbourhood of the point $u_1 = f_1(p), \cdots, u_m = f_m(p)$. Moreover we have

$$f_i^*(g_1(\mathfrak{u}), \cdots, g_n(\mathfrak{u})) = u_i \qquad (1 \leqslant i \leqslant m)$$
$$g_i(f_1^*(\mathfrak{x}), \cdots, f_m^*(\mathfrak{x})) = x_i \qquad (1 \leqslant i \leqslant n)$$

where $\mathfrak{u} = (u_1, \cdots, u_m)$, $\mathfrak{x} = (x_1, \cdots, x_n)$ are points of R^m, R^n respectively, belonging to sufficiently small neighbourhoods of $\mathfrak{u}_0 = (f_1(p), \cdots, f_m(p))$, $\mathfrak{x}_0 = (x_1(p), \cdots, x_n(p))$. We have

$$\Sigma_j \left(\frac{\partial f_i^*}{\partial x_j}\right)_{\mathfrak{x}_0} \left(\frac{\partial g_j}{\partial u_k}\right)_{\mathfrak{u}_0} = \delta_{ik}, \qquad 1 \leqslant i \leqslant m,$$
$$\Sigma_j \left(\frac{\partial g_k}{\partial u_j}\right)_{\mathfrak{u}_0} \left(\frac{\partial f_j^*}{\partial x_i}\right)_{\mathfrak{x}_0} = \delta_{ik}, \qquad 1 \leqslant k \leqslant n.$$

Let us set $a_{ij} = \left(\dfrac{\partial f_i^*}{\partial x_j}\right)_{\mathfrak{x}_0}$. From the first set of equations it follows that the linear equations $\Sigma_j a_{ij} y_i = b_j$ have a solution whatever the right-hand sides, $b_1, \cdots, b_n$, may be. Therefore we must have $m \leqslant n$, and the matrix (a_{ij}) is of rank m. Similarly the second set of equations gives $n \leqslant m$; therefore $m = n$ and $\boxed{(a_{ij})} \neq 0$, which proves that the conditions are necessary.

2) Conversely, let us assume that the conditions 1), 2) are satisfied. Let V be a cubic neighbourhood of p with respect to the system $\{x_1, \cdots, x_n\}$, and let a be the breadth of V. Taking a small enough, we may assume that the functions $f_1, \cdots, f_n$ are defined on V and analytic at every point of V. The implicit function theorem gives the following: there exist two numbers, $a_1 > 0$, $b > 0$ such that if $y_1, \cdots, y_n$ are n real numbers satisfying

$$(2) \qquad |y_i - f_i(p)| < b \qquad (1 \leqslant i \leqslant n)$$

the equations

$$f_i^*(x_1, \cdots, x_n) = y_i$$

have one and only one solution $(x_1, \cdots, x_n)$ which satisfies the conditions

$$|x_i - x_i(p)| < a_1.$$

Moreover, this solution is given by equations of the form

$$x_i = g_i(y_1, \cdots, y_n)$$

where the functions $g_1, \cdots, g_n$ are analytic in the cube Q defined by the inequalities (2).

We may assume without loss of generality that $a_1 < a$. If we assign to every point $\mathfrak{y} = (y_1, \cdots, y_n) \varepsilon Q$ the point $\Phi(\mathfrak{y}) \varepsilon V$ whose coordinates are $x_i = g_i(y_1, \cdots, y_n)$ $(1 \leqslant i \leqslant n)$ we have

$$f_i(\Phi(\mathfrak{y})) = y_i \qquad (1 \leqslant i \leqslant n).$$

It follows that Φ is a homeomorphism of Q with a subset W of V. We can find a number a_2 such that $0 < a_2 < a_1$ and such that the conditions $|x_i - x_i(p)| < a_2$ $(1 \leqslant i \leqslant n)$ imply $|f_i^*(x_1, \cdots, x_n) - f_i(p)| < b$. Therefore W contains all the points $q \varepsilon V$ for which the inequalities $|x_i(q) - x_i(p)| < a_2$ hold, which proves that W is a neighbourhood of p. If $r \varepsilon W$, each of the functions $x_1, \cdots, x_n$ depends analytically on $f_1, \cdots, f_n$ around r, since we have

$$x_i(q) = g_i(f_1(q), \cdots, f_n(q)) \qquad (q \varepsilon W; 1 \leqslant i \leqslant n),$$

It follows that any function $f \varepsilon \mathfrak{A}(r)$ depends analytically on $f_1, \cdots, f_n$ around r. We see that properties 1), 2), 3) of condition III hold for the system $(f_1, \cdots, f_n)$, the neighbourhood W, and the number b. In other words, $\{f_1, \cdots, f_n\}$ is a system of coordinates around p, and W is a cubic neighbourhood of p with respect to this system.

Corollary. *If $\mathfrak{V}$ is a manifold and $p \varepsilon \mathfrak{V}$, the number of functions in a system of coordinates at p is the same for all systems of coordinates at p.*

This number of functions is called the *dimension* of $\mathfrak{V}$ at p. This number does not depend on p. In fact, it follows immediately from Remark 1, p. 70, that p has a neighbourhood V such that the dimension of $\mathfrak{V}$ is the same at all points belonging to V. For every integer $n > 0$, let U_n be the set of points of $\mathfrak{V}$ at which the dimension of $\mathfrak{V}$ is n. Then the sets U_n are all open; they are mutually disjoint, and every point of $\mathfrak{V}$ belongs to one of them. Since $\mathfrak{V}$ is a connected topological space the sets are all empty except one; this proves our assertion.

The common dimension of $\mathfrak{V}$ at all its points is called the *dimension* of $\mathfrak{V}$.

§II. Examples of Manifolds

Let V be a set on which are defined n real valued functions, $f_1, \cdots, f_n$ with the following property: if we assign to every element $p \varepsilon V$ the point $\Phi(p) \varepsilon R^n$ whose coordinates are $f_1(p), \cdots, f_n(p)$, then the mapping $p \rightarrow \Phi(p)$ is a univalent mapping of V onto an open connected subset of R^n.

Under these conditions there exists a manifold $\mathcal{V}$ whose set of points is V and which is determined by the property that the functions $f_1, \cdots, f_n$ form a system of coordinates on $\mathcal{V}$ at each point of V. In fact, since Φ is univalent, there exists a topological space $\mathfrak{V}$, whose set of points is V, and such that Φ is a homeomorphism of $\mathfrak{V}$ with the subspace $\Phi(V)$ of R^n: the open sets of $\mathfrak{V}$ are the sets which are mapped by Φ onto open subsets of R^n. The topological space $\mathfrak{V}$ is connected. If $p \varepsilon \mathfrak{V}$ let $\mathfrak{A}(p)$ be the class of real valued functions defined on neighbourhoods of p and depending analytically on $f_1, \cdots, f_n$ around p. It is a trivial matter to verify that the assignment $p \rightarrow \mathfrak{A}(p)$ satisfies conditions I, II and III of §I. Therefore this assignment defines a manifold $\mathcal{V}$, and $f_1, \cdots, f_n$ obviously form a system of coordinates at any point of $\mathcal{V}$.

If the set V is equipped *a priori* with a topology, and if Φ is a homeomorphism of V, the topological space defined above coincides with the one given *a priori;* the latter space is the underlying space of the manifold we have constructed.

For example, if we take $V = R^n$ with its usual topology, and take for $f_1, \cdots, f_n$ the coordinates in R^n, we obtain a manifold whose underlying space is R^n. This manifold will also be denoted by R^n. A function f, defined in a neighbourhood of a point $p \varepsilon R^n$, is analytic at p on the manifold R^n if, when expressed as a function of the coordinates, it is analytic at the point $x_1 = x_1(p), \cdots, x_n = x_n(p)$.

A manifold which can be obtained by the above procedure has the property that there exists a set of real valued functions, defined on the whole manifold, and forming a system of coordinates at every point of the manifold. There exist, however, manifolds which do not have this property. In this connection we mention the following problem, which seems to be of the utmost difficulty: If $\mathcal{V}$ is a manifold, does there exist a finite set of real valued functions, $f_1, \cdots, f_N$, defined and analytic at all points of $\mathcal{V}$, and having the property that at each $p \varepsilon \mathcal{V}$ some subset of the set $\{f_1, \cdots, f_N\}$ is a coordinate system at p? In fact, it is not even known whether there always exists on a manifold a non-constant function which is everywhere analytic.

We shall now construct a manifold whose underlying space is the one-dimensional torus T^1, i.e. the factor group of the additive group R of real numbers by the group Z of integers. Let $\mathfrak{x}$ be any point of T^1; $\mathfrak{x}$ is a residue class of R modulo Z, i.e. it consists of a real number x and all other real numbers which can be obtained from x by addition of arbitrary integers. If $f(x)$ is a periodic function of period 1 it takes the same value at all points x of the residue class $\mathfrak{x}$; we may denote this value by $f(\mathfrak{x})$, and then f becomes the symbol of a function defined on T^1. In particular, the functions $\sin 2\pi\mathfrak{x}$, and $\cos 2\pi\mathfrak{x}$ are real valued functions defined on T^1. Let $\mathfrak{A}(\mathfrak{x}_0)$ be the class of functions defined in neighbourhoods of $\mathfrak{x}_0$ in T^1 and depending analytically around $\mathfrak{x}_0$ on $\sin 2\pi\mathfrak{x}$, $\cos 2\pi\mathfrak{x}$. It is easy to see that the assignment $\mathfrak{x}_0 \rightarrow \mathfrak{A}(\mathfrak{x}_0)$ satisfies conditions I, II and III of §I. Hence it defines a manifold, which we shall also denote by T^1. If the residue class $\mathfrak{x}_0$ does not contain $\frac{1}{4}$ or $\frac{3}{4}$ the function $\sin 2\pi\mathfrak{x}$ is a system of coordinates at $\mathfrak{x}_0$; if it does not contain 0 or $\frac{1}{2}$ the function $\cos 2\pi\mathfrak{x}$ is a system of coordinates. It is easy to see, however, that no function can be a system of coordinates at every point of T^1.

Let $\mathcal{V}$ be a manifold, U a connected open subset of $\mathcal{V}$, and $\mathfrak{A}(p)$ the class of analytic functions at p on $\mathcal{V}$. If we assign to every $p \varepsilon U$ the class of functions of the form $f \circ I$, where f is any function in $\mathfrak{A}(p)$ and I is the identity mapping of U into $\mathcal{V}$, we clearly obtain a manifold $\mathcal{U}$ whose underlying space is U. Such a manifold is called an *open submanifold* of $\mathcal{V}$.

Definition 1. *Let $\mathcal{V}$, $\mathcal{W}$ be manifolds and let Φ be a mapping of V into $\mathcal{W}$, where V is some neighbourhood of the point $p \varepsilon \mathcal{V}$. The mapping Φ is said to be analytic at p if the following condition is satisfied: if g is any function on $\mathcal{W}$ which is analytic at $\Phi(p)$ then $g \circ \Phi$ is analytic at p on $\mathcal{V}$.*

Suppose furthermore that Φ is a homeomorphism of $\mathcal{V}$ with $\mathcal{W}$. Then Φ is called an analytic isomorphism of $\mathcal{V}$ with $\mathcal{W}$ if both Φ and its reciprocal mapping $\overset{-1}{\Phi}$ are everywhere analytic.

Suppose that $\mathcal{W}$ is a given manifold, that $\mathfrak{B}$ is some topological space and that Φ is a homeomorphism of $\mathfrak{B}$ with some connected subset U of $\mathcal{W}$. Then U is the underlying space of an open submanifold $\mathcal{U}$ of $\mathcal{W}$. We may define a manifold $\mathcal{V}$, whose underlying space is $\mathfrak{B}$, by the condition that Φ shall be an analytic isomorphism of $\mathcal{V}$ with $\mathcal{U}$. To do this we merely assign to each $p \varepsilon \mathfrak{B}$ the class $\mathfrak{A}(p)$ of functions of the form $f \circ \Phi$, with f any function which is analytic at $\Phi(p)$ on $\mathcal{U}$.

Remark. A homeomorphism Φ of a manifold $\mathcal{V}$ with a manifold $\mathcal{W}$ may be everywhere analytic without being an analytic isomorphism.

In fact, let us take for $\mathcal{V}$ the manifold R of real numbers, as defined above, and for $\mathcal{W}$ the manifold which has the same underlying space as $\mathcal{V}$, but which is characterized by the fact that the mapping $x \to \psi(x) = x^3$ is an analytic isomorphism of R with $\mathcal{W}$. Let Φ be the identity mapping of R into $\mathcal{W}$; Φ is clearly an everywhere analytic homeomorphism of R with $\mathcal{W}$. Since the function x is analytic at 0 on R but not on $\mathcal{W}$, Φ is not an analytic isomorphism. This example also shows that distinct manifolds may have the same underlying space.

Proposition 1. *Let Φ be a mapping of a manifold $\mathcal{V}$ into a manifold $\mathcal{W}$. In order for Φ to be an analytic isomorphism of $\mathcal{V}$ with $\mathcal{W}$ it is necessary and sufficient that the following conditions be satisfied:* 1) *Φ is a homeomorphism of $\mathcal{V}$ with $\mathcal{W}$,* 2) *if p is any point of $\mathcal{V}$, and if $\{y_1, \cdots, y_n\}$ is a system of coordinates on $\mathcal{W}$ at $\Phi(p)$ the functions $y_1 \circ \Phi, \cdots, y_n \circ \Phi$ form a system of coordinates at p on $\mathcal{V}$.*

1) Suppose that Φ is an analytic isomorphism and let Ψ be its reciprocal mapping. If f is analytic at p on $\mathcal{V}$, $f \circ \Psi$ is analytic at $\Phi(p)$ on $\mathcal{W}$, and hence depends analytically on $y_1, \cdots, y_n$ around $\Phi(p)$. Since $f = (f \circ \Psi) \circ \Phi$, f depends analytically around p on the functions $y_1 \circ \Phi, \cdots, y_n \circ \Phi$. It follows that, the functions $y_1 \circ \Phi, \cdots, y_n \circ \Phi$ form a system of coordinates at p and that if W is a cubic neighbourhood of $\Phi(p)$ with respect to the system $\{y_1, \cdots, y_n\}$, $\Psi(W)$ is a cubic neighbourhood of p with respect to this system.

2) Suppose that the conditions 1), 2) are satisfied. If g is analytic on $\mathcal{W}$ at $\Phi(p)$, i.e. if g depends analytically around $\Phi(p)$ on $y_1, \cdots, y_n$ then $g \circ \Phi$ depends analytically on $y_1 \circ \Phi, \cdots, y_n \circ \Phi$ around p, and hence $g \circ \Phi$ is analytic at p. This shows that Φ is everywhere analytic. If f is any function which is analytic at p on $\mathcal{V}$, f depends analytically around p on $y_1 \circ \Phi, \cdots, y_n \circ \Phi$. Hence $f \circ \Psi$ depends analytically around $\Phi(p)$ on the functions $(y_1 \circ \Phi) \circ \Psi = y_1, \cdots, (y_n \circ \Phi) \circ \Psi = y_n$, which shows that $f \circ \Psi$ is analytic at $\Phi(p)$ on $\mathcal{W}$. Therefore Ψ is everywhere analytic, and hence Φ is an analytic isomorphism.

§III. Products of Manifolds

Let $\mathcal{V}$ and $\mathcal{W}$ be manifolds of dimensions m and n respectively, and let $\mathfrak{V}$, $\mathfrak{W}$ be their underlying topological spaces. The cartesian product $\mathfrak{V} \times \mathfrak{W}$ is a connected topological space which we shall now make the underlying space of a manifold.

Let (p, q) be a point of $\mathfrak{V} \times \mathfrak{W}(p \in \mathfrak{V}, q \in \mathfrak{W})$. We denote by $\mathfrak{A}(p)$, $\mathfrak{B}(q)$ the classes of analytic functions at p and q on $\mathfrak{V}$ and $\mathfrak{W}$ respectively.

We denote by $\bar{\omega}_1$, $\bar{\omega}_2$ the projections of $\mathfrak{V} \times \mathfrak{W}$ onto $\mathfrak{V}$ and $\mathfrak{W}$

respectively ($\bar{\omega}_1(p, q) = p$, $\bar{\omega}_2(p, q) = q$). Let $\mathcal{C}(p, q)$ be the class consisting of the functions $f \circ \bar{\omega}_1 (f \varepsilon \mathfrak{A}(p))$, $g \circ \bar{\omega}_2 (g \varepsilon \mathfrak{B}(g))$ and of all functions which depend analytically on these around (p, q). The assignment, $(p, q) \to \mathcal{C}(p, q)$, defines a manifold whose underlying space is $\mathfrak{V} \times \mathfrak{W}$. In fact, the class $\mathcal{C}(p, q)$ obviously satisfies conditions I and II of §I. In order to verify that condition III holds we choose systems of coordinates $\{x_1, \cdots, x_m\}$ on $\mathcal{V}$ at p and $\{y_1, \cdots, y_n\}$ on $\mathcal{W}$ at q. If a is a sufficiently small positive number we can find a cubic neighbourhood V of p with respect to the system $\{x_1, \cdots, x_m\}$ and a cubic neighbourhood W of q with respect to the system $\{y_1, \cdots, y_n\}$, both of breadth a.

We set $z_1 = x_1 \circ \bar{\omega}_1, \cdots, z_m = x_m \circ \bar{\omega}_1, z_{m+1} = y_1 \circ \bar{\omega}_2, \cdots, z_{m+n} = y_n \circ \bar{\omega}_2$. The functions $z_1, \cdots, z_{m+n}$ are defined on $V \times W$ and belong to $\mathcal{C}(p', q')$ for every $(p', q') \varepsilon V \times W$. Every function of the form $f \circ \bar{\omega}_1$, $f \varepsilon \mathfrak{A}(p')$, depends analytically on $z_1, \cdots, z_m$ around (p', q'), and every function $g \circ \bar{\omega}_2$, $g \varepsilon \mathfrak{B}(q')$, depends analytically on $z_{m+1}, \cdots, z_{m+n}$ around (p', q'). Hence any function in $\mathcal{C}(p', q')$ depends analytically on $z_1, \cdots, z_{m+n}$ around (p', q').

Finally, if we assign to a point $(p_1, q_1) \varepsilon V \times W$ the point of R^{m+n} whose coordinates are $z_1(p_1, q_1) = x_1(p_1), \cdots, z_m(p_1, q_1) = x_m(p_1)$, $z_{m+1}(p_1, q_1) = y_1(q_1), \cdots, z_{m+n}(p_1, q_1) = y_n(q_1)$, we clearly obtain a homeomorphism of $V \times W$ with a cube of sidelength a in R^{m+n}. Therefore condition III holds.

The manifold obtained in this way is called the product of the manifolds $\mathcal{V}$, $\mathcal{W}$ and denoted by $\mathcal{V} \times \mathcal{W}$. We may, in the same way, define the product of any finite number of manifolds.

If $\mathcal{V}$, $\mathcal{W}$, $\mathcal{X}$ are manifolds, then, strictly speaking, the manifolds $(\mathcal{V} \times \mathcal{W}) \times \mathcal{X}$, $\mathcal{V} \times (\mathcal{W} \times \mathcal{X})$ and $\mathcal{V} \times \mathcal{W} \times \mathcal{X}$, are not the same. However, between any two of them there is a natural analytic isomorphism. For instance, the mapping $((p, q), r) \to (p, (q, r))$ is an analytic isomorphism of $(\mathcal{V} \times \mathcal{W}) \times \mathcal{X}$ onto $\mathcal{V} \times (\mathcal{W} \times \mathcal{X})$ while the mapping $((p, q), r) \to (p, q, r)$ is an analytic isomorphism of $(\mathcal{V} \times \mathcal{W}) \times \mathcal{X}$ onto $\mathcal{V} \times \mathcal{W} \times \mathcal{X}$.

The manifold R^n defined in §II is obviously the product of n manifolds identical with R. If we construct the product of n manifolds identical with T^1 we obtain a manifold whose underlying space is the n-dimensional torus; we shall denote this manifold by T^n.

§IV. TANGENT VECTORS. DIFFERENTIALS

Let $\mathcal{V}$ be a manifold of dimension n, p a point of $\mathcal{V}$, and $\mathfrak{A}(p)$ the class of analytic functions at p. By a *tangent vector* at p we shall mean a mapping L of $\mathcal{A}(p)$ into the real numbers which satisfies the following two conditions:

1) *L is linear, i.e. for any two functions f, g in $\mathcal{A}(p)$ and real numbers a, b we have* $L(af + bg) = aL(f) + bL(g)$,

2) *L is a differentiation, i.e. for any two functions f, g in $\mathcal{A}(p)$ we have* $L(fg) = (L(f))g(p) + f(p)(L(g))$.

If L is a tangent vector and f is a function in $\mathcal{A}(p)$ the number $L(f)$ is often called the *derivative of f in the direction L.*

If L, L' are tangent vectors to $\mathcal{V}$ at the point p, it is clear that (for any λ, $\lambda' \varepsilon R$) the mapping

$$f \to \lambda L(f) + \lambda' L'(f)$$

is again a tangent vector to $\mathcal{V}$ at p. Hence the tangent vectors at p form a vector space, called the *tangent vector space* to $\mathcal{V}$ at p.

Now let $\{x_1, \cdots, x_n\}$ be any coordinate system at p. If f is analytic at p, f has, in some neighbourhood of p, an expression in terms of these coordinates:

$$f(q) = f^*(x_1(q), \cdots, x_n(q))$$

where $f^*(u_1, \cdots, u_n)$ is a function of n real variables, defined and analytic in a neighbourhood of the point $u_1 = x_1(p), \cdots, u_n = x_n(p)$. To simplify the notation we shall write $\frac{\partial f}{\partial x_i}$ when we mean $\left.\frac{\partial f^*}{\partial u_i}\right]_{u_i = x_i(p)}$. Then it is trivial that for any choice of real numbers $\lambda_1, \cdots, \lambda_n$ the mapping of $\mathcal{A}(p)$ into the real numbers defined by

$$f \to \Sigma_i \frac{\partial f}{\partial x_i} \lambda_i$$

is a tangent vector at p. Now we shall prove that every tangent vector at p is of this form by showing that *if $\{x_1, \cdots, x_n\}$ is any coordinate system at p, and L any tangent vector at p, we have, for all $f \varepsilon \mathcal{A}(p)$,*

$$L(f) = \Sigma_i \frac{\partial f}{\partial x_i} L(x_i) \tag{1}$$

This relation is also significant because it shows that a tangent vector is uniquely determined by the values it assigns to the functions of a coordinate system.

To prove (1) we first remark that it is trivial that every tangent vector maps every constant function into 0. If f is any function in $\mathcal{A}(p)$ then we can express f (in a neighbourhood of p) in the form:

$$f = a_0 + a_1(x_1 - x_1^0) + \cdots + a_n(x_n - x_n^0) + \Sigma_{i,j=1}^n (x_i - x_i^0)(x_j - x_j^0)g_{ij}$$

with the functions g_{ij} in $\mathfrak{A}(p)$, and where $x_1^0 = x_1(p), \cdots, x_n^0 = x_n(p)$. Applying L we find

$$\begin{aligned}(2)\quad Lf &= a_1L(x_1 - x_1^0) + \cdots + a_nL(x_n - x_n^0) + L(\Sigma_{i,j=1}^n(x_i - x_i^0)(x_j - x_j^0)g_{ij}) \\ &= a_1Lx_1 + \cdots + a_nLx_n + \Sigma_{i,j=1}^n L((x_i - x_i^0)(x_j - x_j^0)g_{ij})\end{aligned}$$

Then, making use of the differentiation property of a tangent vector, we have

$$\begin{aligned}L((x_i - x_i^0)(x_j - x_j^0)g_{ij}) &= ((x_i(p) - x_i^0)L(x_j - x_j^0) \\ &\quad + (x_j(p) - x_j^0)L(x_i - x_i^0)) + (Lg_{ij})((x_i(p) - x_i^0)(x_j(p) - x_j^0)) \\ &= 0\end{aligned}$$

Hence (2) $\left(\text{because } a_j = \dfrac{\partial f}{\partial x_j}\right)$ yields (1).

If we set

$$L_i(f) = \frac{\partial f}{\partial x_i} \qquad (1 \leqslant i \leqslant n)$$

we obtain a tangent vector L_i for which $L_i(x_j) = \delta_{ij}$. These n tangent vectors are linearly independent, since $(\Sigma_i\lambda_iL_i)(x_j) = \lambda_j$. Moreover, if L is any tangent vector, we have $L(x_j) = (\Sigma_iL(x_i)L_i)(x_j)$ $(1 \leqslant j \leqslant n)$, and hence $L = \Sigma_iL(x_i)L_i$. *It follows that the tangent space is an n-dimensional vector space.*

Now let $\mathcal{W}$ be a manifold and let Φ be a mapping of $\mathcal{V}$ into $\mathcal{W}$, analytic at the point $p \in \mathcal{V}$. Let, moreover, L be a tangent vector to $\mathcal{V}$ at p, and g be any analytic function on $\mathcal{W}$ at the point $q = \Phi(p)$. If we set

$$(3)\qquad M(g) = L(g \circ \Phi)$$

we clearly obtain a tangent vector M to $\mathcal{W}$ at q. It is also clear that the mapping, $L \to M$, is linear.

Definition 1. *The mapping which assigns to every tangent vector L to $\mathcal{V}$ at p the tangent vector M to $\mathcal{W}$ at q, defined by* (3) *is called the differential of the mapping Φ at p. It is usually denoted by $d\Phi$, or $d\Phi_p$.*

Suppose now that Ψ is a mapping of $\mathcal{W}$ into a third manifold $\mathcal{X}$, and that Ψ is analytic at q. If h is any function on $\mathcal{X}$ analytic at $r = \Psi(q)$, the functions $(h \circ \Psi) \circ \Phi$ and $h \circ (\Psi \circ \Phi)$ coincide in the neighbourhood of p. It follows at once that

$$d(\Psi \circ \Phi)_p = d\Psi_q \circ d\Phi_p$$

Proposition 1. *Let $\mathcal{V}$ and $\mathcal{W}$ be manifolds. Let Φ be an analytic mapping of $\mathcal{V}$ into $\mathcal{W}$, and let p be a point of $\mathcal{V}$. Suppose that $d\Phi_p$ is a univalent mapping of the tangent space to $\mathcal{V}$ into the tangent space to $\mathcal{W}$. Then, if $\{y_1, \cdots, y_m\}$ is a system of coordinates at $q = \Phi(p)$ on $\mathcal{W}$, it is possible to select from the set of functions $y_1 \circ \Phi, \cdots, y_m \circ \Phi$ a subset containing n functions which form a system of coordinates at p on $\mathcal{V}$. Moreover, if $\{x_1, \cdots, x_n\}$ is any system of coordinates at p on $\mathcal{V}$, there exists a system of coordinates $z_1, \cdots, z_m$ at q on $\mathcal{W}$ such that x_j coincides in the neighbourhood of p with $z_j \circ \Phi$ $(1 \leqslant j \leqslant n)$.*

In fact, the function $y_i \circ \Phi$ can be expressed in the neighbourhood of p in the form $\varphi_i(x_1, \cdots, x_n)$, where φ_i is a function of n real variables, analytic at the point $x_1 = x_1(p), \cdots, x_n = x_n(p)$. We shall show that the rectangular matrix $\left(\dfrac{\partial\varphi_i}{\partial x_j}\right)_{x=x(p)}$ is of rank n. Suppose that $0 = \Sigma_{j=1}^n \lambda_j \left(\dfrac{\partial\varphi_i}{\partial x_j}\right)_{x=x(p)}$ $(1 \leqslant i \leqslant m)$ is a linear relation between columns of this matrix. Let L_j be the tangent vector to $\mathcal{V}$ at p defined by $L_j(x_k) = \delta_{kj}$ $(1 \leqslant j,\ k \leqslant n)$, and let L be the vector $\Sigma\lambda_j L_j$. We have $L(y_i \circ \Phi) = \Sigma\lambda_j \left(\dfrac{\partial\varphi_i}{\partial x_j}\right)_{x=x(p)} = 0$ whence $(d\Phi(L))y_i = 0$ $(1 \leqslant i \leqslant m)$. It follows that $d\Phi(L) = 0$; since $d\Phi$ is univalent, we have $L = 0$, $\lambda_1 = \lambda_2 = \cdots = \lambda_n = 0$, which proves our assertion.

We can select n indices, $i_1, \cdots, i_n$, from the set $\{1, \cdots, m\}$ so that the determinant formed from the rows with indices $i_1, \ldots, i_n$ is not 0. It is then clear that $y_{i_1} \circ \Phi, \cdots, y_{i_n} \circ \Phi$ form a system of coordinates at p on $\mathcal{V}$.

We can express x_j in the form $x_j = \psi_j(y_{i_1} \circ \Phi, \cdots, y_{i_n} \circ \Phi)$ in the neighbourhood of p; the ψ_j's are analytic functions of n real variables and their functional determinant does not vanish for $y_{i_j} = y_{i_j}(q)$. We set $z_j = \psi_j(y_{i_1}, \cdots, y_{i_n})$ $(1 \leqslant j \leqslant n)$ and take for $z_{n+1}, \cdots, z_m$ those functions y_i whose indices i do not occur among $i_1, \cdots, i_n$. Clearly $\{z_1, \cdots, z_m\}$ is a system of coordinates at q on $\mathcal{W}$, and $z_j \circ \Phi = x_j$ $(1 \leqslant j \leqslant n)$.

Remark. *We see that, under the assumption of Proposition 1, there exists a neighbourhood of p in $\mathcal{V}$ which is mapped topologically under Φ.*

Definition 2. *A mapping Φ of a manifold $\mathcal{V}$ into a manifold $\mathcal{W}$ is said to be regular at the point $p \in \mathcal{V}$ if Φ is analytic at p and $d\Phi_p$ is a univalent mapping.*

Proposition 2. *The notation being as in Proposition 1, suppose that the image under $d\Phi_p$ of the tangent space to $\mathcal{V}$ (at p) covers the whole tangent space to $\mathcal{W}$ (at $q = \Phi(p)$). Then if $\{y_1, \cdots, y_m\}$ is a system of coordinates at q on $\mathcal{W}$, the functions $y_1 \circ \Phi, \cdots, y_m \circ \Phi$ are part of a system of coordinates at p on $\mathcal{V}$.*

Let $x_1, \cdots, x_n$ be the functions of a system of coordinates at p on $\mathcal{V}$. Here again we may express $y_i \circ \Phi$ in the form $\varphi_i(x_1, \cdots, x_n)$ in the neighbourhood of p, and we shall show that the rank of the matrix $\left(\frac{\partial \varphi_i}{\partial x_j}\right)_{x=x(p)}$ is m. In fact, let $\Sigma_{i=1}^m \mu_i \left(\frac{\partial \varphi_i}{\partial x_j}\right)_{x=x(p)} = 0$ be a relation between the rows of this matrix, $(1 \leqslant j \leqslant n)$. Let M_i be the tangent vector to $\mathcal{W}$ at q defined by $M_i(y_k) = \delta_{ik}$ $(1 \leqslant i, k \leqslant m)$. By assumption, there exists a tangent vector L_i to $\mathcal{V}$ at p for which $d\Phi(L_i) = M_i$; we have $\Sigma_j \left(\frac{\partial \varphi_k}{\partial x_j}\right)_{x=x(p)} L_i(x_j) = L_i(y_k \circ \Phi) = \delta_{ik}$. If we multiply by μ_k and sum for $k = 1$ to m, we obtain $\mu_i = 0$, which proves our assertion. We may assume without loss of generality that the determinant formed from the first m columns of our matrix is not 0. The functions $y_1 \circ \Phi, \cdots, y_m \circ \Phi, x_{m+1}, \cdots, x_n$ form a system of coördinates at p on $\mathcal{V}$.

Remark. *It follows immediately that, under the assumption of Proposition 2, the image under Φ of any neighbourhood of p in $\mathcal{V}$ covers a neighbourhood of q in $\mathcal{W}$.*

Proposition 3. *The notation being as in Proposition 1, suppose that $d\Phi_p$ is a linear isomorphism of the tangent space to $\mathcal{V}$ (at p) with the tangent space to $\mathcal{W}$ (at $q = \Phi(p)$). Then there is a neighbourhood V of p which is mapped topologically by Φ onto a neighbourhood W of q in $\mathcal{W}$; moreover, the reciprocal mapping $\overset{-1}{\Phi}$ of W onto V is analytic at q.*

This is an immediate consequence of Propositions 1 and 2.

Proposition 4. *Let Φ be an analytic mapping of a manifold $\mathcal{V}$ into a manifold $\mathcal{W}$. If the differential of Φ is 0 at every point of $\mathcal{V}$, then Φ is a constant mapping* (i.e. *Φ maps $\mathcal{V}$ onto a single point of $\mathcal{W}$*).

Let p be a point of $\mathcal{V}$. Let $\{y_1, \cdots, y_m\}$ be a coordinate system at the point Φp on $\mathcal{W}$, and let W be a cubic neighbourhood of Φp with respect to these coordinates. Let $\{x_1, \cdots, x_n\}$ be a coordinate system at p on $\mathcal{V}$, and let V be a cubic neighbourhood of p with respect to the coordinates x such that $\Phi(V) \subset W$. If $q \in V$, we may write

$y_j(\Phi q) = F_j(x_1(q), \cdots, x_n(q))$, where the functions F_j are analytic in the cube defined by the inequalities $|x_i - x_i(p)| < a$ (where a is the breadth of V). If $q \varepsilon V$, denote by $X_{i,q}$ the tangent vector to $\mathfrak{V}$ at q which is defined by $X_{i,q}x_k = \delta_{ik}$. We have

$$\frac{\partial F_j}{\partial x_i}(x_1(q), \cdots, x_n(q)) = X_{i,q}(y_j \circ \Phi) = (d\Phi_q(X_{i,q}))y_j = 0$$

It follows that the partial derivatives of the functions F_j are equal to 0 and therefore that the functions F_j are constant. This means that Φ maps V onto the point Φp.

To every point $r \varepsilon \mathfrak{W}$ we associate the set U_r of points $p \varepsilon \mathfrak{V}$ which are mapped on r by Φ. It follows from what we have proved that each U_r is open. On the other hand, the sets U_r are mutually disjoint and the union of all these sets is $\mathfrak{V}$. Since $\mathfrak{V}$ is connected, there can be only one set U_r which is not empty. Proposition 4 is thereby proved.

The differential of a function

A real valued analytic function f, defined on $\mathfrak{V}$, may be considered as a mapping of $\mathfrak{V}$ into the manifold R of real numbers. Its differential at p is a linear mapping of the tangent space, $\mathfrak{L}$, to $\mathfrak{V}$ at p into the tangent space $\mathfrak{M}_{x_0}$ to R at $x_0 = f(p)$. Since $\mathfrak{M}_{x_0}$ is a one dimensional linear space over R, spanned by the vector M_0 defined by $M_0(x) = 1$ (considering x as a real valued function on R), we may identify $\mathfrak{M}_{x_0}$ with R itself by identifying M_0 with the number 1. This makes df into a linear function defined on $\mathfrak{L}$, and with real values. It follows directly from the definitions that

$$df(L) = L(f).$$

If f_1, f_2 are analytic functions at p on $\mathfrak{V}$ we have $d(\lambda_1 f_1 + \lambda_2 f_2)(L) = \lambda_1 df_1(L) + \lambda_2 df_2(L)$. Hence the differentials df, for f in $\mathfrak{A}(p)$, form a linear subspace of the space of all linear functions on $\mathfrak{L}$. If $x_1, \cdots, x_n$ form a coordinate system at p, their differentials $dx_1, \cdots, dx_n$ are obviously linearly independent. Hence the space $\mathfrak{D}$ of differentials is of dimension n, and coincides with the space of all linear functions on $\mathfrak{L}$.

The preceding shows that the spaces $\mathfrak{L}$ and $\mathfrak{D}$ may be considered as dual vector spaces.[1]

[1] The dual space of a vector space $\mathfrak{L}$ over a field K is the set of all linear mappings of $\mathfrak{L}$ into K. If $\mathfrak{L}$ is of finite dimension, it may be identified with the dual of its dual space.

Product manifolds

Let $\mathfrak{V}_1$, $\mathfrak{V}_2$ be manifolds of dimensions n_1, n_2 and let $\mathfrak{V} = \mathfrak{V}_1 \times \mathfrak{V}_2$ be their product. Let p_1 be a point of $\mathfrak{V}_1$, p_2 a point of $\mathfrak{V}_2$, and p be the point (p_1, p_2). We denote by $\mathfrak{L}_1$, $\mathfrak{L}_2$, $\mathfrak{L}$ the tangent spaces to $\mathfrak{V}_1$, $\mathfrak{V}_2$, $\mathfrak{V}$ at p_1, p_2, p respectively.

Let, moreover, $\bar{\omega}_1$ and $\bar{\omega}_2$ be the projections of $\mathfrak{V}$ onto $\mathfrak{V}_1$ and $\mathfrak{V}_2$. To every vector $L \varepsilon \mathfrak{L}$ there corresponds vectors $L_1 = d\bar{\omega}_1(L) \varepsilon \mathfrak{L}_1$, $L_2 = d\bar{\omega}_2(L) \varepsilon \mathfrak{L}_2$. Let $\{x_1, \cdots, x_{n_1}\}$ be a coordinate system at p_1 on $\mathfrak{V}_1$, and let $\{y_1, \cdots, y_{n_2}\}$ be a system of coordinates at p_2 on $\mathfrak{V}_2$. The functions $z_1 = x_1 \circ \bar{\omega}_1, \cdots, z_{n_1} = x_{n_1} \circ \bar{\omega}_1$, $z_{n_1+1} = y_1 \circ \bar{\omega}_2, \cdots, z_{n_1+n_2} = y_{n_2} \circ \bar{\omega}_2$ then form a system of coordinates at p on $\mathfrak{V}$.

Let L_1, L_2 be arbitrary vectors in $\mathfrak{L}_1$, $\mathfrak{L}_2$ respectively. There is a vector $L \varepsilon \mathfrak{L}$ defined by the equalities

$$L(z_1) = L_1(x_1), \cdots, L(z_{n_1}) = L_1(x_{n_1});$$
$$L(z_{n_1+1}) = L_2(y_1), \cdots, L(z_{n_1+n_2}) = L_2(y_{n_2}),$$

and it is clear that $d\bar{\omega}_1(L) = L_1$, $d\bar{\omega}_2(L) = L_2$. Since there can be only one vector L for which these equalities hold we see that we may identify $\mathfrak{L}$ with the product of the spaces $\mathfrak{L}_1$ and $\mathfrak{L}_2$.

We have already identified the tangent space to the manifold R of real numbers at any point with R itself. Hence we may identify R^n with the tangent space to the manifold R^n.

§V. INFINITESIMAL TRANSFORMATIONS

Definition 1. *Let $\mathfrak{V}$ be a manifold. A vector field X on $\mathfrak{V}$, (also called an infinitesimal transformation), is a mapping which assigns to every point p a tangent vector $X(p)$ to $\mathfrak{V}$ at this point.*

Let f be any function defined and analytic at the points of some open subset U of $\mathfrak{V}$. Setting $g(p) = X(p)f$ for $p \varepsilon U$ we obtain a function defined on U, which we shall denote by Xf. If for each analytic f the function Xf is also analytic we shall say that X is an *analytic* infinitesimal transformation.

If U is an open subset of $\mathfrak{V}$ on which there exists a system of coordinates $\{x_1, \cdots, x_n\}$, there always exists an analytic infinitesimal transformation defined on U. In fact, let f be any function analytic at a point $p \varepsilon U$; we can express f in the neighbourhood of p as a function $f^*(x_1, \cdots, x_n)$ of $x_1, \cdots, x_n$. Setting $X_1(p)f = \left(\frac{\partial f^*}{\partial x_1}\right)_p$ we obtain a tangent vector at p, and the mapping $p \rightarrow X_1(p)$ is clearly an analytic infinitesimal transformation defined on U. Moreover, if we set $X_i(p) = \left(\frac{\partial f^*}{\partial x_i}\right)_p$ $(1 \leq i \leq n)$ we obtain n analytic infinitesimal transformations which are linearly independent at every point of U. If X is any other infinitesimal transformation defined on U we can write X in the form $X(p) = \Sigma_i A_i(p) X_i(p)$ where $A_1, \cdots, A_n$ are n functions defined on U. If X is analytic, the functions $A_i(p)$ are also analytic, since $A_i = Xx_i$ $(1 \leq i \leq n)$. Conversely, if $A_1, \cdots, A_n$ are n functions defined and analytic on U, it is clear that $X = \Sigma A_i X_i$ is an analytic infinitesimal transformation on U. Since $(Xf)(p) = \Sigma A_i(p) \left(\frac{\partial f^*}{\partial x_i}\right)_p$, we shall call $\Sigma A_i \frac{\partial}{\partial x_i}$ the *symbol* of the infinitesimal transformation X.

If X and Y are analytic infinitesimal transformations defined on a manifold $\mathcal{V}$, the operation $YX = Y \circ X$ is not in general an infinitesimal transformation. For instance, if $\mathcal{V} = R^n$ and if X and Y are defined by $Xf = \frac{\partial f}{\partial x_1}$, $Yf = \frac{\partial f}{\partial x_2}$, we have $YXf = \frac{\partial^2 f}{\partial x_1 \partial x_2}$, and the mapping $f \to \left(\frac{\partial^2 f}{\partial x_1 \partial x_2}\right)_p$ is not a tangent vector to R^n (here p is a point in R^n). However, the operation $U = YX - XY$ is always an analytic infinitesimal transformation; the proof of this consists in a straightforward verification (which we shall omit) that $U(p)$ satisfies the conditions 1) and 2) in the definition of a tangent vector. In terms of a coordinate system $\{x_1, \cdots, x_n\}$ at a point p we can write (for $f \varepsilon \mathfrak{A}(p)$) Xf and Yf in the neighbourhood of p in the forms $\Sigma A_i^*(x_1, \cdots, x_n) \frac{\partial f^*}{\partial x_i}$, $\Sigma B_i^*(x_1, \cdots, x_n) \frac{\partial f^*}{\partial x_i}$. Then, we find that

$$(1) \qquad (Uf)_p = \Sigma_{ij} \left(B_i^* \frac{\partial A_j^*}{\partial x_i} - A_i^* \frac{\partial B_j^*}{\partial x_i}\right)_p \left(\frac{\partial f^*}{\partial x_j}\right)_p$$

This expression for Uf yields a second proof that $XY - YX$ is an analytic infinitesimal transformation.

Definition 2. *If X, Y are analytic infinitesimal transformations on $\mathcal{V}$, then the infinitesimal transformation $U = YX - XY$ will be denoted by $[X, Y]$.*

This bracket operation, which assigns to every pair of analytic infinitesimal transformations (X, Y) the infinitesimal transformation

$[X, Y]$, is a law of composition for the infinitesimal transformations. We also see immediately that, if a is any number and X an infinitesimal transformation, then aX is again an infinitesimal transformation, and, if X, Y are infinitesimal transformations, then so is $X + Y$.

The bracket operation is distributive with respect to addition:

$$[a_1X_1 + a_2X_2, Y] = a_1[X_1, Y] + a_2[X_2, Y]$$
$$[X, a_1Y_1 + a_2Y_2] = a_1[X, Y_1] + a_2[X, Y_2]$$

(for $a_1, a_2 \varepsilon R$, and X, X_1, X_2, Y, Y_1, Y_2 infinitesimal transformations). However, it is *not* associative: in general we have $[[X, Y], Z] \neq [X, [Y, Z]]$. It is easy to prove that it satisfies the following identities:

$$[X, X] = 0$$
$$[[X, Y], Z] + [[Y, Z], X] + [[Z, X], Y] = 0$$

for any analytic infinitesimal transformations X, Y, Z. The first of these identities gives $[X + Y, X + Y] = 0 = [X, Y] + [Y, X]$, whence

$$[Y, X] = -[X, Y].$$

The second, is called the *Jacobi* identity.

Let Φ be an analytic mapping of the manifold $\mathfrak{V}$ into some manifold $\mathfrak{W}$. Let X be an infinitesimal transformation on $\mathfrak{V}$ and Y be an infinitesimal transformation on $\mathfrak{W}$. We shall say that X and Y are *Φ-related* if, for every point $p \varepsilon \mathfrak{V}$, we have

$$d\Phi_p(X_p) = Y_{\Phi p}.$$

If Φ is everywhere regular, there can exist at most one infinitesimal transformation X on $\mathfrak{V}$ which is Φ-related to a given Y on $\mathfrak{W}$, since then X_p is then entirely determined by $d\Phi_p(X_p)$.

Let $\mathfrak{L}_p$ be the tangent space to $\mathfrak{V}$ at p. Its image under $d\Phi_p$ is a subspace $\tilde{\mathfrak{L}}_p$ of the tangent space $\mathfrak{M}_{\Phi p}$ to $\mathfrak{W}$ at the point Φp. If an infinitesimal transformation Y on $\mathfrak{W}$ is Φ-related to an X on $\mathfrak{V}$, we must necessarily have $Y_{\Phi p} \varepsilon \tilde{\mathfrak{L}}_p$ for every $p \varepsilon \mathfrak{V}$.

Proposition 1. *Let Φ be an everywhere regular mapping of a manifold $\mathfrak{V}$ into a manifold $\mathfrak{W}$. If $p \varepsilon \mathfrak{V}$, let, $\mathfrak{L}_p$ be the tangent space to $\mathfrak{V}$ at p and set $\tilde{\mathfrak{L}}_p = d\phi_p(\mathfrak{L}_p)$. If Y is any analytic infinitesimal transformation on $\mathfrak{W}$ such that $Y_{\Phi p} \varepsilon \tilde{\mathfrak{L}}_p$ for every point $p \varepsilon \mathfrak{V}$, then there exists one and only one analytic infinitesimal transformation X on $\mathfrak{V}$ which is Φ-related to Y.*

Under our assumptions, we can find for every $p \epsilon \mathfrak{V}$ an element $X_p \varepsilon \mathfrak{L}_p$ such that $d\Phi_p(X_p) = Y_{\Phi p}$. We have to prove that the assignment $X: p \rightarrow X_p$ is an analytic infinitesimal transformation. From

Proposition 1, §IV, p. 76 it follows that we can find a system of coordinates $\{y_1, \cdots, y_m\}$ at Φp on $\mathcal{W}$ such that $\{y_1 \circ \Phi, \cdots, y_n \circ \Phi\}$ is a coordinate system at p on $\mathcal{V}$ (m, n being the dimensions of the manifolds $\mathcal{W}$, $\mathcal{V}$). If q is in a sufficiently small neighbourhood of p in $\mathcal{V}$ the equality $d\Phi_q(X_q) = Y_{\Phi q}$ gives

$$X_q(y_i \circ \Phi) = Y_{\Phi q} y_i = (Yy_i)_{\Phi q},$$

i.e., the function $X(y_i \circ \Phi)$ coincides in a neighbourhood of p with $Yy_i \circ \Phi$. Since Y is analytic on $\mathcal{W}$, Yy_i is analytic at Φp; hence $Yy_i \circ \Phi$ is analytic at p, and $X(y_i \circ \Phi)$ is analytic at p, which proves that X is analytic at p.

Proposition 2. *Let Φ be any analytic mapping of a manifold $\mathcal{V}$ into a manifold $\mathcal{W}$. Let X_1, X_2 be analytic infinitesimal transformations on $\mathcal{V}$, and Y_1, Y_2, analytic infinitesimal transformations on $\mathcal{W}$. If X_i is Φ-related to Y_i ($i = 1, 2$), then $[X_1, X_2]$ is Φ-related to $[Y_1, Y_2]$.*

Let p be a point of $\mathcal{V}$, and let g be a function on $\mathcal{W}$, analytic at the point $q = \Phi p$. The fact that X_i, Y_i are Φ-related ($i = 1, 2$) may be expressed by the formula

$$(X_i(g \circ \Phi))_{p'} = (Y_i g)_{\Phi p'}$$

or

$$(X_i(g \circ \Phi))_{p'} = (Y_i g \circ \Phi)_{p'},$$

which holds for any point p' in a suitable neighbourhood of p in $\mathcal{V}$. Hence

$$(Y_2 Y_1 g)_{\Phi p'} = (X_2(Y_1 g \circ \Phi))_{p'} = (X_2 X_1(g \circ \Phi))_{p'}.$$

We obtain a similar formula by interchanging the indices 1, 2; subtracting we then find

$$([Y_1, Y_2]g)_{\Phi p'} = ([X_1, X_2](g \circ \Phi))_{p'}$$

whence $[Y_1, Y_2]_p = d\Phi_p([X_1, X_2]_p)$, which proves Proposition 2.

§VI. Submanifolds. Distributions

Definition 1. *Let $\mathcal{V}$ be a manifold. A manifold $\mathcal{W}$ is called a submanifold of $\mathcal{V}$ if the following conditions are satisfied:* 1) *the set of points of $\mathcal{W}$ is a subset of the set of points of $\mathcal{V}$, and* 2) *the identity mapping of $\mathcal{W}$ into $\mathcal{V}$ is regular at every point of $\mathcal{W}$.*

For example, an open submanifold of $\mathcal{V}$ (as defined in §II, p. 73) is a submanifold in the sense of the present definition. In the case of an open submanifold $\mathcal{W}$ of $\mathcal{V}$ the identity mapping of $\mathcal{W}$ into $\mathcal{V}$ is also a homeomorphism, but it is important to realize that this is not always the case for an arbitrary submanifold of $\mathcal{V}$.[1] It is true, however, that this identity mapping is always continuous.

Let I be the identity mapping of a submanifold $\mathcal{W}$ into a manifold $\mathcal{V}$. If $p \varepsilon \mathcal{W}$, and if f is a function analytic at p on $\mathcal{V}$, the function $f \circ I$ is analytic at p on $\mathcal{W}$. This function will be called the *contraction* of the function f to $\mathcal{W}$. From Proposition 1, §IV, p. 76 it follows that we can find a coordinate system $\{x_1, \cdots, x_n\}$ at p on $\mathcal{V}$ such that the traces $x_1 \circ I, \cdots, x_m \circ I$ of $x_1, \cdots, x_m$ on $\mathcal{W}$ form a coordinate system at p on $\mathcal{W}$ (where m is the dimension of $\mathcal{W}$). Let g be an analytic function at p on $\mathcal{W}$; then g may be expressed, in a neighbourhood of p in $\mathcal{W}$, as a function $g^*(x_1 \circ I, \cdots, x_m \circ I)$ of the coordinates $x_i \circ I$. If we set $f(q) = g^*(x_1(q), \cdots, x_m(q))$, f is analytic at p on $\mathcal{V}$, and $f \circ I$ coincides with g on a neighbourhood of p in $\mathcal{W}$. Therefore, *any function which is analytic at a point $p \varepsilon \mathcal{W}$ coincides in a neighbourhood of p on $\mathcal{W}$ with the contraction of a function which is analytic at p on $\mathcal{V}$.*

However, it is not always true that a function which is everywhere defined and analytic on $\mathcal{W}$ coincides with the contraction of a continuous function on $\mathcal{V}$.

Let $\mathfrak{L}_p$ be the tangent space to $\mathcal{V}$ at a point p which belongs to the submanifold $\mathcal{W}$. The mapping dI_p maps the tangent space to $\mathcal{W}$ at p isomorphically onto a vector subspace $\mathfrak{M}_p$ of $\mathfrak{L}_p$. The space $\mathfrak{M}_p$ is also called (although improperly) the tangent space to $\mathcal{W}$ at p.

Let X be any analytic infinitesimal transformation on $\mathcal{V}$, such that $X_p \varepsilon \mathfrak{M}_p$ for every point $p \varepsilon \mathcal{W}$. Since I is everywhere regular, there exists one and only one analytic infinitesimal transformation Y on W such that $X_p = dI_p(Y_p)$ for all $p \varepsilon \mathcal{W}$. The infinitesimal transformation Y is called the *contraction* of X to $\mathcal{W}$. From Proposition 2, §V, it follows that if X_1, X_2 are analytic infinitesimal transformations on $\mathcal{V}$, and Y_1, Y_2 their contractions to $\mathcal{W}$, $[Y_1, Y_2]$ is the contraction of $[X_1, X_2]$.

Definition 2. *An m-dimensional vector subspace of the tangent space to a manifold $\mathcal{V}$ at a point p is called an element of contact of dimension m of $\mathcal{V}$. The point p is called the origin of this element of contact. A law which assigns to every point $p \varepsilon \mathcal{V}$ an element of contact of dimension m and of origin p is called an m-dimensional distribution.*

[1] For instance, a non-compact one parametric sub-group of T^2 may be considered as the underlying set of points of a manifold which is analytically isomorphic with R^1. The identity mapping of this manifold into T^2 is analytic and regular everywhere, but is not a homeomorphism with a subspace of T^2.

Let us denote by $\mathfrak{M}$ a distribution of dimension m on $\mathfrak{V}$, and by $\mathfrak{M}_p$ the m-dimensional subspace which is assigned to the point p by $\mathfrak{M}$.

Definition 3. *We shall say that the distribution $\mathfrak{M}$ is analytic at the point p if the following conditions are satisfied; there exists a neighbourhood V of the point p and a system of m infinitesimal transformations $X_1, \cdots, X_m$, defined and analytic on V, such that, for every point $q \varepsilon V$, the vectors $(X_1)_q, \cdots, (X_m)_q$ form a base of the space $\mathfrak{M}_q$. The system $\{X_1, \cdots, X_m\}$ is then called a local base for the distribution around the point p.*

Remark. There do not always exist analytic distributions on a manifold. For instance, it can be proved that there is no analytic distribution of dimension 1 on a sphere of dimension 4.

Since there always exists on a manifold of dimension n an obvious distribution of dimension n, it follows from the preceeding remark that, given an analytic distribution, it is not always possible to find a system of analytic infinitesimal transformations which forms a base of the distribution at every point.

Definition 4. *Let $\mathfrak{M}$ be an analytic distribution on a manifold $\mathfrak{V}$. A submanifold $\mathfrak{W}$ of $\mathfrak{V}$ is called an integral manifold of $\mathfrak{M}$ if, for every point $p \varepsilon \mathfrak{W}$, $\mathfrak{M}_p$ coincides with the tangent space to $\mathfrak{W}$ at p.*

Let $\mathfrak{M}$ be a distribution, and let X_1, X_2 be infinitesimal transformations, defined in a neighbourhood V of a point p_0, and such that $(X_1)_p$, $(X_2)_p$ both belong to $\mathfrak{M}_p$ for all points $p \varepsilon$V. If p_0 belongs to an integral manifold $\mathfrak{W}$ of $\mathfrak{M}$, X_1 and X_2 have contractions Y_1, Y_2 to $\mathfrak{W}$; therefore $[X_1, X_2]$ has the contraction $[Y_1, Y_2]$; it follows that $[X_1, X_2]_{p_0} \varepsilon \mathfrak{M}_{p_0}$. This shows that certain conditions must be satisfied if a distribution is to have an integral manifold through p_0.

We shall say that an infinitesimal transformation X, defined and analytic in a neighbourhood of a point $p_0 \varepsilon \mathfrak{V}$, *belongs* to the distribution $\mathfrak{M}$ if we have $X_p \varepsilon \mathfrak{M}_p$ for all points p of this neighbourhood. For instance, any infinitesimal transformation of a base of $\mathfrak{M}$ around p belongs to $\mathfrak{M}$.

Definition 5. *We shall say that the analytic distribution $\mathfrak{M}$ is involutive if the following condition is satisfied: if two analytic infinitesimal transformations X_1, X_2, defined on the same open set, both belong to $\mathfrak{M}$, the infinitesimal transformation $[X_1, X_2]$ also belongs to $\mathfrak{M}$.*

From the preceding remarks, it follows that, if every point of $\mathfrak{V}$ belongs to an integral manifold of $\mathfrak{M}$, $\mathfrak{M}$ is necessarily involutive. In the following sections we shall be concerned mainly with the proof of the converse of this proposition.

We shall conclude this section with the proof of:

Proposition 1. *Let Σ be a set of analytic infinitesimal transformations defined on a manifold $\mathcal{V}$, with the following properties:* 1) *the space $\mathfrak{M}_p$ spanned by the vectors $X_p(p \in \mathcal{V}, X \in \Sigma)$ has the same dimension, m, at all points $p \in \mathcal{V}$; and* 2) *if $X, Y \in \Sigma$, the infinitesimal transformation $[X, Y]$ may be expressed as a linear combination of a finite number of elements of Σ, the coefficients being functions on $\mathcal{V}$. Then the assignment $p \to \mathfrak{M}_p$ is an analytic involutive distribution.*

Let p_0 be a point of $\mathcal{V}$; we can find m elements $X_1, \cdots, X_m$ of Σ such that $(X_1)_{p_0}, \cdots, (X_m)_{p_0}$ are linearly independent, and hence span $\mathfrak{M}_{p_0}$. Let $\{x_1, \cdots, x_n\}$ be a coordinate system at p_0. The rectangular matrix whose coefficients are the functions X_ix_j ($1 \leq i \leq m$, $1 \leq j \leq n$) is of rank m at p_0; being continuous, it is also of rank m at all points p of some cubic neighbourhood V of p_0. It follows that the distribution $\mathfrak{M}$ is analytic, the elements $X_1, \cdots, X_m$ forming a base of $\mathfrak{M}$ around p_0.

Let X be any analytic infinitesimal transformation defined on a neighbourhood of p_0 and belonging to $\mathfrak{M}$. We may assume that X is defined on V, and that $X_p = \Sigma_{j=1}^m g_j(p)(X_j)_p$. We assert that the functions $g_j(p)$ are analytic on V. In fact, they satisfy the linear equations

$$(1) \qquad (Xx_k)_p = \Sigma_j g_j(p)(X_jx_k)_p \qquad (1 \leq k \leq n),$$

whose coefficients are analytic on V; moreover, the matrix of the coefficients of $g_1(p), \cdots, g_m(p)$ has rank m at all points of V. In the neighbourhood of any point $p \in V$, the values of the functions $g_j(p)$ may be found by solving a suitably selected system of m equations from (1), and this proves their analyticity.

Let $Y = \Sigma h_jX_j$ be another analytic infinitesimal transformation defined on V and belonging to $\mathfrak{M}$. We want to prove that $[X, Y]$ belongs to $\mathfrak{M}$. It is obviously sufficient to carry out the proof in the case where $X = gX_i$, $Y = hX_j$, for g, h any two analytic functions on V and i, j any two indices between 1 and m. We have

$$\begin{aligned}[X, Y] &= YX - XY = h(X_jg)X_i + ghX_jX_i - ghX_iX_j - g(X_ih)X_j \\ &= (h(X_jg))X_i - (g(X_ih))X_j + gh[X_i, X_j].\end{aligned}$$

But $[X_i, X_j]$ is a linear combination of elements in Σ and hence belongs to $\mathfrak{M}$; it follows that $[X, Y]$ belongs to $\mathfrak{M}$, which proves Proposition 1.

§VII. INTEGRAL MANIFOLDS OF AN INVOLUTIVE DISTRIBUTION (LOCAL THEORY)

Let $\mathcal{V}$ be a manifold, and let $\{x_1, \cdots, x_n\}$ be a coordinate system at a point $p \in \mathcal{V}$. Denote by V a cubic neighbourhood of p with respect to this system, and by a the breadth of V. Let m be any integer $<n$, and let $\xi_{m+1}, \cdots, \xi_n$ be $n - m$ numbers such that $|\xi_{m+h} - x_{m+h}(p)| < a$ $(1 \leqslant h \leqslant n - m)$. Denote by S_ξ the set of points $q \varepsilon V$ whose coordinates satisfy the conditions

$$x_{m+h}(q) = \xi_{m+h} \qquad (1 \leqslant h \leqslant n - m).$$

We can define a manifold $\mathfrak{S}_\xi$ whose set of points is S_ξ by the condition that the contractions of $x_1, \cdots, x_m$ to $\mathfrak{S}_\xi$ shall form a system of coordinates at every point of $\mathfrak{S}_\xi$. It is obvious that $\mathfrak{S}_\xi$ is a submanifold of $\mathfrak{V}$. We shall say that $\mathfrak{S}_\xi$ is the *slice* of V defined by the equations $x_{m+h} = \xi_{m+h}$ $(1 \leqslant h \leqslant n - m)$.

Theorem 1. *Let $\mathfrak{M}$ be an analytic involutive distribution of dimension m on a manifold of dimension n. If p is any point of $\mathfrak{V}$, there exist a coordinate system $\{x_1, \cdots, x_n\}$ at p and a cubic neighbourhood V of p with respect to this system which satisfy the following conditions:* 1) $x_i(p) = 0$ $(1 \leqslant i \leqslant n)$; 2) *let a be the breadth of V and let $\xi_{m+1}, \cdots, \xi_n$ be any $n - m$ numbers such that $|\xi_{m+h}| < a$ $(1 \leqslant h \leqslant n - m)$; then the slice of V which is defined by the equations $x_{m+h} = \xi_{m+h}$ $(1 \leqslant h \leqslant n - m)$ is an integral manifold of $\mathfrak{M}$.*

We first prove

Lemma 1. *Let X be an infinitesimal transformation which is defined and analytic in a neighbourhood of a point $p \varepsilon \mathfrak{V}$ and which is such that $X_p \neq 0$. Then there exist a coordinate system $\{y_1, \cdots, y_n\}$ at p and a cubic neighbourhood W of p with respect to this system which satisfy the following conditions: $y_i(p) = 0$ $(1 \leqslant i \leqslant n)$; X is defined on W and coincides on W with the infinitesimal transformation whose symbol (with respect to the y-coordinates) is $\partial/\partial y_1$.*

We can find a coordinate system $\{z_1, \cdots, z_n\}$ at p such that $X_p z_1 \neq 0$. Let Z be a cubic neighbourhood of p with respect to this system on which X is defined and analytic, and let c be the breadth of Z. If $q \varepsilon Z$, we have $X_q z_i = F_i(z_1(q), \cdots, z_n(q))$ $(1 \leqslant i \leqslant n)$, where the functions F_i are defined and analytic in the cube defined by the inequalities $|z_i - z_i(p)| < c$. We consider the following system of differential equations:

$$\frac{dz_i}{dt} = F_i(z_1, \cdots, z_n) \qquad (1 \leqslant i \leqslant n) \tag{1}$$

Making use of the existence theorem for systems of analytic differential equations, we obtain the following result: there exist a number b_1 such that $0 < b_1 < c$ and a system of n functions $\varphi_i(y_1, \cdots, y_n)$ $(1 \leqslant i \leqslant n)$, defined and analytic in the cube Q_1 specified by the equalities $|y_i| < b_1$ $(1 \leqslant i \leqslant n)$ such that: 1) $|\varphi_i(y_1, \cdots, y_n)| < c$ whenever $(y_1, \cdots, y_n) \varepsilon Q_1$; 2) the equations $z_i = \varphi_i(t, y_2, \cdots, y_n)$ represent a solution of the system (1) with the initial conditions $\varphi_1(0, y_2 \cdots, y_n) = z_1(p)$, $\varphi_i(0, y_2, \cdots, y_n) = y_i + z_i(p)$ for $i > 1$.

We shall prove that the functional determinant $D(\varphi_1, \cdots, \varphi_n)/D(y_1, \cdots, y_n)$ is $\neq 0$ when $y_1 = \cdots = y_n = 0$. We have

$$\left(\frac{\partial \varphi_1}{\partial y_1}\right)_0 = F_1(z_1(p), \cdots, z_n(p)) = X_p z_1 \neq 0$$

On the other hand, if $j > 1$, $(\partial \varphi_i / \partial y_j)_0 = \delta_{ij}$ because $\varphi_i(0, y_2, \cdots, y_n) = (1 - \delta_{1i})y_i + z_i(p)$. Our assertion follows immediately from these formulas.

It follows that there exist a coordinate system $\{y_1, \cdots, y_n\}$ at p and a cubic neighbourhood W of p with respect to this system such that $W \subset Z$ and $z_i(q) = \varphi_i(y_1(q), \cdots, y_n(q))$ for every $q \varepsilon W$. If $q \epsilon W$, we have

$$X_q z_i = F_i(z_1(q), \cdots, z_n(q)) = \frac{\partial \varphi_i}{\partial y_1}(y_1(q), \cdots, y_n(q))$$

whence $X = \partial/\partial y_1$ in W.

Now, we proceed to prove Theorem 1. Let $\{X_1, \cdots, X_m\}$ be a base of $\mathfrak{M}$ around p. We have $(X_1)_p \neq 0$, and we may apply Lemma 1 to X_1. Let $\{y_1, \cdots, y_n\}$ be a coordinate system at p and W a neighbourhood of p which satisfy the conditions of Lemma 1 for X_1, W being furthermore taken so small that $(X_1)_q, \cdots, (X_m)_q$ form a base of $\mathfrak{M}_q$ at every point $q \varepsilon W$. It is clear that, if $m = 1$, the slice of W which is defined by the equations $y_2 = \xi_2, \cdots, y_n = \xi_n$ (where $\xi_2, \cdots, \xi_n$ are any numbers which are smaller in absolute value than the breadth of W) is an integral manifold of $\mathfrak{M}$, which proves Theorem 1 in the case where $m = 1$. To prove Theorem 1 in the general case, we proceed by induction on m. Assume that $m > 1$ and that Theorem 1 is true for distributions of dimension $m - 1$. It is clear that we can find $m - 1$ functions $A_2, \cdots, A_m$, analytic on V, such that $(X_i - A_i X_1)y_1 = 0$ $(2 \leqslant i \leqslant m)$. We set $X_i' = X_i - A_i X_1$; then $(X_1)_q, (X_2')_q, \cdots, (X_m')_q$ form a base of $\mathfrak{M}_q$ at every point $q \varepsilon W$. Let $\mathfrak{X}$ be the slice of W defined by the equation $y_1 = 0$. If $q \varepsilon \mathfrak{X}$, the vectors $(X_i')_q$ $(2 \leqslant i \leqslant m)$ are tangent to $\mathfrak{X}$ at q; it follows that $X_2', \cdots, X_m'$ have contractions $\bar{X}_2, \cdots, \bar{X}_m$ to $\mathfrak{X}$. If $q \varepsilon \mathfrak{X}$, the space spanned by the vectors $(\bar{X}_2)_q, \cdots, (\bar{X}_m)_q$ is the intersection of $\mathfrak{M}_q$ with the tangent space to $\mathfrak{X}$ at q. The distribution $\bar{\mathfrak{M}}$, $q \to \bar{\mathfrak{M}}_q$, is clearly analytic on $\mathfrak{X}$. On the other hand, it follows immediately from Proposition 2, §V, p. 82 that $\mathfrak{M}$ is involutive. Since $\mathfrak{M}$ is of dimension $m - 1$, we may apply our induction assumption to $\mathfrak{M}$. We can find a system of coordinates $\bar{x}_2, \cdots, \bar{x}_m$ at p on $\mathfrak{X}$ and a cubic neighbourhood $\bar{V}$ of p with respect to this system with the following properties: $\bar{x}_i(p) = 0$ $(2 \leqslant i \leqslant n)$; if $\xi_{m+1}, \cdots, \xi_n$ are numbers smaller in absolute value than the breadth a of $\bar{V}$, then the slice $\mathfrak{S}_\xi$ of $\bar{V}$ which is defined by the equations $\bar{x}_{m+h} = \xi_{m+h}$ $(1 \leqslant h \leqslant n - m)$ is an integral manifold of $\mathfrak{M}$. We may furthermore assume that a is at most equal to the breadth of W. Let V be the set of points $q \varepsilon W$ which satisfy the following conditions: the point q' whose coordinates are $(0, y_2(q), \cdots, y_n(q))$ lies in $\bar{V}$ and $|y_1(q)| < a$. If $q \varepsilon V$, we set $x_1(q) = y_1(q)$, $x_i(q) = \bar{x}_i(q')$ for $i > 1$. It is clear that $\{x_1, \cdots, x_n\}$ is a coordinate system at p and that V is a cubic neighbourhood of p with respect to this system. Moreover, since $x_2(q), \cdots, x_n(q)$ depend only upon $y_2(q), \cdots, y_n(q)$ and $x_1(q) = y_1(q)$, the symbol of X_1 (with respect to the coordinates $x_1, \cdots, x_n$) is $\partial/\partial x_1$.

We have $X_1 x_{m+h} = 0$ $(1 \leqslant h \leqslant n - m)$, whence

$$\frac{\partial}{\partial x_1}(X'_i x_{m+h}) = X_1 X'_i x_{m+h} = [X'_i, X_1] x_{m+h}$$

Since $\mathfrak{M}$ is involutive, we have $[X'_i, X_1] = g_{i1} X_1 + \Sigma_{j=2}^m g_{ij} X'_j$ where the functions g_{ij} $(1 \leqslant i, j \leqslant m)$ are analytic on V. Therefore,

$$\frac{\partial}{\partial x_1}(X'_i x_{m+h}) = \Sigma_{j=1}^m g_{ij}(X'_j x_{m+h}) \tag{2}$$

Since any $\bar{\mathfrak{S}}_\xi$ is an integral manifold of $\mathfrak{M}$, we have $X'_i x_{m+h} = 0$ on $\mathfrak{X}$, i.e. for $x_1 = 0$. Considered as functions of x_1, the functions $X'_i x_{m+h}$ $(1 \leqslant i \leqslant m)$ satisfy the linear homogeneous differential system (2). It follows from the uniqueness theorem for systems of differential equations that $X'_i x_{m+h} = 0$ identically on V. This means that any slice of V defined by a system of equations of the form $x_{m+h} = \xi_{m+h}$ $(1 \leqslant h \leqslant n - m)$ is an integral manifold of $\mathfrak{M}$. Theorem 1 is thereby proved for distributions of dimension m.

Proposition 1. *Let $\mathfrak{M}$ be an analytic involutive distribution on a manifold $\mathcal{V}$. If two integral manifolds $\mathcal{W}$ and $\mathcal{W}'$ of $\mathfrak{M}$ have a point p in common, there exists an integral manifold of $\mathfrak{M}$, containing p, which is an open submanifold of both $\mathcal{W}$ and $\mathcal{W}'$.*

We use the notation of Theorem 1. Let $\mathfrak{S}_0$ be the slice of V which is defined by the equations $x_{m+h} = 0$ $(1 \leqslant h \leqslant n - m)$. It will be sufficient to prove that any integral manifold $\mathcal{W}$ of $\mathfrak{M}$ which contains p has an open submanifold which is also an open submanifold of $\mathcal{S}_0$.

Since the identity mapping of $\mathcal{W}$ into $\mathcal{V}$ is continuous, the set $V \cap \mathcal{W}$ is relatively open in $\mathcal{W}$. Because $\mathcal{W}$ is locally connected, the connected component C of p in $V \cap \mathcal{W}$ (in the topology of $\mathcal{W}$) is a relatively open subset of $\mathcal{W}$.[1] Therefore, C is the set of points of an open submanifold $\mathcal{C}$ of $\mathcal{W}$, which is an integral manifold of $\mathfrak{M}$.

Denote by X_i the infinitesimal transformation whose symbol (with respect to the coordinates $x_1, \cdots, x_n$) is $\partial/\partial x_i$. Then, if $q \varepsilon V$, the vectors $(X_1)_q, \cdots, (X_m)_q$ form a base of $\mathfrak{M}_q$. On the other hand, if $q \varepsilon \mathcal{C}$ we know by Proposition 1, §IV, p. 75 that we can select m of the functions $x_1, \cdots, x_n$ whose contractions to $\mathcal{C}$ form a system of coordinates at q. Since $X_i x_{m+h} = 0$ $(1 \leqslant i \leqslant m,\ 1 \leqslant h \leqslant n - m)$, none of these functions can be of index $> m$, which proves that the contractions of $x_1, \cdots, x_m$ to $\mathcal{C}$ form a system of coordinates at any point of $\mathcal{C}$.

If $q \varepsilon \mathcal{C}$, the vectors $(X_1)_q, \cdots, (X_m)_q$ form a base of the tangent space to $\mathcal{C}$ at q. The equations $X_i x_{m+h} = 0$ imply that the differential of the contraction of x_{m+h} to $\mathcal{C}$ is 0; therefore each function x_{m+h} is constant on $\mathcal{C}$ $(1 \leqslant h \leqslant n - m)$ (Cf. Proposition 4, §IV, p. 75). This means that $\mathcal{C}$ is a subset of $\mathcal{S}_0$. Because the contractions of $x_1, \cdots, x_m$ to $\mathcal{C}$ form a system of coordinates at every point of $\mathcal{C}$, $\mathcal{C}$ is an open submanifold of $\mathcal{S}_0$. Proposition 1 is thereby proved.

§VIII. MAXIMAL INTEGRAL MANIFOLDS OF AN INVOLUTIVE DISTRIBUTION

Let $\mathcal{V}$ be a manifold, and let $\mathfrak{M}$ be an analytic involutive distribution on $\mathcal{V}$.

We shall now study the integral manifold of $\mathfrak{M}$ in the large, instead of limiting ourselves to the consideration of a neighbourhood of a point of $\mathcal{V}$.

Let V be the set of points of $\mathcal{V}$. We shall define a new topology on the set V. Let $\mathcal{O}$ be the family of those subsets of V which may be represented as unions of collections of integral manifolds of $\mathfrak{M}$; $\mathcal{O}$ may be taken as the family of open sets in a topology on V. In fact

1) Any union of sets of $\mathcal{O}$ obviously belongs again to $\mathcal{O}$.

2) Let O_1, O_2 be any two sets of $\mathcal{O}$, and let p be a point of $O_1 \cap O_2$. Then there exist two integral manifolds $\mathcal{W}_1, \mathcal{W}_2$ of $\mathfrak{M}$ both containing p, and such that $\mathcal{W}_1 \subset O_1$, $\mathcal{W}_2 \subset O_2$. Proposition 2, §VII, p. 88

[1] In fact, if $q \varepsilon C$, there exists a connected neighbourhood of q contained in $\mathcal{W} \cap V$, and therefore also in C.

shows that $\mathfrak{W}_1$, $\mathfrak{W}_2$ have in common an integral manifold $\mathfrak{W}$ which contains p. We have $p \varepsilon \mathfrak{W} \subset O_1 \cap O_2$, which shows that $O_1 \cap O_2 \varepsilon \mathfrak{O}$.

3) Any open subset U of $\mathfrak{V}$ belongs to $\mathfrak{O}$. In fact, let p be a point of U; there exists an integral manifold $\mathfrak{W}$ of $\mathfrak{M}$ such that $p \varepsilon \mathfrak{W}$. Since $\mathfrak{W}$ is locally connected, the component of p in $\mathfrak{W} \cap U$ is an open subset of $\mathfrak{W}$[1]; as such, it is the underlying space of an open submanifold $\mathfrak{W}_1$ of $\mathfrak{W}$, and $\mathfrak{W}_1$ is clearly an integral manifold of $\mathfrak{M}$ with $p \varepsilon \mathfrak{W}_1 \subset U$. This proves that $U \varepsilon \mathfrak{O}$. It follows in particular that $V \varepsilon \mathfrak{O}$. Furthermore, we see that if $p_1 \neq p_2$, we can find sets O_1, O_2 in $\mathfrak{O}$ such that $p_1 \varepsilon O_1$, $p_2 \varepsilon O_2$, $O_1 \cap O_2 = \phi$.

Let $\mathfrak{V}^*$ be the topological space defined by the family of open sets $\mathfrak{O}$. We shall prove that, if $\mathfrak{W}$ is any integral manifold of $\mathfrak{M}$, $\mathfrak{W}$ is a subspace of $\mathfrak{V}^*$. Let p be any point of $\mathfrak{W}$. Then:

1) Any neighbourhood of p with respect to $\mathfrak{W}$ contains an open submanifold of $\mathfrak{W}$; this submanifold being an integral manifold of $\mathfrak{M}$, it is a set of $\mathfrak{O}$ and is therefore a neighbourhood of p with respect to $\mathfrak{V}^*$.

2) A neighbourhood of p with respect to $\mathfrak{V}^*$ contains a set O such that $p \varepsilon O$, $O \varepsilon \mathfrak{O}$; taking into account the definition of $\mathfrak{O}$, we see that O contains an integral manifold $\mathfrak{W}_1$ of $\mathfrak{M}$ containing p. According to Proposition 2, §VII, p. 88 $\mathfrak{W}_1 \cap \mathfrak{W}$ contains a neighbourhood of p in $\mathfrak{W}$.

From 1) and 2) it follows at once that $\mathfrak{W}$ is an open subspace of $\mathfrak{V}^*$.

Let $\mathfrak{W}$ be any connected component of $\mathfrak{V}^*$, this connected component being considered as a subspace of $\mathfrak{V}^*$. We shall prove that $\mathfrak{W}$ is the underlying space of an integral manifold of $\mathfrak{M}$.

In order to do this, we first select for every point $p \varepsilon V$ some integral manifold $\mathfrak{W}'(p)$ of $\mathfrak{M}$ containing p. If $p \varepsilon \mathfrak{W}$, $\mathfrak{W}'(p)$ is an open connected subset of $\mathfrak{V}^*$, whence, $\mathfrak{W}'(p) \subset \mathfrak{W}$. We denote by $\mathfrak{a}(p)$ the class of real valued functions which are analytic at p on $\mathfrak{W}'(p)$. These functions may be considered as functions defined on neighbourhoods of p in $\mathfrak{W}$. We assert that the assignment $p \to \mathfrak{a}(p)$ defines a manifold on $\mathfrak{W}$. The conditions I, II of §I, p. 68 are obviously fulfilled; the neighbourhood V which occurs in condition III being selected as a neighbourhood W' in $\mathfrak{W}'(p)$ which satisfies condition III for $\mathfrak{W}'(p)$ the conditions III, 1), 2), are fulfilled. As for III, 13), we observe that if $q \varepsilon W'$, the manifolds $\mathfrak{W}'(p)$, $\mathfrak{W}'(q)$ have in common a submanifold which is an open submanifold of both, from which it follows immediately that condition III, 13) is satisfied.

Let $\mathfrak{W}$ be the manifold defined on $\mathfrak{W}$ by the assignment $p \to \mathfrak{a}(p)$. If $p \varepsilon \mathfrak{W}$, $\mathfrak{W}(p)$ is clearly an open submanifold of $\mathfrak{W}$, and therefore $\mathfrak{W}$

[1] Cf. footnote 1), p. 92.

is an integral manifold of $\mathfrak{W}$. It is obviously independent of the choice of the manifolds $\mathcal{W}'(p)$. Since we have $\mathcal{W}'(p) \subset \mathcal{W}$ we see that any integral manifold of $\mathfrak{W}$ which has a point in common with $\mathcal{W}$ is an open submanifold of $\mathcal{W}$.

We have now proved

Theorem 2. *Let $\mathcal{V}$ be a manifold, and let $\mathfrak{M}$ be an involutive distribution on $\mathcal{V}$. Through every point $p \varepsilon \mathcal{V}$ there passes a maximal integral manifold $\mathcal{W}(p)$ of $\mathfrak{M}$, i.e. an integral manifold which is not a subset of any larger integral manifold. Any integral manifold containing p is an open submanifold of $\mathcal{W}(p)$.*

Remark. *The maximal integral manifolds are obviously uniquely characterized by the properties stated in Theorem 2.*

§IX. THE COUNTABILITY AXIOM

Thus far we have not required that the underlying space of a manifold $\mathcal{V}$ should satisfy the Hausdorff countability axioms. These axioms, however, hold for the manifolds which we shall consider later, and this fact has certain important consequences.

Let us call a subset of a manifold $\mathcal{V}$ which is a cubic neighbourhood of one of its points, with respect to a suitable coordinate system at the point, *a cubic subset* of $\mathcal{V}$. It is clear that the countability axioms hold in $\mathcal{V}$ if and only if $\mathcal{V}$ can be covered by a countable collection of cubic subsets. An equivalent form of this condition is given by the following:

The countability axioms hold in a manifold $\mathcal{V}$ if and only if $\mathcal{V}$ can be represented as the union of a countable family of compact subsets of $\mathcal{V}$.

In fact:

1) Suppose that the countability axioms hold in $\mathcal{V}$. Then $\mathcal{V}$ can be represented as the union of a countable family of cubic subsets. Each of them is homeomorphic to a cube in R^n, where n is the dimension of $\mathcal{V}$; since a cube can be represented as the union of a countable family of compact sets, the same holds for $\mathcal{V}$.

2) Suppose that $\mathcal{V}$ is the union of a countable family $(K_1, \cdots, K_m, \cdots)$ of compact subsets. Every point of K_m has a neighbourhood in $\mathcal{V}$ which is a cubic set. Since K_m is compact, it can be covered by a finite number of these cubic sets. This being true for every m, $\mathcal{V}$ can be covered by a countable family of cubic sets.

Now let us consider an involutive distribution $\mathfrak{M}$ on a manifold $\mathcal{V}$. As we have seen (Theorem 1, §VII, p. 88), every point $p \varepsilon \mathcal{V}$ has a neighbourhood V which can be decomposed into slices, each slice being an integral manifold of $\mathfrak{M}$. Let $\mathcal{W}$ be the maximal integral manifold containing p. Then, the intersection $V \cap \mathcal{W}$ is the union of a certain set of these slices. Since two different slices are disjoint, a compact subset of $\mathcal{W}$ can meet at most a finite number of the slices. If the countability axioms hold on $\mathcal{W}$, it follows immediately that the intersection $V \cap \mathcal{W}$ is the union of at most countably many of these slices.

Let S_0 be the slice containing p. Then, obviously, S_0 coincides with the connected component of p in $\mathcal{W} \cap V$, this connected component being taken in the sense of the topology of $\mathcal{W}$. But, if $V \cap \mathcal{W}$ contains at most countably many slices, S_0 is also the connected component of p in the set $V \cap \mathcal{W}$, in the sense of the topology of $\mathcal{V}$. In order to prove this, we may assume that V is a cubic neighbourhood of p with respect to a coordinate system $\{x_1, \cdots, x_n\}$ such that each slice of V is represented by equations of the form

$$x_{r+1} = x_{r+1}(p_1), \quad \cdots, \quad x_n = x_n(p_1),$$

p_1 being a point of the slice. Let $\bar{\omega}$ be the mapping of R^n into R^{n-r} defined by $\bar{\omega}(x_1, \cdots, x_n) = (x_{r+1}, \cdots, x_n)$. Under our assumption, $\bar{\omega}$ maps $V \cap \mathcal{W}$ onto a countable subset of R^{n-r}. Since $\bar{\omega}$ is continuous, it maps every connected component of $V \cap \mathcal{W}$ (in the sense of the topology of $\mathcal{V}$) onto a connected subset of R^{n-r}. Now any connected countable subset of R^{n-r} must clearly consist of a single point; this proves that any connected component of $V \cap \mathcal{W}$ coincides with a slice.

Proposition 1. *Let $\mathfrak{M}$ be an involutive distribution on a manifold $\mathcal{V}$. Let $\mathcal{W}$ be an integral manifold of $\mathfrak{M}$. Suppose that φ is an analytic mapping into $\mathcal{V}$ of a manifold $\mathcal{U}$ and that the image under φ of the set of points of $\mathcal{U}$ is a subset of $\mathcal{W}$. If the countability axioms hold on $\mathcal{W}$, φ is an analytic mapping of $\mathcal{U}$ into $\mathcal{W}$.*

Let s be a point of $\mathcal{U}$, and $p = \varphi(s)$ be its image in $\mathcal{W}$. We select a system of coordinates $\{x_1, \cdots, x_n\}$ at p on $\mathcal{V}$ and a cubic neighbourhood V of p with respect to this system, with the same properties as above. Since φ is continuous, there exists a cubic neighbourhood U of s (with respect to some system of coordinates at s on $\mathcal{U}$) which is mapped by φ onto a subset of V. Moreover, since U is connected, the same thing is true of $\varphi(U)$. Therefore $\varphi(U)$ is a connected subset of $V \cap \mathcal{W}$ (in the topology of V). It follows that $\varphi(U)$ is contained in the slice S_0 of the intersection $V \cap \mathcal{W}$ which contains p.

If f is any function which is analytic at p on $\mathcal{W}$, then f coincides on a neighbourhood of p in $\mathcal{W}$ with the contraction of some function f_1 which is analytic at p on $\mathcal{V}$. The function $f_1 \circ \varphi$ is analytic at s on $\mathcal{U}$, since φ is an analytic mapping of $\mathcal{U}$ into $\mathcal{V}$. If we choose U so that $\varphi(U)$ is in the domain of definition of f_1, $\varphi(U)$ will also be in the domain of definition of f (since $\varphi(U) \subset S_0$), and the functions $f \circ \varphi$, $f_1 \circ \varphi$ will be defined, and will coincide, on U. Therefore $f \circ \varphi$ is analytic at s on U, which proves that φ is an analytic mapping of $\mathcal{U}$ into $\mathcal{W}$.

Proposition 2. *If the countability axioms hold in a manifold $\mathcal{V}$, they also hold in any submanifold of $\mathcal{V}$.*

To prove Proposition 2 we shall first establish the following lemmas.

Lemma 1. *Let $\mathfrak{B}$ be a connected space. Assume that there exists a family $\{V_\alpha\}$ of open subsets of $\mathfrak{B}$ with the following properties: a) the countability axioms hold in every V_α, considered as a subspace of $\mathfrak{B}$, b) there are at most countably many indices β such that V_β meets a given V_α of the family, and c) we have $\bigcup_\alpha V_\alpha = \mathfrak{B}$. Then the countability axioms hold in $\mathfrak{B}$.*

Let α_0 be any index such that $V_{\alpha_0} \neq \phi$; we shall say that an index α is attainable in h steps from α_0 if there exists a sequence $(\alpha_0, \cdots, \alpha_h)$ of $h + 1$ indices, beginning with α_0 and ending with $\alpha_h = \alpha$, such that $V_{\alpha_{i-1}} \cap V_{\alpha_i} \neq \phi$ $(1 \leqslant i \leqslant h)$. Let A_h be the set of indices α which have this property. We shall prove, by induction on h, that A_h is countable. This statement is true, by assumption, for $h = 1$. Now assume it to hold for h; if $\alpha \varepsilon A_{h+1}$, there exists an index $\beta \varepsilon A_h$ such that $V_\alpha \cap V_\beta \neq \phi$. There are only countably many indices β in A_h and, for each of them, there are only countably many indices α with $V_\alpha \cap V_\beta \neq \phi$; this proves our statement for $h + 1$. Let A be the set $\bigcup_{h=1}^{\infty} A_h$; A is then a countable set. We set $V = \bigcup_{\alpha \varepsilon A} V_\alpha$; V is an open subset of $\mathfrak{B}$ and the countability axioms hold in V. Let p be any point adherent to V; then p belongs to some V_α. Since V_α is open, we have $V \cap V_\alpha \neq \phi$, whence $V_\alpha \cap V_\beta \neq \phi$ for some $\beta \varepsilon A$; if $\beta \varepsilon A_h$, we have $\alpha \varepsilon A_{h+1}$, $V_\alpha \subset V$, $p \varepsilon V$, which proves that V is also closed. Since $\mathfrak{B}$ is connected we have $V = \mathfrak{B}$: Lemma 1 is proved.

Lemma 2. *Let $\mathfrak{B}$ be a connected and locally connected space. Assume that $\mathfrak{B}$ can be covered by the union of a countable family of open subsets V_k $(k = 1, \cdots)$ which have the following property: any component of any one of the sets V_k satisfies the countability axioms. Then the countability axioms hold in $\mathfrak{B}$.*

Let $V_{k,\alpha}$ be the components of V_k, α running over a set of indices A_k. Taking Lemma 1 into account, it will be sufficient to prove that, k, m and $\alpha \epsilon A_k$ being given, there are only countably many indices $\beta \epsilon A_m$ for which $V_{k,\alpha} \cap V_{m,\beta} \neq \phi$. The set $V_m \cap V_{k,\alpha}$ is an open subset of $V_{k,\alpha}$; since the countability axioms hold in $V_{k,\alpha}$, the set $V_m \cap V_{k,\alpha}$ has only countably many components K_ρ $(\rho = 1, \cdots)$. Each K_ρ is a connected subset of V_m, and therefore belongs to a uniquely determined component, $V_{m,\beta(\rho)}$ of V_m. Let β be any index such that $V_{k,\alpha} \cap V_{m,\beta} \neq \phi$; if p is a point of $V_{k,\alpha} \cap V_{m,\beta}$, p belongs to one of the sets K_ρ. This set K_ρ, being a connected subset of V_m and having a point in common with $V_{m,\beta}$ is contained in $V_{m,\beta}$, whence $\beta = \beta(\rho)$, which proves our assertion.

Lemma 3. *Let $\mathfrak{V}$ be a connected space. Assume that there exists a continuous mapping φ of $\mathfrak{V}$ into R^d with the following property: p being any point of $\mathfrak{V}$, there exists an open subset V of $\mathfrak{V}$, containing p, which is mapped topologically under φ onto an open subset of R^d. Then the countability axioms hold in $\mathfrak{V}$.*

We can find a countable set $\{U_1, \cdots\}$ of open subsets of R^d with the following property: r being any point of R^d, any neighbourhood of r contains some set U_k. We may furthermore assume that the sets U_k are connected. If k is any integer >0, we consider the family of those open subsets $V_{k,\alpha}$ of $\mathfrak{V}$ which are mapped topologically onto U_k under φ; α runs over a set of indices A_k which may be empty for some k. We assume that $V_{k,\alpha} \neq V_{k,\beta}$ for $\alpha \neq \beta$. If $V_{k,\alpha}$ has a point in common with $V_{k,\beta}$, we have $\alpha = \beta$. In fact let ψ_α, ψ_β be the mappings of U_k onto $V_{k,\alpha}$, $V_{k,\beta}$ respectively which are reciprocal to the mappings induced by φ on these two sets, and let W be the set $V_{k,\alpha} \cap V_{k,\beta}$. Assume for a moment that W has a boundary point p in $V_{k,\alpha}$; then we have $p = \lim p_n$, where (p_n) is a sequence of points of W. We have therefore, $\varphi(p) = \lim \varphi(p_n)$ and $\psi_\beta(\varphi(p)) = \lim \psi_\beta(\varphi(p_n))$. For any point $q \varepsilon W$ we clearly have $\psi_\beta(\varphi(q)) = q$, whence $\psi_\beta(\varphi(p)) = \lim p_n = p$; on the other hand the point $\psi_\beta(\varphi(p))$ is in $V_{k,\beta}$ and hence is in W. This shows that $p \varepsilon W$, which, because W is open, is a contradiction. Therefore W has no boundary point in $V_{k,\alpha}$; this latter set being connected (because it is homeomorphic to U_k), we have $W = V_{k,\alpha}$. The same argument would show that $W = V_{k,\beta}$, whence $\alpha = \beta$. Because the sets $V_{k,\alpha}$ are open it now follows that they are the components of $V_k = \bigcup_{\alpha \epsilon A_k} V_{k,\alpha}$. Furthermore, it is clear that the countability axioms hold in every $V_{k,\alpha}$ and that every point of $\mathfrak{V}$ belongs to one of the sets V_k. Therefore, Lemma 3 follows from Lemma 2.

Lemma 4. *If $\mathcal{V}$ is a d-dimensional submanifold of R^n, the countability axioms hold in $\mathcal{V}$.*

Let x_i^* $(1 \leqslant i \leqslant n)$ be the contractions to $\mathcal{V}$ of the coordinates in R^n. Let $I = \{i_1, \cdots, i_d\}$ be any set of distinct numbers between 1 and n, and let V_I be the set of points of $\mathcal{V}$ at which the functions $x_{i_1}^*, \cdots, x_{i_d}^*$ form a coordinate system on $\mathcal{V}$; V_I is an open subset of $\mathcal{V}$, and every point of $\mathcal{V}$ belongs to one of the sets V_I. Let V' be any component of V_I and let φ be the mapping of V' into R^d defined by $\varphi(p) = (x^*_{i_1}(p), \cdots, x^*_{i_d}(p))(p \in V')$; it follows immediately from Lemma 3, applied to the space V' and the mapping φ, that the countability axioms hold in V'. Lemma 4 then follows from Lemma 2.

We can now prove Proposition 1. Since the countability axioms hold in $\mathcal{V}$, we may cover $\mathcal{V}$ by a countable number of open subsets V_k $(1 \leq k < \infty)$, each of which is a cubic neighbourhood of some point of $\mathcal{V}$ with respect some system of coordinates at the point. Set $W_k = V_k \cap \mathcal{W}$, where $\mathcal{W}$ is a submanifold of $\mathcal{V}$. The sets W_k are open in $\mathcal{W}$. If W'_k is a component of W_k (in the topology of $\mathcal{W}$), W'_k, considered as a subspace of $\mathcal{W}$, is the underlying space of an open submanifold $\mathcal{W}'_k$ of $\mathcal{W}$. Considering V_k as a subspace of $\mathcal{V}$, it is the underlying space of an open submanifold $\mathcal{V}_k$ of $\mathcal{V}$, and $\mathcal{W}'_k$ is a submanifold of $\mathcal{V}_k$. Since V_k is a cubic neighbourhood with respect to some system of coordinates, $\mathcal{V}_k$ is analytically isomorphic to an open submanifold of R^n (where $n = \dim \mathcal{V}$). It follows that $\mathcal{W}'_k$ is analytically isomorphic to a submanifold of R^n. By Lemma 4, the countability axioms hold in $\mathcal{W}_k$. By Lemma 2, the countability axioms hold in $\mathcal{W}$.

CHAPTER IV

Analytic Groups. Lie Groups

Summary. Chapter II was concerned with the study of groups which are at the same time topological spaces. Now, we shall consider groups which are at the same time manifolds; this leads to the notion of an analytic group, which is defined in §I.

The most important new concept brought about by the introduction of analytic groups is the concept of the Lie algebra, which is defined in §II. To every analytic group there is associated a Lie algebra, and the relationships which may hold between analytic groups have their counterpart in the corresponding relationships between Lie algebras. Thus, the analytic subgroups of an analytic group $\mathcal{G}$ correspond to the subalgebras of the Lie algebra of $\mathcal{G}$ (Theorem 1, §IV, p. 107), and the analytic homomorphisms of an analytic group $\mathcal{G}$ into an analytic group $\mathcal{H}$ correspond to the homomorphisms of the Lie algebra of $\mathcal{G}$ into the Lie algebra of $\mathcal{H}$ (Theorem 2, §VI, p. 111). The analytic groups which are considered here are not necessarily closed subgroups and are not necessarily topological subgroups, although they are submanifolds. It is shown in §V that, if an analytic subgroup is closed as a set of points, it is necessarily a topological subgroup. If $\mathfrak{H}$ is a topological subgroup of an analytic group $\mathcal{G}$ such that the connected component of the neutral element in $\mathfrak{H}$ is the underlying topological group of a closed analytic subgroup of $\mathcal{G}$, the homogeneous space $\mathcal{G}/\mathfrak{H}$ has a structure of manifold which is defined in §V. If $\mathfrak{H}$ is distinguished, $\mathcal{G}/\mathfrak{H}$ is an analytic group. It is shown in §VII that the Lie algebra of $\mathcal{G}/\mathfrak{H}$ is a factor algebra of the Lie algebra of $\mathcal{G}$.

The notion of exponential mapping for matrices can be generalized to the case of an arbitrary analytic group. Thus, the elements of the Lie algebra of an analytic group $\mathcal{G}$ can be used to represent parametrically the elements of a neighbourhood of the neutral element in $\mathcal{G}$. The definition of this generalized exponential mapping is the object of §VIII. The exponential mapping is used in §IX to complete the indications which were given in §VII on the homomorphisms of analytic groups.

In §X, it is shown that the addition and bracket operation in the Lie algebra correspond (approximately) to multiplication and the building of commutators in the group.

In §XI, it is shown that an analytic group $\mathcal{G}$ has a representation by linear transformations operating on the Lie algebra of the group. From this, one can deduce that an analytic group whose Lie algebra has no center $\neq \{0\}$ is at least locally isomorphic with a subgroup of the linear group. Furthermore, it is proved that a given Lie algebra can be represented as the Lie algebra of some analytic group provided its center contains only 0. That this *proviso* is not really necessary will be proved much later (in volume 2).

In section XII, it is proved that the commutator subgroup of an analytic group is the underlying group of an analytic subgroup, the derived group.

An analytic group bears *ipso facto* the structure of a topological group; it is shown in §XIII that the analytical structure (and in particular the Lie algebra) is uniquely determined in terms of the topological structure (Theorem 3, p. 129). This remarkable fact holds only for *real* analytic groups, not for groups with complex parameters (Cf. volume 2). To avoid the condition of connectedness which was involved in the notion of manifold, we define a *Lie group* to be a locally connected topological group $\mathfrak{G}$ such that the component of the neutral element in $\mathfrak{G}$ is the underlying topological group of some analytic group (which is then uniquely determined). It follows from this last fact that we may use freely the concept of Lie algebra when dealing with Lie groups.

In section XIV, we derive a sufficient condition for a topological group to be a Lie group. Our result is a slight generalization of the result of Cartan according to which any closed subgroup of a Lie group is a Lie Group.

In section XV, we show that the group of automorphisms of any algebra whatsoever on the field of real numbers is a Lie group. From this, we deduce that the group of automorphisms of a connected Lie group is a Lie group.

§I. DEFINITION OF THE NOTION OF ANALYTIC GROUP. EXAMPLES

Definition 1. *An analytic group is a pair* $(\mathcal{V}, G)$ *formed by a manifold* $\mathcal{V}$ *and a group* G *which satisfy the following conditions:* 1) *the set of points of* $\mathcal{V}$ *coincides with the set of elements of* G; 2) *the mapping* $(\sigma, \tau) \to \sigma\tau^{-1}$ *of the manifold* $\mathcal{V} \times \mathcal{V}$ *into* $\mathcal{V}$ *is everywhere analytic.*

The manifold $\mathcal{V}$ is called the *underlying manifold* of the analytic group. The underlying topological space of $\mathcal{V}$ is also called the underlying space of the group. Since every analytic mapping is continuous, the pair formed by the underlying space and the group G is a topological group, which is called the *underlying topological group* of the analytic group. This topological group is obviously connected and locally simply connected.

The additive group of R^n, associated with the manifold R^n which was defined in Chapter III, §II, p. 73 is an analytic group which we shall again denote by R^n.

We now consider the group $GL(n, C)$. If $\sigma = (x_{ij}(\sigma))$ is a matrix in this group, we denote by $x'_{ij}(\sigma)$ and $x''_{ij}(\sigma)$ the real and imaginary parts of $x_{ij}(\sigma)$. If we assign to σ the point $\Phi(\sigma) \varepsilon R^{2n^2}$ whose coordinates are the numbers $x'_{ij}(\sigma)$, $x''_{ij}(\sigma)$ (arranged in some fixed order), we obtain a homeomorphism, Φ, of $GL(n, C)$ with the subset of R^{2n^2} composed of the points for which

$$\left| x'_{ij} + \sqrt{-1}\, x''_{ij} \right| \neq 0$$

This set is open. On the other hand, the group $GL(n, C)$ is connected. Hence we may define a manifold $\mathcal{V}$ whose underlying space is the underlying space of $GL(n, C)$ by the requirement that the $2n^2$ functions x'_{ij}, x''_{ij} shall form a system of coordinates at every point of $\mathfrak{V}$ (Cf. Chapter III, §II, p. 73).

The quantities $x'_{ij}(\sigma\tau^{-1})$, $x''_{ij}(\sigma\tau^{-1})$ may be expressed as rational functions of the $x'_{ij}(\sigma)$, $x'_{ij}(\tau)$, $x''_{ij}(\sigma)$, $x''_{ij}(\tau)$, and the denominators of these rational functions are $\neq 0$ on $GL(n, C)$. It follows that the mapping $(\sigma, \tau) \to \sigma\tau^{-1}$ is analytic, which proves that the pair formed by $\mathfrak{V}$ and $GL(n, C)$ is an analytic group. This analytic group will again be denoted by $GL(n, C)$.

Finally, let us consider the torus group T^1, i.e. the factor group of R by the group of integers. We have defined the topological space T^1 as the underlying space of a manifold T^1, and we have seen that the functions $\cos 2\pi\mathfrak{x}$, $\sin 2\pi\mathfrak{x}$ are everywhere analytic on T^1 and that, at each point, one of them is a coordinate on T^1. The formulas

$$\cos 2\pi(\mathfrak{x} - \mathfrak{y}) = \cos 2\pi\mathfrak{x} \cos 2\pi\mathfrak{y} + \sin 2\pi\mathfrak{x} \sin 2\pi\mathfrak{y}$$
$$\sin 2\pi(\mathfrak{x} - \mathfrak{y}) = \sin 2\pi\mathfrak{x} \cos 2\pi\mathfrak{y} - \sin 2\pi\mathfrak{y} \cos 2\pi\mathfrak{x}$$

show immediately that the manifold T^1 and the group T^1, considered together, make up an analytic group, which we shall also denote by T^1.

Now, let $\mathcal{G}$ and $\mathcal{H}$ be two analytic groups, $\mathfrak{V}$ and $\mathfrak{W}$ their underlying manifolds and G and H their underlying groups. Then the product $G \times H$ is a group, and the set of elements of this group is also the set of points of $\mathfrak{V} \times \mathfrak{W}$. Let $((\sigma, \tau), (\sigma_1, \tau_1))$ be a pair of elements of $G \times H$, with $\sigma, \sigma_1 \varepsilon G$ and $\tau, \tau_1 \varepsilon H$. The mapping $((\sigma, \tau), (\sigma_1, \tau_1)) \to ((\sigma, \sigma_1), (\tau, \tau_1))$ of $(\mathfrak{V} \times \mathfrak{W}) \times (\mathfrak{V} \times \mathfrak{W})$ into $(\mathfrak{V} \times \mathfrak{V}) \times (\mathfrak{W} \times \mathfrak{W})$ is obviously analytic. The mappings $(\sigma, \sigma_1) \to \sigma\sigma_1^{-1}$, $(\tau, \tau_1) \to \tau\tau_1^{-1}$ of $\mathfrak{V} \times \mathfrak{V}$ into $\mathfrak{V}$ and of $\mathfrak{W} \times \mathfrak{W}$ into $\mathfrak{W}$ are analytic by assumption. It follows easily that the mapping $((\sigma, \sigma_1), (\tau, \tau_1)) \to (\sigma\sigma_1^{-1}, \tau\tau_1^{-1})$ of $(\mathfrak{V} \times \mathfrak{V}) \times (\mathfrak{W} \times \mathfrak{W})$ into $\mathfrak{V} \times \mathfrak{W}$ is analytic. Hence the same is true of the mapping $((\sigma, \tau), (\sigma_1, \tau_1)) \to (\sigma, \tau)(\sigma_1, \tau_1)^{-1} = (\sigma\sigma_1^{-1}, \tau\tau_1^{-1})$ of $(\mathfrak{V} \times \mathfrak{W}) \times (\mathfrak{V} \times \mathfrak{W})$ into $\mathfrak{V} \times \mathfrak{W}$. This means that the pair formed by the manifold $\mathfrak{V} \times \mathfrak{W}$ and the group $G \times H$ is an analytic group, which we shall call the *product* of the analytic groups $\mathcal{G}$ and $\mathcal{H}$ and denote by $\mathcal{G} \times \mathcal{H}$. We can define in a similar way the product of any finite number of analytic groups.

§II. The Lie Algebra

Let $\mathcal{G}$ be an analytic group. We denote by $\mathfrak{g}_\sigma$ the tangent space to $\mathcal{G}$ at the point σ. If σ and τ are any elements of $\mathcal{G}$, there exists a unique element $\rho(\rho = \tau\sigma^{-1})$ such that the left translation associated with ρ maps σ on τ: we denote this left translation by $\Phi_\rho = \Phi_{\tau\sigma^{-1}}$.

It follows immediately from the definition of an analytic group that the mapping $\xi \to \xi^{-1}$ of $\mathcal{G}$ into itself is analytic. Hence the mapping $\xi \to \rho\xi$, which is obtained by performing successively the mappings $\xi \to \xi^{-1}$ and $\xi \to \rho\xi^{-1}$, is also analytic: every left translation in an analytic group is an analytic isomorphism of the underlying manifold with itself.

It follows that the mapping Φ_ρ has a differential $d\Phi_\rho$ which maps $\mathfrak{g}_\sigma$ isomorphically onto $\mathfrak{g}_\tau$.

Definition 1. *An infinitesimal transformation X on $\mathcal{G}$ is said to be left invariant if, for any σ, $\tau \in \mathcal{G}$, we have*

$$d\Phi_{\tau\sigma^{-1}} X_\sigma = X_\tau$$

Let ϵ be the neutral element of $\mathcal{G}$. In order for X to be left invariant it is sufficient for the equality $d\Phi_\tau X_\epsilon = X_\tau$ to hold for every $\tau \in \mathcal{G}$. In fact, assume that such is the case; since $\Phi_{\sigma^{-1}}$ is the reciprocal mapping of Φ_σ, $d\Phi_{\sigma^{-1}}$ is the reciprocal mapping of $d\Phi_\sigma$ and we have $X_\epsilon = d\Phi_{\sigma^{-1}} X_\sigma$, whence

$$X_\tau = d\Phi_\tau(d\Phi_{\sigma^{-1}} X_\sigma) = d(\Phi_\tau \circ \Phi_{\sigma^{-1}}) X_\sigma = d\Phi_{\tau\sigma^{-1}} X_\sigma.$$

It follows immediately that, given an element $X_\epsilon \in \mathfrak{g}_\epsilon$, there exists one and only one left invariant infinitesimal transformation X which takes the value X_ϵ at ϵ.

We shall now prove that *every left invariant infinitesimal transformation X is analytic.* Let σ_0 be any element of $\mathcal{G}$. We select a system of coordinates $\{x_1, \cdots, x_n\}$ on $\mathcal{G}$ at σ_0 and a cubic neighbourhood V_1 of σ_0 with respect to this system. There exists a cubic neighbourhood V_2 of σ_0 such that the conditions $\sigma \in V_2$, $\tau \in V_2$ imply $\sigma\sigma_0^{-1}\tau \in V_1$. Let σ be any element of V_2; we have $X_\sigma x_i = (d\Phi_{\sigma\sigma_0^{-1}} X_{\sigma_0}) x_i = X_{\sigma_0}(x_i \circ \Phi_{\sigma\sigma_0^{-1}})$. The functions $x_i'(\sigma, \tau) = x_i(\sigma\sigma_0^{-1}\tau)$ are defined and analytic on $V_2 \times V_2$; we have $x_i'(\sigma, \tau) = f_i(x_1(\sigma), \cdots, x_n(\sigma), x_1(\tau), \cdots, x_n(\tau))$, where the functions $f_i(y_1, \cdots, y_n, z_1, \cdots, z_n)$ are analytic in their $2n$ arguments in the neighbourhood of the system of values $y_k = x_k(\sigma_0)$, $z_k = x_k(\sigma_0)$ $(1 \leqslant k \leqslant n)$. We have

$$X_\sigma x_i = \Sigma_{j=1}^n (X_{\sigma_0} x_j) \left(\frac{\partial f_i}{\partial z_j}\right)_{\sigma, \sigma_0}$$

where the indices σ, σ_0 mean that the partial derivatives are taken for $y_k = x_k(\sigma)$, $z_k = x_k(\sigma_0)$ $(1 \leqslant k \leqslant n)$. The quantities $X_{\sigma_0} x_j$ are constants, and $(\partial f_i / \partial z_j)_{\sigma, \sigma_0}$, considered as a function of σ, is analytic at σ_0. Therefore, the functions $X_\sigma x_i$ are analytic at σ_0, which proves that X is analytic at σ_0.

Let X and Y be any two left invariant infinitesimal transformations. We have

$$d\Phi_{\tau\sigma^{-1}}([X, Y]_\sigma) = [d\Phi_{\tau\sigma^{-1}}(X), d\Phi_{\tau\sigma^{-1}}(Y)]_\tau = [X, Y]_\tau$$

which proves that $[X, Y]$ is also left invariant.

The left invariant infinitesimal transformations of $\mathcal{G}$ form a vector space $\mathfrak{g}$ of dimension n over the field of real numbers (where n is the dimension of $\mathcal{G}$). Moreover, the conditions $X \varepsilon \mathfrak{g}$, $Y \varepsilon \mathfrak{g}$ imply $[X, Y] \varepsilon \mathfrak{g}$.

Definition 2. *Let K be a field, and let $\mathfrak{g}$ be a vector space of finite dimension over K. Suppose moreover that there is given a law of composition $(X, Y) \rightarrow [X, Y]$ in $\mathfrak{g}$ with the following properties:*

1. *It is bilinear*, i.e.

$$\begin{aligned} [a_1X_1 + a_2X_2, Y] &= a_1[X_1, Y] + a_2[X_2, Y] \\ [X, a_1Y_1 + a_2Y_2] &= a_1[X_1 Y_1] + a_2[X, Y_2] \end{aligned} \qquad \begin{pmatrix} a_1, a_2 \varepsilon K; X, Y, X_1, Y_1, \\ X_2, Y_2 \varepsilon \mathfrak{g} \end{pmatrix}$$

2. *It satisfies the following conditions:* $[X, X] = 0$, $[[X, Y], Z] + [[Y, Z], X] + [[Z, X], Y] = 0$ (*for any* $X, Y, Z \varepsilon \mathfrak{g}$). *Then* $\mathfrak{g}$, *equipped with this law of composition, is called a Lie algebra over* K.

Remark. It follows immediately from the definition that

$$[X, Y] + [Y, X] = 0 \qquad (X, Y \varepsilon \mathfrak{g});$$

because

$$\begin{aligned} 0 = [X + Y, X + Y] &= [X, X] + [X, Y] + [Y, X] + [Y, Y] \\ &= [X, Y] + [Y, X] \end{aligned}$$

In terms of this definition, we see that the left invariant infinitesimal transformations of an analytic group $\mathcal{G}$ form a Lie algebra (whose dimension is equal to the dimension of $\mathcal{G}$) over the field of real numbers. This Lie algebra is called the Lie algebra of $\mathcal{G}$.

Instead of the left invariant infinitesimal transformations, we might have considered the right invariant infinitesimal transformations. If Ψ_τ is the right translation associated with an element τ, the right invariant infinitesimal transformations Y are characterized by the condition that $Y_\tau = d\Psi_\tau Y_\epsilon$ for any $\tau \varepsilon \mathcal{G}$. Let J be the mapping $\xi \rightarrow \xi^{-1}$ of $\mathcal{G}$ into itself. It is clear that J is an analytic isomorphism of the underlying manifold of $\mathcal{G}$ with itself. Let X be any left invariant infinitesimal transformation; we assert that the infinitesimal transformation Y defined by $Y_\sigma = dJ(X_{\sigma^{-1}})$ is right invariant. In fact, we have

$$d\Psi_\tau Y_\epsilon = d\Psi_\tau(dJ(X_\epsilon)) = d(\Psi_\tau \circ J)X_\epsilon$$

But $\Psi_\tau \circ J$ maps an arbitrary element ξ into $\xi^{-1}\tau = (\tau^{-1}\xi)^{-1}$, whence

$\Psi_\tau \circ J = J \circ \Phi_{\tau^{-1}}$. Therefore,

$$d\Psi_\tau Y_\epsilon = dJ(d\Phi_{\tau^{-1}} X_\epsilon) = dJ(X_{\tau^{-1}}) = Y_\tau$$

which proves our assertion.

The right invariant infinitesimal transformations also form a Lie algebra. But, since $dJ([X_1, X_2]) = [dJ(X_1), dJ(X_2)]$, this new Lie algebra is isomorphic with the Lie algebra of left invariant infinitesimal transformations, and hence does not give any further information about the structure of $\mathcal{G}$.

§III. EXAMPLES OF LIE ALGEBRAS

Let us first consider the additive group R of real numbers. If we denote by x the coordinate in R, there exists an infinitesimal transformation X on R defined by $X_a x = 1$ (for all $a \varepsilon R$). This infinitesimal transformation is left invariant. In fact, let Φ_a be the translation corresponding to an element a; then $\Phi_a b = b + a$, $(d\Phi_a X_0)x = X_0(x \circ \Phi_a) = X_0(x + a) = 1 = X_a x$, which proves our assertion. It follows that X is a basic element of the Lie algebra of R, which consists of all multiples λX of $X (\lambda \varepsilon R)$.

Now, let $\mathcal{G}$ and $\mathcal{H}$ be analytic groups, and let $\mathfrak{g}$ and $\mathfrak{h}$ be their Lie algebras. We know that the tangent space to $\mathcal{G} \times \mathcal{H}$ at a point (σ, τ) of this manifold may be identified with the product $\mathfrak{g}_\sigma \times \mathfrak{h}_\tau$ of the tangent spaces $\mathfrak{g}_\sigma$ and $\mathfrak{h}_\tau$ to $\mathcal{G}$ and $\mathcal{H}$ at σ and τ respectively. Let X be a left invariant infinitesimal transformation on $\mathcal{G}$ and let Y be a left invariant infinitesimal transformation on $\mathcal{H}$. If we assign to every $(\sigma, \tau) \varepsilon \mathcal{G} \times \mathcal{H}$ the tangent vector (X_σ, Y_τ) to $\mathcal{G} \times \mathcal{H}$ at (σ, τ), we obtain an infinitesimal transformation Z on $\mathcal{G} \times \mathcal{H}$. We shall prove that Z is left invariant. Let $\bar{\omega}_1$ and $\bar{\omega}_2$ denote the projections of $\mathcal{G} \times \mathcal{H}$ on $\mathcal{G}$ and on $\mathcal{H}$. The vector $Z_{\sigma,\tau}$ is determined by the conditions

$$d\bar{\omega}_1 Z_{\sigma,\tau} = X_\sigma \qquad d\bar{\omega}_2 Z_{\sigma,\tau} = Y_\tau$$

Let Φ_σ, Ψ_τ, $\Theta_{\sigma,\tau}$ be the left translations associated with σ, τ, (σ, τ) in $\mathcal{G}$, $\mathcal{H}$ and $\mathcal{G} \times \mathcal{H}$ respectively. We have

$$\bar{\omega}_1 \circ \Theta_{\sigma,\tau} = \Phi_\sigma \circ \bar{\omega}_1 \qquad \bar{\omega}_2 \circ \Theta_{\sigma,\tau} = \Psi_\tau \circ \bar{\omega}_2$$

whence, if ϵ, η represent the neutral elements in $\mathcal{G}$, $\mathcal{H}$,

$$d\bar{\omega}_1(d\Theta_{\sigma,\tau} Z_{\epsilon,\eta}) = d\Phi_\sigma(d\bar{\omega}_1 Z_{\epsilon,\eta}) = d\Phi_\sigma X_\epsilon = X_\sigma$$
$$d\bar{\omega}_2(d\Theta_{\sigma,\tau} Z_{\epsilon,\eta}) = d\Psi_\tau(d\bar{\omega}_2 Z_{\epsilon,\eta}) = d\Psi_\tau Y_\eta = Y_\tau$$

This proves that $Z_{\sigma,\tau} = d\Theta_{\sigma,\tau} Z_{\epsilon,\eta}$, and therefore that Z is left invariant. On the other hand, if Z and Z' are two analytic infinitesimal transformations on $\mathcal{G} \times \mathcal{H}$, we have $d\bar{\omega}_i[Z, Z'] = [d\bar{\omega}_i Z, d\bar{\omega}_i Z']$ $(i = 1, 2)$. This leads us to the following definition:

Definition 1. *Let $\mathfrak{g}$ and $\mathfrak{h}$ be two Lie algebras over the same field K. Let us define the following law of composition in $\mathfrak{g} \times \mathfrak{h}$:*

$$[(X, Y), (X', Y')] = ([X, X'], [Y, Y'])$$

We then obtain a Lie algebra whose underlying vector space is the product of the underlying vector spaces of $\mathfrak{g}$ and $\mathfrak{h}$. This Lie algebra is called the product of $\mathfrak{g}$ and $\mathfrak{h}$ and is denoted by $\mathfrak{g} \times \mathfrak{h}$.

We have proved that *the Lie algebra of the product of two analytic groups is the product of the Lie algebras of these groups.*

The definition and the result may easily be extended to the case of the product of several Lie algebras or analytic groups.

In particular, we see that the Lie algebra $\mathfrak{r}^n$ of the analytic group R^n is the product of n Lie algebras identical with the Lie algebra $\mathfrak{r}$ of R. Now, if X, X' are any two elements of $\mathfrak{r}$, we have $[X, X'] = 0$; hence the same is true of the Lie algebra of R^n.

Since T^1 is of dimension 1, its Lie algebra is also of dimension 1, and is therefore identical with $\mathfrak{r}$. If we call T^n the product of n analytic groups identical with T^1, we see that the Lie algebra of T^n is $\mathfrak{r}^n$.

We shall now find the Lie algebra of $GL(n, C)$. The coefficients $x_{ij}(\sigma)$ of a matrix $\sigma \varepsilon GL(n, C)$ are complex valued functions, whose real and imaginary parts $x'_{ij}(\sigma)$ and $x''_{ij}(\sigma)$ are analytic functions on $GL(n, C)$. If X is an analytic infinitesimal transformation on $GL(n, C)$, we set $Xx_{ij} = Xx'_{ij} + \sqrt{-1}\, Xx''_{ij}$.

With every left invariant infinitesimal transformation X on $GL(n, C)$ we associate the matrix $(a_{ij}(X))$, where $a_{ij}(X) = X_\epsilon x_{ij}$ $(1 \leqslant i, j \leqslant n)$ and where ϵ is the neutral element of $GL(n, C)$. We obtain in this way a linear mapping of the Lie algebra $\mathfrak{gl}(n, C)$ of $GL(n, C)$ into the vector space $\mathfrak{M}_n(C)$ consisting of all matrices of degree n with real or complex coefficients. ($\mathfrak{M}_n(C)$ is a vector space of dimension $2n^2$ over R). If $a_{ij}(X) = 0$ for all (i, j), we have $X_\epsilon = 0$, because: he $2n^2$ functions x'_{ij}, x''_{ij} form a system of coordinates at ϵ. Since X is left invariant, it follows that $X = 0$. Our linear mapping is therefore a linear isomorphism of $\mathfrak{gl}(n, C)$ with a subspace of $\mathfrak{M}_n(C)$. Since $\mathfrak{gl}(n, C)$ and $\mathfrak{M}_n(C)$ are both of dimension $2n^2$, the image of $\mathfrak{gl}(n, C)$ is the whole of $\mathfrak{M}_n(C)$.

It remains to compute the matrix $(a_{ij}([X, Y]))$ when $(a_{ij}(X))$ and $(a_{ij}(Y))$ are known, X and Y being two left invariant infinitesimal transformations. We have $X_\sigma x_{ij} = d\Phi_\sigma X_\epsilon x_{ij} = X_\epsilon(x_{ij} \circ \Phi_\sigma)$, where Φ_σ

is the left translation associated with an element $\sigma \varepsilon GL(n, C)$. Hence

$$X_\sigma x_{ij} = \Sigma_k x_{ik}(\sigma) X_\epsilon x_{kj} = \Sigma_k x_{ik}(\sigma) a_{kj}(X)$$

If we consider $X_\sigma x_{ij}$ as a function of σ, we have

$$Y_\epsilon(X_\sigma x_{ij}) = \Sigma_k a_{ik}(Y) a_{kj}(X).$$

Let us denote by $\tilde{X}$, $\tilde{Y}$ the matrices $(a_{ij}(X))$, $(a_{ij}(Y))$; then the matrix whose coefficients are the numbers $Y_\epsilon(X_\sigma x_{ij})$ is $\tilde{Y}\tilde{X}$, and the matrix whose coefficients are the numbers $X_\epsilon(Y_\sigma x_{ij})$ is $\tilde{X}\tilde{Y}$. It follows that the matrix which corresponds to $[X, Y]$ is $\tilde{Y}\tilde{X} - \tilde{X}\tilde{Y}$.

Therefore, *the Lie algebra of $GL(n, C)$ is isomorphic with the space of all complex matrices of degree n, this space being equipped with the following law of composition:* $[\tilde{X}, \tilde{Y}] = \tilde{Y}\tilde{X} - \tilde{X}\tilde{Y}$.

More generally, if K is any field, the set of all matrices of degree n with coefficients in K can be turned into a Lie algebra over K by defining the bracket operation by the formula $[\tilde{X}, \tilde{Y}] = \tilde{Y}\tilde{X} - \tilde{X}\tilde{Y}$.

Let $\mathfrak{g}$ be a vector space of finite dimension over a field K, and let $\{X_1, \cdots, X_n\}$ be a base of $\mathfrak{g}$ over K. If we want to define a law of composition which will define a structure of Lie algebra on $\mathfrak{g}$, it will obviously be sufficient to give the expressions for the elements $[X_i, X_j]$ $(1 \leqslant i, j \leqslant n)$, say

$$[X_i, X_j] = \Sigma_k c_{ijk} X_k$$

The constants c_{ijk} are called the *constants of structure* of the Lie algebra with respect to the base $\{X_1, \cdots, X_n\}$. These constants cannot be chosen arbitrarily. In fact, we must have

$$(1) \qquad [X_i, X_j] + [X_j, X_i] = 0$$

$$(2) \qquad [[X_i, X_j], X_k] + [[X_j, X_k], X_i] + [[X_k, X_i], X_j] = 0$$

and hence

$$(3) \qquad c_{ijk} + c_{jik} = 0$$

$$(4) \qquad \Sigma_h(c_{ijh}c_{hkl} + c_{jkh}c_{hil} + c_{kih}c_{hjl}) = 0 \qquad (1 \leqslant i, j, k, l \leqslant n)$$

If K is of characteristic $\neq 2$, these conditions are not only necessary but also sufficient to insure that the law of composition defined by the constants c_{ijk} will have the properties described in Definition 1. In fact, assume that conditions (3) and (4) are satisfied, and let $X = \Sigma_i x_i X_i$, $Y = \Sigma_i y_i X_i$, $Z = \Sigma_i z_i X_i$ be any three elements of $\mathfrak{g}$. We have $[X, X] = \Sigma_{ij} x_i x_j [X_i, X_j] = 0$ because $[X_i, X_j] + [X_j, X_i] = 0$

and

$$[[X, Y], Z] + [[Y, Z], X] + [[Z, X], Y] = \Sigma_{ijk} x_i y_j z_k([[X_i, X_j], X_k] + [[X_j, X_k], X_i] + [[X_k, X_i], X_j]) = 0$$

which proves our assertion.

Moreover, it will obviously be sufficient that the conditions (1) hold for $i \leqslant j$ and that the conditions (2) hold for $i < j < k$.

The construction of all Lie algebras of a given dimension over say the field of real numbers is thus reduced to a purely algebraic problem. The proof of the fact that to every Lie algebra there corresponds an analytic group is very difficult and must be postponed to the second volume.

§IV. ANALYTIC SUBGROUPS

Definition 1. *Let* $\mathcal{G}$ *be an analytic group. An analytic group* $\mathcal{H}$ *is called an analytic subgroup of* $\mathcal{G}$ *if the following conditions are satisfied:* 1) *the underlying manifold of* $\mathcal{H}$ *is a submanifold of the underlying manifold of* $\mathcal{G}$; 2) *the underlying group of* $\mathcal{H}$ *is a subgroup of the underlying group of* $\mathcal{G}$.(3)

Let $\mathcal{H}$ be an analytic subgroup of $\mathcal{G}$. We denote by $\mathfrak{g}$ the Lie algebra of $\mathcal{G}$ and by $\mathfrak{g}_\sigma$ the tangent vector space to $\mathcal{G}$ at a point $\sigma \varepsilon \mathcal{G}$. Let $\mathfrak{h}$ be the set of elements $X \varepsilon \mathfrak{g}$ such that X_ϵ belongs to the tangent space $\mathfrak{h}_\epsilon$ to $\mathcal{H}$ at the neutral element ϵ. Let σ be any element of $\mathcal{H}$; if Φ_σ is the left translation associated with σ, Φ_σ induces an analytic isomorphism of the underlying manifold of $\mathcal{H}$ with itself. It follows that $d\Phi_\sigma(\mathfrak{h}_\epsilon)$ is the tangent space $\mathfrak{h}_\sigma$ to $\mathcal{H}$ at σ. If $X \varepsilon \mathfrak{h}$, we have $X_\sigma \varepsilon \mathfrak{h}_\sigma$ for all $\sigma \varepsilon \mathcal{H}$, and therefore X has a contraction to the submanifold $\mathcal{H}$, and this contraction is a left invariant infinitesimal transformation on $\mathcal{H}$. If X and Y are in $\mathfrak{h}$, we know that $[X, Y]$ also has a contraction to $\mathcal{H}$; in particular, we have $[X, Y] \varepsilon \mathfrak{h}$.

Definition 2. *Let* $\mathfrak{g}$ *be a Lie algebra. A subset* $\mathfrak{h}$ *of* $\mathfrak{g}$ *is called a subalgebra of* $\mathfrak{g}$ *if the following conditions are satisfied:* 1) $\mathfrak{h}$ *is a vector subspace of* $\mathfrak{g}$; 2) *the conditions* $X \epsilon \mathfrak{h}$, $Y \epsilon \mathfrak{h}$ *imply* $[X, Y] \varepsilon \mathfrak{h}$.

With this terminology, we see that the set $\mathfrak{h}$ introduced above is a subalgebra of the Lie algebra $\mathfrak{g}$ of $\mathcal{G}$, and that this subalgebra is isomorphic to the Lie algebra of $\mathcal{H}$ (an isomorphism being obtained by assigning to every $X \varepsilon \mathfrak{h}$ its contraction to $\mathcal{H}$). We may therefore identify the Lie algebra of $\mathcal{H}$ with a subalgebra of $\mathfrak{g}$.

Conversely, let $\mathfrak{h}$ be any subalgebra of $\mathfrak{g}$. To every $\sigma \varepsilon \mathcal{G}$ we assign the subspace $\mathfrak{h}_\sigma$ of $\mathfrak{g}_\sigma$ composed of the elements X_σ for $X \varepsilon \mathfrak{h}$. We obtain in this way a distribution $\mathfrak{M}$ on $\mathcal{G}$ which is obviously analytic.

If $\{X_1, \cdots, X_m\}$ is a base of $\mathfrak{h}$, the infinitesimal transformations $X_1, \cdots, X_m$ form a base of $\mathfrak{M}$ around every point of $\mathcal{G}$. Remembering that $\mathfrak{h}$ is a subalgebra, we see immediately that the distribution $\mathfrak{M}$ is involutive. Let $\mathcal{H}$ be the maximal integral manifold of the distribution $\mathfrak{M}$ containing the neutral element ϵ. If σ is any element of $\mathcal{H}$, the left translation Φ_σ is an analytic isomorphism of the underlying manifold of $\mathcal{H}$ with itself. Since $d\Phi_\sigma\mathfrak{h}_\tau = \mathfrak{h}_{\sigma\tau}$, this analytic isomorphism leaves the distribution $\mathfrak{M}$ invariant. It follows immediately that Φ_σ permutes among themselves the maximal integral manifolds of $\mathfrak{M}$. If $\sigma \varepsilon \mathcal{H}$, $\Phi_{\sigma^{-1}}$ maps $\mathcal{H}$ on a maximal integral manifold which contains $\Phi_{\sigma^{-1}}\sigma = \epsilon$, whence $\Phi_{\sigma^{-1}}\mathcal{H} = \mathcal{H}$. It follows that the set of points of $\mathcal{H}$ is a subgroup H of the underlying group of $\mathcal{G}$, and that, if $\sigma \varepsilon H$, Φ_σ induces an analytic isomorphism of $\mathcal{H}$ with itself. We wish to prove furthermore that the mapping $(\sigma, \tau) \to \sigma\tau^{-1}$ of $\mathcal{H} \times \mathcal{H}$ onto $\mathcal{H}$ is analytic.

We know that the mapping $(\sigma, \tau) \to \sigma\tau^{-1}$ is an analytic mapping of $\mathcal{H} \times \mathcal{H}$ into $\mathcal{G}$ (because $\mathcal{H} \times \mathcal{H}$ is clearly a submanifold of $\mathcal{G} \times \mathcal{G}$). Therefore, making use of Proposition 1, §IX, Chapter III, p. 94, we see that it will be sufficient to prove that the countability axiom holds in $\mathcal{H}$. Since $\mathcal{H}$ is a submanifold of $\mathcal{G}$, it will be enough to prove that the countability axiom holds in $\mathcal{G}$ (Cf. Proposition 2, §IX, Chap. III, p. 94). Let V be a cubic neighbourhood of the neutral element ϵ in $\mathcal{G}$ with respect to some system of coordinates at ϵ. Since V is homeomorphic to a cube in some cartesian space, it contains a countable dense subset E. Let D be the group generated by the elements of E; then D is countable. We shall prove that $\mathcal{G}$ is the union of the sets δV, $\delta \varepsilon D$; this will clearly prove that the countability axiom holds in $\mathcal{G}$. Let σ be any element of $\mathcal{G}$. Since $\mathcal{G}$ is connected, the elements of V form a set of generators of $\mathcal{G}$, and we may write σ in the form $\sigma_1^{a_1} \cdots \sigma_h^{a_h}$ with $\sigma_k \varepsilon V$, $a_k = \pm 1$ $(1 \leqslant k \leqslant h)$. For each k we can find a sequence $(\theta_{k,n})$ of elements of E which converges to σ_k. Set $\delta_n = \theta_{1,n}^{a_1} \cdots \theta_{h,n}^{a_h}$; then we have $\lim_{n\to\infty} \delta_n = \sigma$. Since σV^{-1} is a neighbourhood of σ, there exists an integer n such that $\delta_n \varepsilon \sigma V^{-1}$, whence $\sigma \varepsilon \delta_n V$. Since $\delta_n \varepsilon D$, our assertion is proved.

We have therefore proved that $\mathcal{H}$ is an analytic subgroup of $\mathcal{G}$. The subalgebra of $\mathfrak{g}$ which is associated with $\mathcal{H}$ by the construction which was indicated at the beginning of this section is obviously $\mathfrak{h}$.

Conversely, let $\mathcal{H}'$ be any analytic subgroup of $\mathcal{G}$ whose Lie algebra is $\mathfrak{h}$ Then it is clear that $\mathcal{H}'$ is an integral manifold of the distribution $\mathfrak{M}$ and therefore that $\mathcal{H}'$ is an open submanifold of $\mathcal{H}$. Since $\mathcal{H}'$ contains ϵ, it contains a neighbourhood of ϵ with respect to $\mathcal{H}$. Since

$\mathfrak{H}$ is connected, the elements of any neighbourhood of ϵ form a set of generators of $\mathfrak{H}$, whence $\mathfrak{H}' = \mathfrak{H}$. We have proved:

Theorem 1. *Let $\mathcal{G}$ be an analytic group. If $\mathfrak{H}$ is an analytic subgroup of $\mathcal{G}$, the Lie algebra of $\mathfrak{H}$ may be considered as a subalgebra of the Lie algebra of $\mathcal{G}$. Every subalgebra of the Lie algebra of $\mathcal{G}$ is the Lie algebra of one and only one analytic subgroup of $\mathcal{G}$.*

Remark. The maximal integral manifolds of the distribution $\mathfrak{M}$ which was introduced in the course of the proof are clearly the cosets $\sigma\mathfrak{H}$ obtained from $\mathfrak{H}$ by the left translations in $\mathcal{G}$.

§V. Closed Analytic Subgroups

Let $\mathcal{G}$ be an analytic group of dimension m, and let $\mathfrak{H}$ be an analytic subgroup of $\mathcal{G}$ of dimension n. We assume furthermore that the set of points of $\mathfrak{H}$ is closed in the underlying space of $\mathcal{G}$. We shall prove that, under this condition, the underlying space of $\mathfrak{H}$ is a subspace of the underlying space of $\mathcal{G}$, and therefore that the underlying topological group of $\mathfrak{H}$ is a topological subgroup of the underlying topological group of $\mathcal{G}$.

Since $\mathfrak{H}$ and its cosets are the maximal integral manifolds of an involutive distribution on $\mathcal{G}$ (cf. §IV, p. 107), we can find a system of coordinates $\{x_1, \cdots, x_m\}$ at the neutral element ϵ on $\mathcal{G}$ and a cubic neighbourhood V of ϵ with respect to this system which have the following property: a being the breadth of V, if $\xi_{n+1}, \cdots, \xi_m$ are any $m - n$ numbers such that $|\xi_{n+j}| < a$ $(1 \leqslant j \leqslant m - n)$, the slice S_ξ of V characterized by the equations $x_{n+j} = \xi_{n+j}$ $(1 \leqslant j \leqslant m - n)$ is contained in some coset modulo $\mathfrak{H}$ (cf. Theorem 1, §VII, Chapter III, p. 88). The set $\mathfrak{H} \cap V$ consists of at most countably many slices S_ξ because the axiom of countability holds in $\mathfrak{H}$ (cf. §IX, Chapter III, p. 94). Let Ξ be the set of points $(\xi_{n+1}, \cdots, \xi_m) \epsilon R^{m-n}$ such that $S_\xi \subset \mathfrak{H}$. Since $\mathfrak{H}$ is closed, Ξ is relatively closed in the cube defined by $|x_{n+j}| < a$ $(1 \leqslant j \leqslant m - n)$ (the x_{n+j}'s being taken as coordinates in R^{m-n}). Since Ξ is countable, it has at least one isolated point ξ^0. Let σ_0 be a point of S_{ξ^0}. If W is a sufficiently small neighbourhood of ϵ with respect to $\mathfrak{H}$, $\sigma_0 W$ is a neighbourhood of σ_0 with respect to S_{ξ^0}. Since ξ^0 is isolated in Ξ, there exists a neighbourhood V' of ϵ in $\mathcal{G}$ such that $\sigma_0 W = \mathfrak{H} \cap \sigma_0 V'$. It follows immediately that, σ being any point of $\mathfrak{H}$, the neighbourhoods of σ with respect to $\mathfrak{H}$ are the intersections with $\mathfrak{H}$ of the neighbourhoods of σ with respect to $\mathcal{G}$. This proves that $\mathfrak{H}$ is a subspace of $\mathcal{G}$.

At the same time, we see that every point of Ξ is isolated, and therefore that $V \cap \mathfrak{H} \subset S_0$ provided a is small enough, where S_0 is the slice which contains ϵ.

Now, let $\mathfrak{H}_1$ be a closed topological subgroup of $\mathfrak{G}$ which contains $\mathfrak{H}$ as a relatively open subgroup (which implies that the component of the neutral element in $\mathfrak{H}_1$ is $\mathfrak{H}$). If a is small enough, we shall have $V \cap \mathfrak{H}_1 \subset \mathfrak{H}$, and therefore $V \cap \mathfrak{H}_1 = S_0$. We assert that there exists a number $a_1 > 0 (a_1 < a)$ with the following property: if the numbers $|\xi_{n+j}|$, $|\xi'_{n+j}|$ $(1 \leqslant j \leqslant m - n)$ are all $\leqslant a_1$, and if the points $\xi = (\xi_{n+1}, \cdots, \xi_m)$, $\xi' = (\xi'_{n+1}, \cdots, \xi'_m) \varepsilon R^{m-n}$ are distinct, then the slices S_ξ, $S_{\xi'}$, belong to distinct cosets modulo $\mathfrak{H}_1$. In fact, the neighbourhood V having been so chosen that $V \cap \mathfrak{H}_1 = S_0$, let V_2, V_1 be cubic neighbourhoods of ϵ such that $V_2 V_2 \subset V$, $V_1^{-1} V_1 \subset V_2$; we assert that the half-breadth a_1 of V_1 has the stated property. In fact, let σ and τ be two elements of V_1 which belong to the same coset modulo $\mathfrak{H}_1$. We have $\tau^{-1}\sigma \varepsilon V_2 \cap \mathfrak{H}_1 = V_2 \cap S_0$, whence $\sigma \varepsilon \tau (V_2 \cap S_0)$. But $V_2 \cap S_0$ is connected and $\tau(V_2 \cap S_0) \subset V_2 V_2 \subset V$; therefore σ and τ belong to the same component of $\tau \mathfrak{H}_1 \cap V$, i.e. to the same slice S_ξ, which proves our assertion.

Since $\mathfrak{H}_1$ is a closed subgroup of $\mathfrak{G}$, the homogeneous space $\mathfrak{G}/\mathfrak{H}_1$ has a natural topology (Cf. §III, Chapter II, p. 29). We denote by $\bar{\omega}$ the natural mapping of $\mathfrak{G}$ onto $\mathfrak{G}/\mathfrak{H}_1$. Let F be the subset of V defined by the conditions

$$x_1 = 0, \quad \cdots, \quad x_n = 0, \quad |x_{n+j}| \leqslant a_1 \quad (1 \leqslant j \leqslant m - n)$$

Then $\bar{\omega}$ maps F in a continuous univalent way; since F is compact, this mapping is topological. Since $\bar{\omega}(F)$ contains the image under $\bar{\omega}$ of the cubic neighbourhood of breadth a_1 of ϵ, $\bar{\omega}(F)$ is a neighbourhood of $\bar{\omega}(\epsilon)$ in $\mathfrak{G}/\mathfrak{H}_1$. We have proved

Proposition 1. *Let $\mathfrak{H}$ be a closed analytic subgroup of the analytic group $\mathfrak{G}$. Then $\mathfrak{H}$ is a topological subgroup of $\mathfrak{G}$. Let $\mathfrak{H}_1$ be a closed topological subgroup of $\mathfrak{G}$ which contains $\mathfrak{H}$ as a relatively open subgroup. Then there exists a subset F of $\mathfrak{G}$ which contains the neutral element ϵ, is homeomorphic to a closed cube in a cartesian space and is mapped topologically onto a neighbourhood of $\bar{\omega}(\epsilon)$ in $\mathfrak{G}/\mathfrak{H}_1$ by the natural mapping $\bar{\omega}$ of $\mathfrak{G}$ onto $\mathfrak{G}/\mathfrak{H}_1$.*

We shall now define the notion of *analytic function* at a point p of the space $\mathfrak{G}/\mathfrak{H}_1$. Let f be a function defined on a neighbourhood of p, and let σ be any point of the set $\overset{-1}{\bar{\omega}}(p)$. The function $f \circ \bar{\omega} = g$ is defined on a subset of $\mathfrak{G}$ which can be represented as a union of cosets modulo $\mathfrak{H}_1$, and we have $g(\rho\tau) = g(\rho)$ if $\tau \varepsilon \mathfrak{H}_1$. Therefore, if g is analytic at σ, it is also analytic at every point of the form $\sigma\tau$, $\tau \varepsilon \mathfrak{H}_1$, i.e. at every point of $\overset{-1}{\bar{\omega}}(p)$. If this is the case, we shall say that f is analytic at p on $\mathcal{G}/\mathfrak{H}_1$. Let $\mathfrak{A}(p)$ be the class of functions analytic at p. The classes $\mathfrak{A}(p)$ have obviously the properties I, II of §I, Chapter III, p. 68. The set F being constructed as above, denote by W the set of interior points of $\bar{\omega}(\sigma F)$. If $q \varepsilon W$, we set $y_i(q) = x_{n+i}(\rho)$, where ρ is the point of F such that $\bar{\omega}(\sigma\rho) = q$. The mapping $q \rightarrow (y_1(q), \cdots, y_{m-n}(q))$ maps W topologically onto a cube in R^{m-n}; the functions y_i are analytic at every point of W and conversely, every function which is analytic at a point $q \varepsilon W$ is analytically dependent around q on the functions $y_1, \cdots, y_{m-n}$. It follows easily that we can define a manifold whose underlying space is $\mathcal{G}/\mathfrak{H}_1$ and such that $\mathfrak{A}(p)$ is the class of analytic functions at p on this manifold. It is this manifold which we shall from now on denote by $\mathcal{G}/\mathfrak{H}_1$. If I is the set of points of F whose coordinates are $< a_1$ in absolute value, the contraction of $\bar{\omega}$ to I is an analytic isomorphism of I (considered as a submanifold of $\mathcal{G}$) with an open submanifold of $\mathcal{G}/\mathfrak{H}_1$.

To every $\rho \varepsilon \mathcal{G}$ there corresponds a homeomorphism of $\mathcal{G}/\mathfrak{H}_1$ with itself which maps a point $p = \bar{\omega}(\sigma)$ on $\rho p = \bar{\omega}(\rho\sigma)$. *The mapping* $(\rho, p) \rightarrow \rho p$ *is an analytic mapping of* $\mathcal{G} \times (\mathcal{G}/\mathfrak{H}_1)$ *into* $\mathcal{G}/\mathfrak{H}_1$. In fact, $p_0 = \bar{\omega}(\sigma_0)$ being any point of $\mathcal{G}/\mathfrak{H}_1$, let $\bar{\omega}^*$ be the reciprocal mapping of the contraction of $\bar{\omega}$ to $\sigma_0 I$. Then $\bar{\omega}^*$ is an analytic mapping of a neighbourhood of p_0 into $\mathcal{G}$ and we have $\rho p = \bar{\omega}(\rho\bar{\omega}^*(p))$ if p is in this neighbourhood, which proves our assertion.

§VI. Analytic Homomorphisms

Definition 1. *A homomorphism H of an analytic group $\mathcal{G}$ into an analytic group $\mathcal{H}$ is called an analytic homomorphism if it is an analytic mapping of the underlying manifold of $\mathcal{G}$ into the underlying manifold of $\mathcal{H}$.*

Let H be an analytic homomorphism of $\mathcal{G}$ into $\mathcal{H}$. Let X be any left invariant infinitesimal transformation on $\mathcal{G}$. Denote by ϵ and η the neutral elements of $\mathcal{G}$ and $\mathcal{H}$ respectively; then $dH_\epsilon X_\epsilon$ is a tangent vector to $\mathcal{H}$ at η. Denote by Y the left invariant infinitesimal transformation on $\mathcal{H}$ such that $Y_\eta = dH_\epsilon X_\epsilon$. Then we assert that

$$Y_{H\sigma} = dH_\sigma X_\sigma \tag{1}$$

for any $\sigma \varepsilon \mathcal{G}$. Denote by Φ_σ the left translation in $\mathcal{G}$ which maps ϵ on σ and by $\Psi_{H\sigma}$ the left translation in $\mathcal{H}$ which maps η on $H\sigma$. From the fact that H is a homomorphism, it follows immediately that $H \circ \Phi_\sigma$

$= \Psi_{H\sigma} \circ H$. Therefore

$$dH_\sigma X_\sigma = d(H \circ \Phi_\sigma)X_\epsilon = d(\Psi_{H\sigma} \circ H)X_\epsilon = d\Psi_{H\sigma}(dH_\epsilon X_\epsilon) = d\Psi_{H\sigma}Y_\eta = Y_{H\sigma}$$

which proves our assertion.

The formula (1) means that Y is H-related to X (Cf. §V, Chapter III, p. 82). We set $Y = dH(X)$. We know that, if X and X' are any two left invariant infinitesimal transformations on $\mathcal{G}$, then

$$dH([X, X']) = [dH(X), dH(X')]$$

(cf. Proposition 1, §V, Chapter III, p. 82).

Definition 2. *Let $\mathfrak{g}$ and $\mathfrak{h}$ be Lie algebras over the same field K. A mapping Δ of $\mathfrak{g}$ into $\mathfrak{h}$ is called a homomorphism if* a) *it is a linear mapping;* b) *we have* $\Delta([X, X']) = [\Delta(X), \Delta(X')]$ *for any* $X, X' \varepsilon \mathfrak{g}$.

We see that to every analytic homomorphism H of $\mathcal{G}$ into $\mathcal{H}$ there corresponds a homomorphism dH of the Lie algebra $\mathfrak{g}$ of $\mathcal{G}$ into the Lie algebra $\mathfrak{h}$ of $\mathcal{H}$. We shall now examine whether the converse of this statement holds true.

Let Δ be a homomorphism of $\mathfrak{g}$ into $\mathfrak{h}$, and let $\mathfrak{e}$ be the subset of $\mathfrak{g} \times \mathfrak{h}$ composed of all elements of the form $(X, \Delta X)(X \varepsilon \mathfrak{g})$. Since Δ is a homomorphism, we see easily that $\mathfrak{e}$ is a subalgebra of $\mathfrak{g} \times \mathfrak{h}$. We know that $\mathfrak{g} \times \mathfrak{h}$ is the Lie algebra of $\mathcal{G} \times \mathcal{H}$ (Cf. §III, p. 104). Let $\mathcal{E}$ be the analytic subgroup of $\mathcal{G} \times \mathcal{H}$ whose Lie algebra is $\mathfrak{e}$ (Theorem 1, §IV, p. 107). Let $\bar{\omega}_1$ be the projection of $\mathcal{G} \times \mathcal{H}$ on $\mathcal{G}$, and let $\bar{\omega}_1'$ be the contraction of $\bar{\omega}_1$ to $\mathcal{E}$. The mapping $\bar{\omega}_1'$ is clearly an analytic homomorphism of $\mathcal{E}$ into $\mathcal{G}$. Moreover, if $X \varepsilon \mathfrak{g}$, $d\bar{\omega}_1'$ maps $(X_\epsilon, (\Delta X)_\eta)$ on X_ϵ, which shows that $d\bar{\omega}_1'$ maps the tangent space to $\mathcal{E}$ at (ϵ, η) in a univalent way onto $\mathfrak{g}_\epsilon$ (the tangent space to $\mathcal{G}$ at ϵ). It follows that we can find a cubic neighbourhood U of (ϵ, η) with respect to $\mathcal{E}$ (relative to some system of coordinates on $\mathcal{E}$) which is mapped topologically by $\bar{\omega}_1'$ onto a neighbourhood of ϵ in $\mathcal{G}$ (cf. Proposition 3, §IV, Chapter III, p. 76). Let λ be the reciprocal mapping of the contraction of $\bar{\omega}_1'$ to U; then λ is a local homomorphism of $\mathcal{G}$ into $\mathcal{E}$. *If the group $\mathcal{G}$ is simply connected*, this local homomorphism may be extended to a homomorphism (which we also denote by λ) of $\mathcal{G}$ into $\mathcal{E}$ (Theorem 2, §IX, Chapter II, p. 54). Since $\lambda \circ \bar{\omega}_1'$ coincides with the identity on U and since $\mathcal{E}$ is connected, $\lambda \circ \bar{\omega}_1'$ is the identity mapping and therefore $\bar{\omega}_1'$ is a homeomorphism of $\mathcal{E}$ with $\mathcal{G}$. Let $\bar{\omega}_2$ be the projection of $\mathcal{G} \times \mathcal{H}$ on $\mathcal{H}$; then $H = \bar{\omega}_2 \circ \lambda$ is clearly an analytic homomorphism of $\mathcal{G}$ into $\mathcal{H}$. If X is a left invariant infinitesimal transformation on $\mathcal{G}$, we have $d\lambda(X_\epsilon) = (X_\epsilon, (\Delta X)_\eta)$, $d\bar{\omega}_2(X_\epsilon, (\Delta X)_\eta) = (\Delta X)_\eta$, whence $dH(X) = \Delta X$.

Assuming that there exists an analytic homomorphism H of $\mathcal{G}$ into $\mathcal{H}$ such that $dH = \Delta$ (we have seen that this is the case if $\mathcal{G}$ is simply connected), we shall prove that H is uniquely determined. In fact, let H' be any analytic homomorphism of $\mathcal{G}$ into $\mathcal{H}$ such that $dH' = \Delta$. The mapping $\sigma \to \theta(\sigma) = (\sigma, H'\sigma)$ maps isomorphically the underlying group of $\mathcal{G}$ onto a subgroup E' of the underlying group of $\mathcal{G} \times \mathcal{H}$. Furthermore, θ is a regular analytic mapping of $\mathcal{G}$ into $\mathcal{G} \times \mathcal{H}$. We can define an analytic group $\mathcal{E}'$ whose underlying group is E' by the condition that θ shall be an analytic isomorphism of $\mathcal{G}$ with $\mathcal{E}'$. The identity mapping of $\mathcal{E}'$ into $\mathcal{G} \times \mathcal{H}$ may be represented in the form $\theta \circ \overset{-1}{\theta}$ and is therefore analytic and regular. We conclude that $\mathcal{E}'$ is an analytic subgroup of $\mathcal{G} \times \mathcal{H}$. The mapping $\bar{\omega}_1 \circ \theta$ is the identity mapping of $\mathcal{G}$ onto itself; the mapping $\bar{\omega}_2 \circ \theta$ coincides with H'. It follows that, if $X \in \mathfrak{g}$, we have $d\bar{\omega}_1(d\theta(X)) = X$, $d\bar{\omega}_2(d\theta(X)) = dH'(X) = \Delta(X)$, whence $d\theta(X) = (X, \Delta(X))$. This shows that the Lie algebra of $\mathcal{E}'$ coincides with the Lie algebra $\mathfrak{e}$ constructed above and depends only on Δ. Therefore the group $\mathcal{E}'$ is also uniquely determined by Δ, and the same is true of H'.

We have therefore proved

Theorem 2. *Let $\mathcal{G}$ and $\mathcal{H}$ be analytic groups, and let $\mathfrak{g}$ and $\mathfrak{h}$ be their Lie algebras. If H is an analytic homomorphism of $\mathcal{G}$ into $\mathcal{H}$, there corresponds to H a homomorphism dH of $\mathfrak{g}$ into $\mathfrak{h}$ such that, X being any element of $\mathfrak{g}$, the infinitesimal transformation X and $dH(X)$ are H-related. If H and H' are analytic homomorphisms such that $dH = dH'$, then $H = H'$. If $\mathcal{G}$ is simply connected, every homomorphism of $\mathfrak{g}$ into $\mathfrak{h}$ can be represented in the form dH, where H is an analytic homomorphism of $\mathcal{G}$ into $\mathcal{H}$.*

The restriction that $\mathcal{G}$ be simply connected cannot be removed. In fact, we have seen that T^1 and R have the same Lie algebra $\mathfrak{r}$. It is clear that the identity mapping of $\mathfrak{r}$ into itself cannot be obtained from any homomorphism of T^1 into R, since the only homomorphism of T^1 into R is the one which maps every element upon 0.

However, the proof has shown that the following result holds in every case: *to every homomorphism Δ of $\mathfrak{g}$ into $\mathfrak{h}$ there corresponds a local homomorphism H of a neighbourhood U of ϵ in $\mathcal{G}$ into $\mathcal{H}$ which is analytic at every point $\sigma \in U$ and which satisfies the condition $dH(X_\sigma) = (\Delta X)_{H\sigma}$ for every $X \in \mathfrak{g}$ and $\sigma \in U$.*

To conclude this section, we observe that a homomorphism H of an analytic group $\mathcal{G}$ into an analytic group $\mathcal{H}$ is certainly everywhere analytic if it is analytic at ϵ. In fact, we have $H(\sigma_0\sigma) = H(\sigma_0)H(\sigma)$ and our statement follows from the fact that the left translations in $\mathcal{G}$ and $\mathcal{H}$ are analytic isomorphisms of these manifolds with themselves.

§VII. FACTOR GROUPS OF AN ANALYTIC GROUP

Let $\mathcal{G}$ be an analytic group, and let $\mathfrak{H}$ be a topological subgroup of $\mathcal{G}$ which satisfies the following conditions: $\mathfrak{H}$ is closed in $\mathcal{G}$; the component of the neutral element in $\mathfrak{H}$ is the underlying topological group of an analytic subgroup $\mathcal{H}$ of $\mathcal{G}$ and is relatively open in $\mathfrak{H}$. Then we have already defined a structure of manifold on $\mathcal{G}/\mathfrak{H}$. Now, let us assume furthermore that $\mathfrak{H}$ is distinguished. Then $\mathcal{G}/\mathfrak{H}$ has a structure of group. We shall prove that the group $\mathcal{G}/\mathfrak{H}$ and the manifold $\mathcal{G}/\mathfrak{H}$, taken together, form an analytic group.

The mapping $(\rho, p) \to \rho^{-1}p$ of $\mathcal{G} \times (\mathcal{G}/\mathfrak{H})$ into $\mathcal{G}/\mathfrak{H}$ is analytic (cf. §V, p. 109). Moreover, since $\mathfrak{H}$ is distinguished, $\rho^{-1}p$ depends only upon the coset q modulo $\mathfrak{H}$ which contains ρ, and $\rho^{-1}p = q^{-1}p$. Let $\bar{\omega}$ be the natural mapping of $\mathcal{G}$ onto $\mathcal{G}/\mathfrak{H}$. We have seen in §V that, given a point $q_0^{-1} \varepsilon \mathcal{G}/\mathfrak{H}$, there exists an analytic mapping $\bar{\omega}^*$ into $\mathcal{G}$ of a neighbourhood of q_0^{-1} such that $\bar{\omega}(\bar{\omega}^*(q^{-1})) = q^{-1}$ when q^{-1} is in this neighbourhood. We have $q^{-1}p = \bar{\omega}^*(q^{-1})p$, which proves that the mapping $(p, q) \to q^{-1}p$ is analytic, and therefore that $\mathcal{G}/\mathfrak{H}$ is an analytic group.

Denote by $\mathcal{K}$ the group $\mathcal{G}/\mathfrak{H}$ and by $\mathfrak{g}$, $\mathfrak{h}$ and $\mathfrak{k}$ the Lie algebras of $\mathcal{G}$, $\mathcal{H}$ and $\mathcal{K}$ respectively. By Theorem 2, §VI, p. 111 there corresponds to $\bar{\omega}$ a homomorphism $d\bar{\omega}$ of $\mathfrak{g}$ into $\mathfrak{k}$. We know that there exists a submanifold I of $\mathcal{G}$, containing the neutral element and such that the contraction of $\bar{\omega}$ to I is an analytic isomorphism of I with an open submanifold of $\mathcal{K}$ (cf. §V, p. 109). It follows that every tangent vector to $\mathcal{K}$ at $\bar{\omega}(\epsilon)$ is the image under $d\bar{\omega}_\epsilon$ of some tangent vector to $\mathcal{G}$ at ϵ. In other words, we have $d\bar{\omega}(\mathfrak{g}) = \mathfrak{k}$.

Let $\mathfrak{h}_1$ be the set of elements $X \varepsilon \mathfrak{g}$ such that $d\bar{\omega}(X) = 0$; $\mathfrak{h}_1$ is a vector subspace of $\mathfrak{g}$ of dimension $\leqslant m - (m - n) = n$ (where m and n are the dimensions of $\mathcal{G}$ and $\mathcal{H}$ respectively). On the other hand, $\bar{\omega}$ maps $\mathcal{H}$ onto $\bar{\omega}(\epsilon)$. It follows immediately that $d\bar{\omega}(X_\epsilon) = 0$ for any $X \varepsilon \mathfrak{h}$, whence $\mathfrak{h} \subset \mathfrak{h}_1$. Since $\mathfrak{h}$ is of dimension n, we have $\mathfrak{h} = \mathfrak{h}_1$. From the definition of $\mathfrak{h}_1$ it follows immediately that the conditions $X \varepsilon \mathfrak{h}$, $Y \varepsilon \mathfrak{g}$ imply $[X, Y] \varepsilon \mathfrak{h}$.

Definition 1. *A vector subspace $\mathfrak{h}$ of a Lie algebra $\mathfrak{g}$ is called an ideal in $\mathfrak{g}$ if the conditions $X \varepsilon \mathfrak{h}$, $Y \varepsilon \mathfrak{g}$ imply $[X, Y] \varepsilon \mathfrak{h}$.*

Let $\mathfrak{h}$ be an ideal in the Lie algebra $\mathfrak{g}$. Let Y_1^* and Y_2^* be two elements of the vector space $\mathfrak{g}/\mathfrak{h}$. If Y_1 and Y_2 are elements of $\mathfrak{g}$ which belong to the cosets Y_1^* and Y_2^* respectively, the element $[Y_1, Y_2]$ belongs to a coset which depends only upon Y_1^* and Y_2^*, as follows from the formula

$$[Y_1 + X_1, Y_2 + X_2] = [Y_1, Y_2] + [Y_1, X_2] - [Y_2, X_1] + [X_1, X_2] \equiv [Y_1, Y_2] \pmod{\mathfrak{h}}$$

(if X_1 and X_2 belong to $\mathfrak{h}$). If we denote by $[Y_1^*, Y_2^*]$ the coset which contains $[Y_1, Y_2]$, we see immediately that the law of composition $(Y_1^*, Y_2^*) \to [Y_1^*, Y_2^*]$ defines a structure of Lie algebra on $\mathfrak{g}/\mathfrak{h}$. The Lie algebra defined in this way will again be denoted by $\mathfrak{g}/\mathfrak{h}$.

Returning to the consideration of the group $\mathcal{G}/\mathfrak{H}$, we observe that, if $X \varepsilon \mathfrak{g}$, $d\bar{\omega}(X)$ depends only upon the residue class of X modulo $\mathfrak{h}$, and therefore that $d\bar{\omega}$ defines in a natural way a linear isomorphism δ of $\mathfrak{g}/\mathfrak{h}$ onto $\mathfrak{k}$. Since $d\bar{\omega}$ is a homomorphism, δ is an isomorphism of the Lie algebra $\mathfrak{g}/\mathfrak{h}$ with $\mathfrak{k}$. We have therefore proved:

Proposition 1. *Let $\mathcal{G}$ be an analytic group, and let $\mathfrak{H}$ be a closed distinguished topological subgroup of $\mathcal{G}$. Assume that the component of the neutral element in $\mathfrak{H}$ is relatively open in $\mathfrak{H}$ and is the underlying topological group of an analytic subgroup $\mathcal{H}$ of $\mathcal{G}$. Let $\mathfrak{g}$ and $\mathfrak{h}$ be the Lie algebras of $\mathcal{G}$ and $\mathcal{H}$ respectively; then $\mathfrak{h}$ is an ideal in $\mathfrak{g}$ and the Lie algebra of $\mathcal{G}/\mathfrak{H}$ is isomorphic with $\mathfrak{g}/\mathfrak{h}$.*

We shall prove later that, if $\mathcal{H}$ is an analytic subgroup of $\mathcal{G}$ (not necessarily closed), the Lie algebra of $\mathcal{H}$ is an ideal in the Lie algebra of $\mathcal{G}$ if and only if $\mathcal{H}$ is distinguished in $\mathcal{G}$.

§VIII. The Exponential Mapping. Canonical Coordinates

Let $\mathcal{G}$ be an analytic group, and let $\mathfrak{g}$ be the Lie algebra of $\mathcal{G}$. We shall consider the analytic homomorphisms into $\mathcal{G}$ of the additive group R of real numbers. If we denote by t the coordinate in R, the Lie algebra $\mathfrak{r}$ of R is spanned by the infinitesimal transformation L defined by $Lt = 1$. If f is an analytic function on R at a point t_0, we have $L_{t_0}f = (df/dt)_{t_0}$.

If X is any element of $\mathfrak{g}$, there exists a homomorphism of $\mathfrak{r}$ into $\mathfrak{g}$ which maps L on X. Since R is simply connected, there corresponds to this homomorphism an analytic homomorphism $t \to \theta(t, X)$ of R into $\mathcal{G}$ (cf. Theorem 2, §VI, p. 111). If f is an analytic function on $\mathcal{G}$ at the point $\sigma_0 = \theta(t_0, X)$, we have

$$X_{\sigma_0}f = \left(\frac{df(\theta(t, X))}{dt}\right)_{t_0} \qquad (1)$$

Suppose for instance that $\mathcal{G}$ is the group $GL(n, C)$. Let $x_{ij}(\sigma)$ be the coefficients of a matrix $\sigma \varepsilon GL(n, C)$. We may represent an element X of the Lie algebra of $\mathcal{G}$ by the matrix $(X_\epsilon x_{ij}) = \tilde{X}$ (cf. §III, p. 104). The matrix $(X_\sigma x_{ij})$ is then equal to $\sigma\tilde{X}$ for any $\sigma \varepsilon GL(n, C)$. The coefficients of the matrix $\theta(t, X)$ are analytic functions of t; if we denote by $d\theta(t, X)/dt$ the matrix whose coefficients are the derivatives of the coefficients of $\theta(t, X)$, Formula (1) gives

$$\frac{d\theta(t, X)}{dt} = \theta(t, X)\tilde{X} \tag{2}$$

On the other hand, we have $\theta(O, X) = \epsilon$ (the unit matrix). The solution of the differential equation (2) with the initial condition $\theta(O, X) = \epsilon$ is

$$\theta(t, X) = \exp t\tilde{X}$$

(Cf. §II, Chapter I, p. 5). This leads us to the following generalization:

Definition 1. *Let X be an arbitrary element of the Lie algebra $\mathfrak{g}$ of an analytic group $\mathcal{G}$. Let L be the element of the Lie algebra $\mathfrak{r}$ of the group R which is defined by $Lt = 1$ (where t is the coordinate in R). Let $t \to \theta(t, X)$ be the analytic homomorphism of R into $\mathcal{G}$ whose differential is the homomorphism of $\mathfrak{r}$ into $\mathfrak{g}$ which maps L upon X. We shall denote the element $\theta(1, X)$ by* exp X.

The exponential mapping is therefore a mapping of $\mathfrak{g}$ into $\mathcal{G}$. We have clearly

$$\exp (t_1 + t_2)X = (\exp t_1X)(\exp t_2X)$$
$$\exp (-tX) = (\exp tX)^{-1}$$

Let $\{X_1, \cdots, X_n\}$ be a base in $\mathfrak{g}$. We may express each $X \varepsilon \mathfrak{g}$ in the form $X = \Sigma_i^n u_i(X)X_i$. We can define a manifold on the set $\mathfrak{g}$ by the condition that $u_1, \cdots, u_n$ shall form a coordinate system at every point of this manifold. The manifold obtained in this way is clearly independent of the choice of the base $\{X_1, \cdots, X_n\}$.

Now we shall prove that the exponential mapping is an analytic mapping of the manifold which has just been defined. Let $\{y_1, \cdots, y_n\}$ be a system of coordinates on $\mathcal{G}$ at the neutral element ϵ, such that $y_i(\epsilon) = 0$ $(1 \leqslant i \leqslant n)$, and let W be a cubic neighbourhood of ϵ with respect to this system. Let $u_1, \cdots, u_n$ be any real numbers; there exists a number $t_1 > 0$ such that $\exp t\Sigma_i^n u_iX_i \varepsilon W$ for $|t| < t_1$. Let $T(u_1, \cdots, u_n)$ be the least upper bound of the numbers t_1

satisfying this condition. If $|t| < T(\mathfrak{u})$, we set

$$y_j(\exp t\Sigma_i^n u_i X_i) = F_j(t; \mathfrak{u})$$

(where $\mathfrak{u}$ stands for the n letters $u_1, \cdots, u_n$). If $\sigma \varepsilon W$, we have $(X_i)_\sigma y_j = U_{ij}(y_1(\sigma), \cdots, y_n(\sigma))$, where the functions $U_{ij}(y_1, \cdots, y_n)$ are defined and analytic whenever $|y_k| < a$ $(1 \leqslant k \leqslant n)$ (where a is the breadth of W). It follows from the definition of the exponential mapping that

$$\frac{dF_j}{dt}(t; \mathfrak{u}) = \Sigma_{i=1}^n u_i U_{ij}(F_1(t; \mathfrak{u}), \cdots, F_n(t; \mathfrak{u}))$$

In other words, the equations $y_i = F_i(t; \mathfrak{u})$ $(1 \leqslant i \leqslant n)$ represent, for $|t| < T(\mathfrak{u})$, a solution of the differential system

$$(1) \qquad \frac{dy_j}{dt} = \Sigma_{i=1}^n u_i U_{ij}(y_1, \cdots, y_n) \qquad (1 \leqslant j \leqslant n)$$

Furthermore, we have $F_i(O; \mathfrak{u}) = 0$ $(1 \leqslant i \leqslant n)$.

Now, we apply the existence theorem to the system (1), and we obtain the following result. There exist two numbers $b > 0$ and $c > 0$ and n functions $F_i^*(t; \mathfrak{u})$, which are defined and analytic in the domain defined by the conditions $|u_k| < b$ $(1 \leqslant k \leqslant n)$, $|t| < c$, and which have the following properties: 1) $F_i^*(0; \mathfrak{u}) = 0$ $(1 \leqslant i \leqslant n)$; 2) $|F_i^*(t; \mathfrak{u})| < a$; 3) the equations $y_i = F_i^*(t; \mathfrak{u})$ represent a solution of the system (1). Making use of the uniqueness theorem, we conclude that the equalities $F_i(t, \mathfrak{u}) = F_i^*(t; \mathfrak{u})$ $(1 \leqslant i \leqslant n)$ hold provided $|u_k| < b$ $(1 \leqslant k \leqslant n)$, $|t| < \min\{c, T(\mathfrak{u})\}$. It follows immediately from the definition of $T(\mathfrak{u})$ that

$$\text{l.u.b.}_{|t|<T(\mathfrak{u})}(\max_{1 \leqslant i \leqslant n}|F_i(t, \mathfrak{u})|) = a$$

Making use of property 2) of the functions F_i^*, we see that $T(\mathfrak{u}) \geqslant c$ provided $|u_k| < b$ $(1 \leqslant k \leqslant n)$. On the other hand, we clearly have

$$F_i(t; \lambda\mathfrak{u}) = F_i(\lambda t; \mathfrak{u})$$

provided $|t| < |\lambda|^{-1}T(\mathfrak{u})$. It follows easily that the functions $F_i(1; \mathfrak{u})$ are defined and analytic provided $|u_k| < bc$ $(1 \leqslant k \leqslant n)$. This shows that the exponential mapping is analytic at the origin of $\mathfrak{g}$.

If X is any element of $\mathfrak{g}$, there exists an integer $M > 0$ such that $M^{-1}X$ is contained in some open neighbourhood of 0 on which the exponential mapping is analytic. Since

$$\exp Y = (\exp M^{-1}Y)^M$$

we see that the exponential mapping is analytic at X.

Now we shall see that the exponential mapping is regular at the origin of $\mathfrak{g}$ (note that it is not true in general that the exponential mapping is everywhere regular). In fact, let X_i be the tangent vector to $\mathfrak{g}$ at the origin which is defined by the conditions $X_i u_j = \delta_{ij}$ $(1 \leqslant j \leqslant n)$. It is clear that the differential of the exponential mapping maps X_i upon $(X_i)_\epsilon$. Since the vectors $(X_i)_\epsilon$ form a base of the tangent space to $\mathcal{G}$ at ϵ, our assertion is proved. It follows immediately that there exists a coordinate system $\{x_1, \cdots, x_n\}$ at ϵ on $\mathcal{G}$ such that the equalities $x_i(\exp \Sigma_k u_k X_k) = u_i$ $(1 \leqslant i \leqslant n)$ hold provided $|u_1|, \cdots, |u_n|$ are sufficiently small.

Definition 2. *Let $\{X_1, \cdots, X_n\}$ be a base of the Lie algebra of an analytic group $\mathcal{G}$. A system of coordinates $\{x_1, \cdots, x_n\}$ at the neutral element ϵ of $\mathcal{G}$ is called a canonical system of coordinates (with respect to the base $\{X_1, \cdots, X_n\}$) if the equalities $x_i(\exp \Sigma_i u_i X_i) = u_i$ $(1 \leqslant i \leqslant n)$ hold whenever $|u_1|, \cdots, |u_n|$ are sufficiently small. Let a be a number such that our equalities hold whenever $|u_i| < a$ $(1 \leqslant i \leqslant n)$. A cubic neighbourhood of breadth $< a$ with respect to the coordinates $x_1, \cdots, x_n$ is called a canonical neighbourhood of ϵ.*

§IX. FIRST APPLICATIONS OF CANONICAL COORDINATES

Proposition 1. *Let $\bar{\omega}$ be an analytic homomorphism of an analytic group $\mathcal{G}$ into an analytic group $\mathcal{H}$. If X is any element of the Lie algebra of $\mathcal{G}$, we have*

$$\bar{\omega}(\exp X) = \exp (d\bar{\omega}(X)).$$

In fact, let L be the infinitesimal transformation of the additive group R which is defined by $dt \cdot L = 1$ (where t is the coordinate in R), and let θ be the mapping $t \rightarrow \exp tX$. We have $d\theta(L) = X$, whence $d(\bar{\omega} \circ \theta)(L) = d\bar{\omega}(X)$. If θ' is the mapping $t \rightarrow \exp t \cdot d\bar{\omega}(X)$ of R into $\mathcal{H}$, we have also $d\theta'(L) = d\bar{\omega}(X)$; it follows that $\theta' = \bar{\omega} \circ \theta$, which proves Proposition 1.

Corollary 1. *If the mapping $\bar{\omega}$ of Proposition* 1 *is univalent, it is also everywhere regular.*

In fact, assume that $d\bar{\omega}(X) = 0$ for some X in the Lie algebra of $\mathcal{G}$. We have $\bar{\omega}(\exp tX) = \bar{\omega}(\epsilon)$ for every $t \varepsilon R$; since $\bar{\omega}$ is univalent, we have $\exp tX = \epsilon$, $X = 0$, which proves that $d\bar{\omega}$ is a univalent mapping. Corollary 1 follows immediately.

Corollary 2. *Let $\mathcal{H}$ be an analytic sub-group of $\mathcal{G}$, and let $\mathfrak{h}$ be the Lie algebra of $\mathcal{H}$. If U is a neighbourhood of* 0 *in $\mathfrak{h}$, the elements* $\exp X$, *for $X \varepsilon U$, form a neighbourhood of the neutral element in $\mathcal{H}$.*

This follows immediately from Proposition 1, applied to the identity mapping of $\mathcal{H}$ into $\mathcal{G}$.

Corollary 3. *The notation being as in Proposition* 1, *let us assume that* $\mathfrak{g}$ *and* $\mathfrak{h}$ *are the Lie algebras of* $\mathcal{G}$ *and* $\mathcal{H}$ *respectively. Then* $\bar{\omega}$ *maps* $\mathcal{G}$ *onto the analytic subgroup of* $\mathcal{H}$ *whose Lie algebra is the image of* $\mathfrak{g}$ *under* $d\bar{\omega}$.

Since $d\bar{\omega}$ is a homomorphism of $\mathfrak{g}$ into $\mathfrak{h}$, $d\bar{\omega}(\mathfrak{g})$ is clearly a subalgebra $\mathfrak{i}$ of $\mathfrak{h}$. Let $\mathfrak{J}$ be the corresponding analytic subgroup of $\mathcal{H}$. The elements of the form $\exp X$, $X \varepsilon \mathfrak{g}$, form a system of generators of $\mathcal{G}$. It follows that the elements $\bar{\omega}(\exp X) = \exp d\bar{\omega}(X)$ form a system of generators of $\bar{\omega}(\mathcal{G})$. But these elements also form a system of generators of $\mathfrak{J}$ by Corollary 2.

Corollary 4. *The notation being as in Proposition* 1, *let* $\mathfrak{n}$ *be the kernel of the homomorphism* $d\bar{\omega}$, *and let* $\mathcal{N}$ *be the analytic subgroup of* $\mathcal{G}$ *whose Lie algebra is* $\mathfrak{n}$. *The group* $\mathcal{N}$ *is the component of the neutral element in the kernel* $\mathfrak{N}_1$ *of* $\bar{\omega}$ *and is relatively open in* $\mathfrak{N}_1$.

If $X \varepsilon \mathfrak{n}$, we have (by Proposition 1), $\bar{\omega}(\exp X) = \exp d\bar{\omega}(X) = \eta$ (the neutral element of $\mathcal{H}$). Therefore, it follows from Corollary 2 that $\mathcal{N} \subset \mathfrak{N}_1$. Let B be a neighbourhood of the zero element in $\mathfrak{h}$ which is mapped topologically under the exponential mapping of $\mathfrak{h}$ into $\mathcal{H}$; let A be a neighbourhood of the zero element in $\mathfrak{g}$ such that $d\bar{\omega}(A) \subset B$ ($d\bar{\omega}$, being a linear mapping, is clearly continuous). If an element $X \varepsilon A$ is such that $\bar{\omega}(\exp X) = \exp d\bar{\omega}(X) = \eta$, we have $d\bar{\omega}(X) = 0$ whence $X \varepsilon \mathfrak{n}$, $\exp X \varepsilon \mathcal{N}$. Let V be the image of A under the exponential mapping of $\mathfrak{g}$ into $\mathcal{G}$. Then V is a neighbourhood of the neutral element in $\mathcal{G}$ and we have $\mathfrak{N}_1 \cap V \subset \mathcal{N}$. It follows that the underlying group of $\mathcal{N}$ is relatively open in $\mathfrak{N}_1$. This group is therefore also relatively closed in $\mathfrak{N}_1$. Since $\mathfrak{N}_1$ is clearly closed in $\mathcal{G}$, the set of points of $\mathcal{N}$ is closed and $\mathcal{N}$ is a subspace of $\mathcal{G}$. Corollary 4 is thereby proved.

In §III, we have defined the set $\mathcal{M}_n(C)$ of all complex matrices of degree n as a Lie algebra. The set M^R of the real matrices of degree n is clearly a subalgebra of $\mathcal{M}_n(C)$; the same holds for the set M^S of matrices of trace O, because, if X and Y are any two matrices in $\mathcal{M}_n(C)$, we have $Sp([X, Y]) = SpYX - SpXY = 0$ according to a well-known property of traces. The set M^h of skew-hermitian matrices is also a sub-algebra. In fact, assume that $\bar{X} + {}^tX = 0$, $\bar{Y} + {}^tY = 0$, and set $Z = YX - XY$: we have

$$ {}^tZ = {}^t(YX) - {}^t(XY) = {}^tX\,{}^tY - {}^tY\,{}^tX = \bar{X}\bar{Y} - \bar{Y}\bar{X} = -\bar{Z}. $$

Proposition 2. *Each of the groups* $SL(n, C)$, $U(n)$, $SU(n)$, $SL(n, R)$, $SO(n)$ *is the underlying topological group of an analytic sub-group of* $GL(n, C)$.

We write down the proof for $SO(n)$ only, since the other groups may be treated similarly. The set $M^R \cap M^s \cap M^h$ is a sub-algebra of $\mathfrak{gl}(n, C)$. Let $\mathcal{H}$ be the corresponding sub-group. By Proposition 4, §II, Chapter I, p. 5, together with Corollary 2 above, we see that there is a set W which is a neighbourhood of the neutral element in both $\mathcal{H}$ and $SO(n)$. Since $\mathcal{H}$ and $SO(n)$ are both connected, W is a set of generators for either one of these groups, which proves that $\mathcal{H}$ and $SO(n)$ have the same elements. In particular, the set of points of $\mathcal{H}$ is closed in $GL(n, C)$, and therefore $\mathcal{H}$ is a sub-space of $GL(n, C)$ (cf. §V, p. 109), which proves Proposition 2.

From now on, the notations $SL(n, C)$, $U(n)$, $SU(n)$, $SL(n, R)$, $SO(n)$ will refer to the analytic groups which have been defined in Proposition 2.

It can be proved in the same way that $Sp(n)$ is the underlying topological group of an analytic group; the same argument could be applied to $Sp(n, C)$ if we knew that the latter group was connected, which we shall prove later.

§X. CANONICAL COORDINATES OF PRODUCTS AND COMMUTATORS

Let $\mathcal{G}$ be an analytic group. We select a canonical system of coordinates $\{x_1, x_2, \cdots, x_n\}$ at the neutral elements ϵ of $\mathcal{G}$. Let V be a cubic neighbourhood of ϵ with respect to this system (whose breadth we denote by a), and let V_1 be a cubic neighbourhood of ϵ, of breadth a_1, such that $V_1V_1 \subset V$.

If $\sigma, \tau \varepsilon V_1$, we have

$$x_i(\sigma\tau) = f_i(x_1(\sigma), \cdots, x_n(\sigma); x_1(\tau), \cdots, x_n(\tau)),$$

where the functions $f_i(y_1, y_2, \cdots, y_n; z_1, \cdots, z_n)$ are analytic in the domain defined by $|y_i| < a_1$, $|z_i| < a_1$. They can be expanded in power-series in $y_1, y_2, \cdots, y_n, z_1, z_2, \cdots, z_n$ and those power series converge for $|y_i| < a_2$, $|z_i| < a_2$ (where $a_2 \leq a_1$). We write

$$f_i = \Sigma_{l=0}^{\infty} P_{il}(y_1, y_2, \cdots, y_r; z_1, z_2, \cdots, z_n)$$

where P_{il} is a polynomial of degree l in $z_1, z_2, \cdots, z_n$ whose coefficients are analytic functions of $y_1, y_2, \cdots, y_n$. We have $\tau = \exp \Sigma x_i(\tau)X_i$; we set $\tau(u) = \exp \Sigma u x_i(\tau)X_i$, and we have

$$(1) \quad x_i(\sigma\tau(u)) = \Sigma_0^{\infty} P_{il}(x_1(\sigma), \cdots, x_n(\sigma); x_1(\tau), \cdots, x_n(\tau))u^l$$

provided $|x_i(\sigma)| < a_2$, $|x_i(\tau)| < a_2$, $|u| \leq 1$.

On the other hand, if f is any function which is analytic on V_1, we have

$$\frac{df(\sigma\tau(u))}{du} = ((\Sigma x_i(\tau)X_i)f)_{\sigma\tau(u)}.$$

Let us set $T = \Sigma x_i(\tau)X_i$. By applying our formula to the functions Tf, T^2f, $\cdots$, we get

$$\frac{d^lf(\sigma\tau(u))}{du^l} = (T^lf)_{\sigma\tau(u)},$$

and therefore the Taylor expansion at $u = 0$ of $f(\sigma\tau(u))$, considered as a function of u, is

$$f(\sigma\tau(u)) = f(\sigma) + \Sigma_1^\infty \frac{u^l}{l!}(T^lf)_\sigma.$$

If $f = x_i$ and $|x_i(\sigma)| < a_2$, $|x_i(\tau)| < a_2$, the formulas (1) show that this series in u converges for $|u| \leq 1$, and that

$$P_{il}(x_1(\sigma), \cdots, x_n(\sigma); x_1(\tau), \cdots, x_n(\tau)) = \frac{1}{l!}(T^lx_i)_\sigma.$$

Now we set $S = \Sigma x_i(\sigma)X_i$, $\sigma(t) = \exp tS$. If g in an analytic function at ϵ, the Taylor expansion in t of $g(\sigma(t))$ is

$$g(\sigma(t)) = g(\epsilon) + \Sigma_1^\infty \frac{t^k}{k!}(S^kg)_\epsilon.$$

Applying this formula to the functions $(T^lx_i)_\sigma$, we finally get

$$x_i(\sigma\tau) = \Sigma_{k,l=0}^\infty \frac{1}{k!l!}(S^kT^lx_i)_\epsilon. \tag{2}$$

and these formulas hold for $|x_i(\sigma)| < a_2$, $|x_i(\tau)| < a_2$ (S^0 and T^0 stand for the identity operators: $S^0f = T^0f = f$).

The quantity $(S^kT^lx_i)_\epsilon$ is a polynomial of degree $k + l$ in the quantities $x_1(\sigma), \cdots, x_n(\sigma), x_1(\tau), \cdots, x_n(\tau)$; therefore the formulas (2) give the Taylor expansions of the functions $x_i(\sigma\tau)$.

Let $f(\sigma, \tau)$ be any function analytic at (ϵ, ϵ) on $\mathcal{G} \times \mathcal{G}$. If the Taylor expansion of $f(\sigma, \tau)$ by means of $x_1(\sigma), \cdots, x_n(\sigma); x_1(\tau), \cdots, x_n(\tau)$ begins with terms of total degree h in these $2n$ variables, we shall say that f is of order h at (ϵ, ϵ).

If we observe that $x_i(\sigma) = x_i(\epsilon) + \Sigma_1^\infty \frac{1}{k!}(S^kx_i)_\epsilon$, $x_i(\tau) = x_i(\epsilon) + \Sigma_1^\infty \frac{1}{l!}(T^lx_i)_\epsilon$, the formulas (2) give

$$x_i(\sigma\tau) = x_i(\sigma) + x_i(\tau) + (STx_i)_\epsilon + R_{1,i}(\sigma, \tau), \tag{3}$$

where $R_{1,i}$ is of order $\geqslant 3$.

The difference $x_i(\sigma\tau) - x_i(\sigma) - x_i(\tau)$ is of order 2 at (ϵ, ϵ); moreover it vanishes if either σ or τ coincides with ϵ.

It follows that

$$(4) \qquad x_i(\sigma\tau) - x_i(\sigma) - x_i(\tau) = \Sigma_{j,j'=1}^{n} x_j(\sigma)x_{j'}(\tau)A_{ijj'}(\sigma, \tau)$$

where $A_{ijj'}$ is analytic at (ϵ, ϵ).

Replacing σ by $\sigma\tau$ and τ by $\tau^{-1}\sigma^{-1}\tau\sigma$, we have
$x_i(\tau\sigma) - x_i(\sigma\tau) - x_i(\tau^{-1}\sigma^{-1}\tau\sigma) = \Sigma_{jj'}x_j(\sigma\tau)x_{j'}(\tau^{-1}\sigma^{-1}\tau\sigma)A_{ijj'}(\sigma\tau, \tau^{-1}\sigma^{-1}\tau\sigma)$.
Since $x_{j'}(\tau^{-1}\sigma^{-1}\tau\sigma)$ vanishes when either τ or σ coincides with ϵ it must be of order 2 at least, which shows that the difference $x_i(\tau\sigma) - x_i(\sigma\tau) - x_i(\tau^{-1}\sigma^{-1}\tau\sigma)$ is of order 3 at least. On the other hand, we have, by (3)

$$x_i(\tau\sigma) - x_i(\sigma\tau) = ([S, T]x_i)_\epsilon + R_{1,i}(\sigma, \tau) - R_{1,i}(\tau, \sigma).$$

whence

$$(5) \qquad x_i(\tau^{-1}\sigma^{-1}\tau\sigma) = ([S, T]x_i)_\epsilon + R_{2,i}(\sigma, \tau)$$

where $R_{2,i}$ is of order 3 at least.

We have $\tau(\tau^{-1}\sigma^{-1}\tau\sigma) = \sigma^{-1}\tau\sigma$; hence, by (4), we get

$$x_i(\sigma^{-1}\tau\sigma) = x_i(\tau) + ([S, T]x_i)_\epsilon + R_{3,i}(\sigma, \tau)$$

where $R_{3,i}$ is of order 3 at least. Changing σ in σ^{-1}, and observing that $[T, S] = -[S, T]$, we get

$$(6) \qquad x_i(\sigma\tau\sigma^{-1}) = x_i(\tau) + ([T, S]x_i)_\epsilon + R_{4,i}(\sigma, \tau),$$

with $R_{4,i}$ of order 3 at least.

§XI. THE ADJOINT REPRESENTATION

Let $\mathcal{G}$ be an analytic group, and let $\mathfrak{g}$ be the Lie algebra of $\mathcal{G}$. By an analytic endomorphism of $\mathcal{G}$ we understand an analytic homomorphism of $\mathcal{G}$ into itself. If an analytic endomorphism is also an analytic isomorphism of the manifold of $\mathcal{G}$ with itself, we say that it is an analytic automorphism of $\mathcal{G}$. It is clear that the analytic automorphisms of $\mathcal{G}$ form a group A.

Let α be any element of A. Then $d\alpha$ is a homomorphism of $\mathfrak{g}$ with itself. Moreover the formula $\overset{-1}{\alpha} \circ \alpha = e$ (the identity mapping) shows that $d(\overset{-1}{\alpha})$ is the reciprocal mapping of $d\alpha$. It follows that $d\alpha$ is an automorphism of $\mathfrak{g}$, i.e. it is a linear mapping of $\mathfrak{g}$ onto itself such that $d\alpha([X, Y]) = [d\alpha(X), d\alpha(Y)]$ for $X, Y \varepsilon \mathfrak{g}$. The mapping $\alpha \to d\alpha$ is obviously a linear representation of A. This representation is faithful; in fact, assume that $d\alpha(X) = X$ for all $X \varepsilon \mathfrak{g}$; by Proposition 1, §IX, p. 118 we have $\alpha(\exp X) = \exp X$: α leaves invariant the elements of a neighbourhood V of the neutral element in $\mathcal{G}$. Since $\mathcal{G}$ is connected, it is generated by the elements of V, and α leaves invariant every element of $\mathcal{G}$, which proves our assertion.

Let now σ be any element of $\mathcal{G}$; we denote by α_σ the mapping $\tau \to \sigma\tau\sigma^{-1}$ of $\mathcal{G}$ into itself. Clearly $\alpha_\sigma \varepsilon A$, and the mapping $\sigma \to \alpha_\sigma$ is a homomorphism of $\mathcal{G}$ into A.

Definition 1. *The mapping $\sigma \to d\alpha_\sigma$ is called the adjoint representation of $\mathcal{G}$.*

Proposition 1. *The adjoint representation of $\mathcal{G}$ is an analytic homomorphism of $\mathcal{G}$ into $GL(n, C)$.*

Let $\{X_1, X_2, \cdots, X_n\}$ be a base of $\mathfrak{g}$ and let $\{x_1, \cdots, x_n\}$ be a canonical system of coordinates relative to this base. If

$$d\alpha_\sigma(X_i) = \Sigma_{j=1}^n a_{ji}(\sigma)X_j,$$

the matrix which represents σ in the adjoint representation is $(a_{ij}(\sigma))$, If $\tau = \exp tX_i$, we have $\sigma\tau\sigma^{-1} = \exp \Sigma_{j=1}^n ta_{ji}(\sigma)X_j$ by Proposition 1. §IX, p. 118. Therefore the quantities $ta_{ji}(\sigma)$ are the canonical coordinates of $\sigma\tau\sigma^{-1}$ (if t is small enough). By formula (6), §X, p. 120, we know that these coordinates are analytic functions of σ at ϵ, which proves that the adjoint representation is analytic at ϵ, and therefore also everywhere (cf. remark at the end of §VI, p. 111).

Let A be the adjoint representation. Then dA is a homomorphism of $\mathfrak{g}$ into $\mathfrak{gl}(n, C)$. Let X be any element of $\mathfrak{g}$; $dA(X)$ is a matrix X^* and the matrix $A(\exp tX)$ is $\exp tX^*$. Therefore

$$X^* = \lim_{t\to 0} \frac{A(\exp tX) - E}{t}$$

where E is the unit matrix. By formula (6), §X, p. 120, we have

$$(A(\exp tX))(X_i) = X_i + t[X_i, X] + t^2\varphi(t)$$

where $\varphi(t)$ remains bounded when $t \to 0$. Therefore X^*, considered as a linear endomorphism of $\mathfrak{g}$, maps any element $Y = \Sigma a_i X_i$ on $[Y, X]$.

Now let $\mathfrak{g}$ be any Lie algebra over a field K. If n is any integer, the matrices of degree n with coefficients in K form a Lie algebra $\mathfrak{gl}(n, K)$, in which the bracket operation is $[X, Y] = YX - XY$.

Definition 2. *If $\mathfrak{g}$ is a Lie algebra over the field K, we call a homomorphism of $\mathfrak{g}$ into $\mathfrak{gl}(n, K)$ a representation of $\mathfrak{g}$ (in K, of degree n).*

Let X be any element of $\mathfrak{g}$, and let $\mathsf{P}(X)$ be the mapping $Y \to [Y, X]$ of $\mathfrak{g}$ into itself. If we select a base in $\mathfrak{g}$, $\mathsf{P}(X)$ may be represented as a matrix of degree $n = \dim \mathfrak{g}$. The mapping $X \to \mathsf{P}(X)$ is clearly linear; moreover

$$[Y, [X_1, X_2]] = -[X_1, [X_2, Y]] - [X_2, [Y, X_1]] = (\mathsf{P}(X_2)\mathsf{P}(X_1) - \mathsf{P}(X_1)\mathsf{P}(X_2))Y,$$

whence $\mathsf{P}([X_1, X_2]) = [\mathsf{P}(X_1), \mathsf{P}(X_2)]$ which proves that P is a representation.

Definition 3. *The mapping which assigns to every $X \varepsilon \mathfrak{g}$ the linear endomorphism $\mathsf{P}(X)$ of $\mathfrak{g}$ defined by $\mathsf{P}(X)Y = [Y, X]$ for all $Y \varepsilon \mathfrak{g}$ is called the adjoint representation of $\mathfrak{g}$.*

Returning now to the case where $\mathfrak{g}$ is the Lie algebra of an analytic group $\mathcal{G}$, we see that the mapping denoted above by dA is the adjoint representation of $\mathfrak{g}$.

Let $\mathcal{H}$ be any analytic sub-group of $\mathcal{G}$, and let $\mathfrak{h}$ be its Lie algebra. It is clear that $\sigma\mathcal{H}\sigma^{-1}$ is also an analytic sub-group of $\mathcal{G}$ whose Lie algebra is the image of $\mathfrak{h}$ under the mapping $d\alpha_\sigma$. Therefore a necessary and sufficient condition for $\mathcal{H}$ to be distinguished is that $\mathfrak{h}$ be transformed into itself by every operation $d\alpha_\sigma$, $\sigma \varepsilon \mathcal{G}$. This condition may be put in another form. Assume that it is satisfied: then, if $X \varepsilon \mathfrak{h}$, we have $(A(\exp tY))(X) \epsilon \mathfrak{h}$ for every t, whence

$$[X, Y] = \lim_{t \to 0} \frac{A(\exp tY) - E}{t} X \varepsilon \mathfrak{h}$$

and $\mathfrak{h}$ is an ideal in $\mathfrak{g}$. Conversely, assume that $\mathfrak{h}$ is an ideal in $\mathfrak{g}$; let Y be any element of $\mathfrak{g}$, and let Y^* be the matrix which represents Y in the adjoint representation of $\mathfrak{g}$. We have $Y^*(\mathfrak{h}) \subset \mathfrak{h}$; if $\sigma = \exp Y$, we have $d\alpha_\sigma = \exp Y^*$, whence $d\alpha_\sigma(\mathfrak{h}) \subset \mathfrak{h}$. But the elements σ for which $d\alpha_\sigma(\mathfrak{h}) \subset \mathfrak{h}$ obviously form a sub-group of $\mathcal{G}$; this subgroup contains all the elements $\exp Y$, $Y \varepsilon \mathfrak{g}$, i.e. it contains a neighbourhood of the neutral element in $\mathcal{G}$. Since $\mathcal{G}$ is connected, our subgroup is the whole of $\mathcal{G}$, which proves that $\mathcal{H}$ is a distinguished sub-group.

We have therefore proved the following result, which was announced in §VII, p. 114:

Proposition 2. *If $\mathcal{H}$ is an analytic sub-group of the analytic group $\mathcal{G}$, a necessary and sufficient condition for $\mathcal{H}$ to be distinguished is that its Lie algebra be an ideal in the Lie algebra of $\mathcal{G}$.*

The kernel of the adjoint representation of the group $\mathcal{G}$ is clearly the center of $\mathcal{G}$. The kernel of the adjoint representation of the Lie algebra $\mathfrak{g}$ (i.e. the set of elements $X \varepsilon \mathfrak{g}$ which are mapped on O under the adjoint representation) is the set of elements $X \epsilon \mathfrak{g}$ such that $[X, Y] = 0$ for every $Y \varepsilon \mathfrak{g}$.

Definition 4. *If* $\mathfrak{g}$ *is a Lie algebra, the center of* $\mathfrak{g}$ *is the set of the elements* $X\varepsilon\mathfrak{g}$ *for which* $[X, Y] = 0$ *holds for every* $Y\varepsilon\mathfrak{g}$.

The center of a Lie algebra is clearly an ideal.

The following result is an immediate consequence of Corollary 4 to Proposition 1, §IX, p. 118:

Proposition 3. *Let* $\mathfrak{C}_1$ *be the center of an analytic group* $\mathfrak{G}$, *and let* $\mathfrak{C}$ *be the connected component in* $\mathfrak{C}_1$ *of the neutral element. Then* $\mathfrak{C}$ *is the underlying group of a closed analytic sub-group of* $\mathfrak{G}$ *whose Lie algebra is the center of the Lie algebra of* $\mathfrak{G}$.

Now let $\mathfrak{g}$ be any Lie algebra over the field R, and let us denote by P the adjoint representation of $\mathfrak{g}$. Then $\mathsf{P}(\mathfrak{g})$ is a sub-algebra of $\mathfrak{gl}(n, C)$, and therefore there exists an analytic sub-group of $GL(n, C)$ whose Lie algebra is $\mathsf{P}(\mathfrak{g})$ (cf. Theorem 1, §IV p. 107). On the other hand, $\mathsf{P}(\mathfrak{g})$ is isomorphic to $\mathfrak{g}/\mathfrak{c}$, where $\mathfrak{c}$ is the center of $\mathfrak{g}$. Therefore we have the following result:

Proposition 4. *If* $\mathfrak{g}$ *is a Lie algebra over the field* R, *there exists an analytic group whose Lie algebra is isomorphic with* $\mathfrak{g}/\mathfrak{c}$ *where* $\mathfrak{c}$ *is the center of* $\mathfrak{g}$.

It follows in particular that if the center of a Lie algebra $\mathfrak{g}$ contains only the zero element, there exists an analytic group whose Lie algebra is isomorphic to $\mathfrak{g}$. The result remains true for any Lie algebra, but is considerably more difficult to prove.

§XII. The Derived Group

Let $\mathfrak{g}$ be a Lie algebra over a field K. The vector-space spanned by the elements of the form $[X, Y]$, with $X\epsilon\mathfrak{g}$, $Y\epsilon\mathfrak{g}$ is obviously an ideal in $\mathfrak{g}$.

Definition 1. *The ideal spanned in a Lie algebra* $\mathfrak{g}$ *by the elements of the form* $[X, Y]$, $(X\varepsilon\mathfrak{g}, Y\varepsilon\mathfrak{g})$ *is called the derived algebra of* $\mathfrak{g}$. *It is generally denoted by* $\mathfrak{g}'$.

A Lie algebra is said to be *abelian* if the element $[X, Y]$ is equal to 0 for any two elements X, Y of the Lie algebra. The derived algebra of any Lie algebra $\mathfrak{g}$ is characterized by the following properties:

Proposition 1. *If* $\mathfrak{g}'$ *is the derived algebra of* $\mathfrak{g}$, *the factor algebra* $\mathfrak{g}/\mathfrak{g}'$ *is abelian. Conversely, if* $\mathfrak{h}$ *is any ideal such that* $\mathfrak{g}/\mathfrak{h}$ *is abelian,* $\mathfrak{h}$ *contains the derived algebra.*

In fact, let X, Y be two elements of $\mathfrak{g}$ and let $\tilde{X}$, $\tilde{Y}$ be their cosets modulo $\mathfrak{h}$. The condition "$[\tilde{X}, \tilde{Y}] = 0$ for all X, Y" is equivalent to the condition "$[X, Y]\varepsilon\mathfrak{h}$ for all X, Y," and this proves proposition 1.

Now let $\mathfrak{G}$ be an analytic group, and let $\mathfrak{g}$ be its Lie algebra. To the derived algebra $\mathfrak{g}'$ of $\mathfrak{g}$ there corresponds an analytic sub-group

$\mathcal{G}'$ of $\mathcal{G}$, which is called the *derived group* of $\mathcal{G}$. It is always a distinguished sub-group, but not necessarily closed.

Let $\mathfrak{a}$ be the factor algebra $\mathfrak{g}/\mathfrak{g}'$. If d is the dimension of $\mathfrak{a}$, $\mathfrak{a}$ is the Lie algebra of the additive group R^d. There exists a natural homomorphism $\bar{\omega}$ of $\mathfrak{g}$ onto $\mathfrak{a}$, which maps any element X onto its coset $\bar{\omega}(X)$ modulo $\mathfrak{g}'$. There exists a neighbourhood U of the neutral element ϵ in $\mathcal{G}$ and a continuous mapping φ of U into R^d such that

$$\begin{aligned} \varphi(\sigma\tau) &= \varphi(\sigma) + \varphi(\tau) && \text{if } \sigma, \tau, \sigma\tau \epsilon U \\ \varphi(\exp X) &= \exp \bar{\omega}(X) && \text{if } X \text{ is sufficiently near } 0. \end{aligned}$$

Therefore $\varphi(\sigma\tau\sigma^{-1}\tau^{-1}) = 0$ if σ, τ are sufficiently near ϵ. But $\sigma\tau\sigma^{-1}\tau^{-1}$ can be written, for σ, τ sufficiently small, in the form $\exp X(\sigma, \tau)$, where $X(\sigma, \tau)$ tends to zero when σ, τ tend to ϵ. Since $\varphi(\exp X(\sigma, \tau))$ is the neutral element of R^d, we have $\bar{\omega}(X(\sigma, \tau)) = 0$, $X(\sigma, \tau) \varepsilon \mathfrak{g}'$ and $\sigma\tau\sigma^{-1}\tau^{-1} \varepsilon \mathcal{G}'$ when σ, τ belong to a sufficiently small neighbourhood U_1 of ϵ in $\mathcal{G}$. Since $\mathcal{G}$ is connected, the elements of U_1 constitute a set of generators of $\mathcal{G}$, and it follows easily that $\mathcal{G}'$ contains the commutator group $\mathfrak{C}$ of $\mathcal{G}$, i.e. the group generated by all the commutators $\sigma\tau\sigma^{-1}\tau^{-1}$ for σ, $\tau \varepsilon \mathcal{G}$.

We shall prove that, conversely, every element of $\mathcal{G}'$ belongs to $\mathfrak{C}$. Since $\mathcal{G}'$ is connected, it is sufficient to prove it for the elements of a neighbourhood of ϵ in $\mathcal{G}'$.

We can find a finite number of pairs $(Y_1, Z_1), \cdots, (Y_r, Z_r)$ of elements in $\mathfrak{g}$ such that the elements $U_1 = [Y_1, Z_1] \cdots, U_r = [Y_r, Z_r]$ form a base of $\mathfrak{g}'$. Let $U_{r+1}, \cdots, U_n$ be $n - r$ elements such that $\{U_1, U_2, \cdots, U_n\}$ is a base of $\mathfrak{g}$. We select a canonical system of coordinates, say $\{u_1, u_2, \cdots, u_n\}$, with respect to this base. Let V be a cubic neighbourhood of ϵ with respect to this system.

We set $\sigma_i(s) = \exp sY_i$; $\tau_i(t) = \exp tZ_i$; $(1 \leqq i \leqq r)$ and $\rho_i(s, t) = \sigma_i(s)\tau_i(t)\sigma_i^{-1}(s)\tau_i^{-1}(t)$. The functions $u_i(\rho_j(s, t))$ are analytic at $s = t = 0$. We write

$$u_i(\rho_j(s, t)) = ta_{ij}(s) + t^2b_{ij}(s) + \cdots$$

where the functions $a_{ij}(s)$, $b_{ij}(s)$, $\cdots$ are analytic at $s = 0$. Moreover, since $U_i = [Y_i, Z_i]$, it follows from formula (5), §X, page 120, that

$$\lim_{s\to 0} \frac{a_{ij}(s)}{s} = -\delta_{ij}.$$

Therefore we can find a value $s_1 \neq 0$ of s such that $\exp sX_i \varepsilon V$ for $|s| \leqq |s_1|$ and that $\boxed{a_{ij}(s_1)} \neq 0$.

We set $\rho_i(t) = \rho_i(s_1, t)$ $(1 \leqq i \leqq r)$, and

$$\rho(t_1, t_2, \cdots, t_r) = \rho_1(t_1)\rho_2(t_2) \cdots \rho_r(t_r).$$

Then we have

$$\left(\frac{\partial u_i(\rho(t_1, \cdots t_r))}{\partial t_j}\right)_{t=0} = a_{ij}(s_1)$$

and $u_{r+h}(\rho(t_1, t_2, \cdots, t_r)) = 0$ if $|t_1| \cdots, |t_r|$ are sufficiently small, $(1 \leqq h \leqq n - r)$. The second formula is true because $\rho(t_1, t_2, \cdots, t_r)$ belongs to the commutator group, and hence also to $\mathfrak{G}'$.

Since $\boxed{a_{ij}(s)} \neq 0$ the theorem on implicit functions shows that every element $\sigma \varepsilon V$ such that $u_{r+1}(\sigma) = \cdots = u_n(\sigma) = 0$ may be written in the form $\rho(t_1, t_2, \cdots, t_r)$ provided $|u_1(\sigma)|, \cdots, |u_r(\sigma)|$ are sufficiently small. But these conditions all hold for the elements σ of a suitable neighbourhood of ϵ in $\mathfrak{G}'$. Since $\rho(t_1, t_2, \cdots, t_r)$ always belongs to the commutator group of $\mathfrak{G}$, our assertion is proved.

Remark. If the group $\mathfrak{G}$ is simply connected, there corresponds to the projection $\bar{\omega}$ of $\mathfrak{g}$ onto $\mathfrak{g}/\mathfrak{g}'$ a continuous homomorphism H of $\mathfrak{G}$ into R^d. The kernel of this homomorphism is a closed sub-group of $\mathfrak{G}$ whose Lie algebra is $\mathfrak{g}'$. The connected component of ε in this group coincides with $\mathfrak{G}'$; on the other hand, this component is obviously a closed sub-group. Hence: *If $\mathfrak{G}$ is a simply connected analytic group, the commutator group of $\mathfrak{G}$ is a closed sub-group.*

The assumption that $\mathfrak{G}$ is simply connected cannot be removed in this statement, as can be shown by examples.

Exactly by the same method as was applied to $\mathfrak{G}'$, it can be proved that *if $\mathfrak{G}$ is a simply connected analytic groups, and if $\mathfrak{h}$ is an ideal in the Lie algebra $\mathfrak{g}$ of $\mathfrak{G}$, the analytic sub-group of $\mathfrak{G}$ whose Lie algebra is $\mathfrak{h}$ is always closed, provided there exists an analytic group whose Lie algebra is isomorphic to $\mathfrak{g}/\mathfrak{h}$.*

§XIII. Topological Invariance of the Lie Algebra

We shall prove in this section that two analytic groups which have the same underlying topological group coincide also as analytic groups and, in particular, have the same Lie algebra.

Let $\mathfrak{G}$ be an analytic group; we select a canonical system of coordinates $\{x_1, x_2, \cdots, x_n\}$ at the neutral element ϵ of $\mathfrak{G}$, and we take a cubic neighbourhood V_1 of ϵ with respect to this system. Let a_1 be the breadth of V_1.

Lemma 1. *Let a be a number such that $0 < a < a_1$, and V be the set of points $\sigma \varepsilon V_1$ such that $|x_i(\sigma)| < a$ $(1 \leqslant i \leqslant n)$. If an element $\sigma \varepsilon V_1$*

is such that $|x_i(\sigma)| \leqslant a_1 - a$ $(1 \leqslant i \leqslant n)$ *and that* $\sigma, \sigma^2, \cdots, \sigma^k$ *all belong to* V, *we have* $|x_i(\sigma)| < \dfrac{a}{k}$ *and* $x_i(\sigma^l) = lx_i(\sigma)$ $(1 \leqslant i \leqslant n, 0 \leqslant l \leqslant k)$.

Let us set $X = \Sigma x_i(\sigma)X_i$, where $\{X_1, \cdots, X_n\}$ is the base of the Lie algebra $\mathfrak{g}$ of $\mathcal{G}$ corresponding to the system of coordinates $\{x_1, \cdots, x_n\}$. We know that $\exp tX \epsilon V_1$ for all values of t such that $|tx_i(\sigma)| < a_1$ $(1 \leqslant i \leqslant n)$. We may assume $\sigma \neq \epsilon$; we then define a number t_0 by the equality $t_0 \cdot \max(|x_i(\sigma)|) = a_1$. We want to prove that $t_0 > k$. There exists an integer h such that $h < t_0 \leqslant h + 1$, and we have $\sigma^h = \exp hX \epsilon V_1$, $x_i(\sigma^h) = hx_i(\sigma)$ $(1 \leqslant i \leqslant n)$. If i is an index such that $t_0|x_i(\sigma)| = a_1$, we have $h|x_i(\sigma)| \geqslant a_1 - |x_i(\sigma)| \geqslant a$ and therefore σ^h does not belong to V, from which it follows that $t_0 > h > k$. Hence, for $1 \leqslant l \leqslant k$, we have $x_i(\sigma^l) = x_i(\exp lX) = lx_i(\sigma)$; if we set $l = k$ we obtain $|x_i(\sigma)| < \dfrac{a}{k}$.

Remark. The equality $x_i(\sigma^l) = lx_i(\sigma)$ obviously holds also for $-k \leqslant l \leqslant k$.

Now let Θ be a continuous homomorphism into $\mathcal{G}$ of the additive group R of real numbers. The notation being as in Lemma 1, we can find an interval $]-t_1, t_1[$ (with $t_1 > 0$) such that $\Theta(t) \epsilon V$ if t belongs to this interval. We set $f_i(t) = x_i(\Theta t)$ $(-t_1 < t < t_1)$; if k is a sufficiently large integer, we have $|f_i(k^{-1}t)| < a_1 - a$ $(1 \leqslant i \leqslant n)$. On the other hand, since Θ is a homomorphism, we have $\Theta(t) = (\Theta(k^{-1}t))^k$, whence

$$f_i(t) = kf_i\left(\frac{t}{k}\right)$$

$$f_i\left(l\frac{t}{k}\right) = lf_i\left(\frac{t}{k}\right) = \frac{l}{k}f_i(t) \qquad \text{for } k \text{ sufficiently large and } |l| \leqslant k.$$

Since the functions $f_i(t)$ are continuous, it follows immediately that $f_i(t't) = t'f_i(t)$ (for $|t'| \leqslant 1$). Let t_2 be any fixed number $\neq 0$ in the interval $]-t_1, t_1[$. We have $\Theta(t) = \exp tt_2^{-1}X$, where $X = \Sigma f_i(t_2)X_i$ and this formula holds for $|t| \leqslant |t_2|$. Since the mappings $t \to \Theta(t)$, $t \to \exp tt_2^{-1}X$ are both homomorphisms, the formula holds for all values of t, which proves that Θ is analytic.

Proposition 1. *Every continuous homomorphism H of an analytic group $\mathcal{G}$ into an analytic group $\mathcal{H}$ is an analytic homomorphism.*

Let $\{X_1, \cdots, X_n\}$ be a base of the Lie algebra $\mathfrak{g}$ of $\mathcal{G}$. The mapping $t \to H(\exp tX_i)$ is a continuous homomorphism of R into $\mathcal{H}$; hence there exists for each i an element Y_i of the Lie algebra of $\mathcal{H}$ such that $H(\exp tX_i) = \exp tY_i$. We have

$$(1)\quad H((\exp t_1X_1)(\exp t_2X_2)\cdots(\exp t_nX_n)) = (\exp t_1Y_1)(\exp t_2Y_2)\cdots(\exp t_nY_n).$$

Let $x_1, x_2, \cdots, x_n$ be the canonical coordinates on $\mathcal{G}$ corresponding to the base $\{X_1, X_2, \cdots, X_n\}$ of $\mathfrak{g}$. If $|t_1|, \cdots, |t_n|$ are sufficiently small, the element $(\exp t_1X_1)(\exp t_2X_2)\cdots(\exp t_nX_n)$ will be in a canonical neighbourhood of the neutral element and its coordinates $\varphi_1(t_1, \cdots, t_n), \cdots, \varphi_n(t_1, \cdots, t_n)$ will be analytic functions of $t_1, \cdots, t_n$. If $t_i = \delta_{ij}t$, we have $\varphi_i = \delta_{ij}t$, whence $\left(\frac{\partial\varphi_i}{\partial t_j}\right)_{0,\ldots,0} = \delta_{ij}$. Therefore the functional determinant $\frac{D(\varphi_1, \varphi_2, \cdots, \varphi_n)}{D(t_1, t_2, \cdots, t_n)}$ is equal to 1 for $t_1 = t_2 = \cdots = t_n = 0$. It follows that there is a neighbourhood V of the neutral element such that every $\sigma \varepsilon V$ may be written in the form $(\exp t_1X_1)(\exp t_2X_2)\cdots(\exp t_nX_n)$, where the numbers $t_1, t_2, \cdots, t_n$ depend analytically on σ. Formula (1) shows at once that the mapping $\sigma \rightarrow H(\sigma)$ is analytic at the neutral element, and therefore everywhere.

The announced result now follows immediately:

Theorem 3. *If two analytic groups $\mathcal{G}$, $\mathcal{G}'$ have the same underlying topological group, they coincide.*

In fact, all we have to do is to apply Proposition 1 to the identity mappings of $\mathcal{G}$ into $\mathcal{G}'$ and of $\mathcal{G}'$ into $\mathcal{G}$: these mappings are analytic, and therefore they are mutually reciprocal analytic isomorphisms.

Definition 1. *A locally connected topological group $\mathfrak{G}$ is said to be a Lie group if the connected component of the neutral element in $\mathfrak{G}$ is the underlying topological group of an analytic group $\mathcal{G}_1$.*

If such is the case, we know by Theorem 3 that $\mathcal{G}_1$ is uniquely determined. Its Lie algebra is also called the Lie algebra of $\mathfrak{G}$.

A Lie group is always locally cartesian (i.e. there exists a neighbourhood of the neutral element which is homeomorphic to R^n). It has been conjectured by Hilbert that, conversely, every locally cartesian group is a Lie group. This conjecture has been proved to be true under some restrictive conditions; for instance we know that it is true for compact groups and also for abelian groups. Although it seems almost certain that it is true in general, the proof will probably require a set of completely new methods of approach.

Every discrete group is of course a Lie group (of dimension 0). The linear groups which were discussed in Chapter I are all Lie groups.

The center of a Lie group is a Lie group. The product of a finite number of Lie groups is a Lie group. In the next section, we shall prove that every closed topological subgroup of a Lie group is a Lie group.

§XIV. A Criterion for Lie Groups

Proposition 1. *Let $\mathfrak{G}$ be a locally compact topological group which admits a continuous univalent homomorphism H into a Lie group $\mathfrak{H}$. Then $\mathfrak{G}$ itself is a Lie group.*

Let $\mathfrak{F}$ be the set of all continuous homomorphisms of the additive group R into $\mathfrak{G}$. To every $\Theta \varepsilon \mathfrak{F}$ there corresponds an element $Y = Y(\Theta)$ in the Lie algebra $\mathfrak{h}$ of $\mathfrak{H}$ such that

$$H(\Theta t) = \exp tY \qquad (t \varepsilon R).$$

The elements $Y(\Theta)$ corresponding to the various elements $\Theta \varepsilon \mathfrak{F}$ form a sub-set $\mathfrak{h}_1$ of $\mathfrak{h}$. We denote by r the maximal number of linearly independent elements which can be found in $\mathfrak{h}_1$ (r may be zero). We construct a base $\{Y_1, Y_2, \cdots, Y_n\}$ of $\mathfrak{h}$ whose r first elements $Y_1, Y_2, \cdots, Y_r$ belong to $\mathfrak{h}_1$. There corresponds to this base a canonical system of coordinates $\{y_1, y_2, \cdots, y_n\}$ at the neutral element η of $\mathfrak{H}$. Let V_1 be a cubic neighbourhood of η with respect to this system.

Since H is continuous, there exists a compact neighbourhood U of the neutral element ϵ in $\mathfrak{G}$ such that $H(U) \subset V_1$. Let B be the boundary of the set U; B is a compact set and does not contain ϵ. Since H is univalent, $H(B)$ does not contain η. Therefore, there exists a number $a > 0$ (smaller than the breadth of V_1) such that the inequality $\max_i |y_i(H\sigma)| > a$ holds at every point $\sigma \varepsilon B$. Let V be the cubic neighbourhood of breadth a of η.

If τ is any element of V_1, we set $d(\tau) = (\Sigma_1^n y_i^2(\tau))^{\frac{1}{2}}$.

Lemma 1. *Suppose that a sequence of elements $\sigma_k \neq \epsilon$ in U converges towards ϵ in such a way that each one of the sequences $\dfrac{y_i(H\sigma_k)}{d(H\sigma_k)}$ converges to a limit u_i when $k \to \infty$. Then $Y = \Sigma u_i Y_i$ belongs to $\mathfrak{h}_1$, and, if Θ is the corresponding element of $\mathfrak{F}$, we have $\Theta(t) \varepsilon U$ for $|t| < a$.*

For every k, there exists a largest integer l_k which has the following two properties: (1) $l_k d(H\sigma_k) < a$; (2) for any integer m such that $0 \leqslant m \leqslant l_k$, we have $\sigma_k^m \varepsilon U$. Let σ' be any cluster point of the sequence $(\sigma_k^{l_k})$; since $l_k d(H\sigma_k) < a$, we have $|l_k y_i(H\sigma_k)| < a$ $(1 \leqslant i \leqslant n)$ whence $(H\sigma_k)^{l_k} = H(\sigma_k^{l_k}) \varepsilon V$; moreover, $y_i(H\sigma')$ is a cluster point of the numerical sequence $(l_k y_i(H\sigma_k))$, which shows that $|y_i(H\sigma')| \leqslant a$ $(1 \leqslant i \leqslant n)$. In particular, no cluster point of the sequence $(\sigma_k^{l_k})$ can belong to B.

For every k, either we have $(l_k + 1)d(H\sigma_k) > a$ or $\sigma_k^{l_k+1}$ does not belong to U. Let K be the set of integers for which the second eventuality occurs. If K were infinite, the elements $\sigma_k^{l_k}$, for $k \varepsilon K$, would form an infinite sequence which would have a cluster point $\sigma' \varepsilon U$ (because U is compact). But, since $\sigma_k^{l_k+1} = \sigma_k^{l_k}\sigma_k$, $\lim \sigma_k = \epsilon$, σ' would also be a cluster point of a sequence of elements belonging to the complement of U. Hence, σ' would belong to B, which is impossible. Therefore, if k is sufficiently large, we have $l_k d(H\sigma_k) < a \leqslant (l_k + 1)d(H\sigma_k)$, whence $\lim l_k d(H\sigma_k) = a$.

Let t be any number such that $0 \leqslant t < a$. For each k, let m_k be the largest integer such that $m_k \leqslant a^{-1}tl_k$. Then $m_k < l_k$, whence $|m_k y_i(H\sigma_k)| < a$ $(1 \leqslant i \leqslant n)$ and $\sigma_k^{m_k} \varepsilon U$. It follows that $(H\sigma_k)^{m_k} \varepsilon V$ and that $y_i(H\sigma_k^{m_k}) = m_k y_i(H\sigma_k)$ $(1 \leqslant i \leqslant n)$. We have $\lim m_k/l_k = ta^{-1}$, $\lim l_k d(H\sigma_k) = a$, whence

$$\lim m_k y_i(H\sigma_k) = tu_i \qquad (1 \leqq i \leqq n)$$

which shows that

$$\lim H(\sigma_k^{m_k}) = \exp \Sigma t u_i Y_i = \exp tY. \tag{1}$$

On the other hand, since U is compact, the mapping H, which is continuous and univalent, maps U homeomorphically. Hence we may conclude from (1) that the sequence $\sigma_k^{m_k}$ has a limit $\sigma(t)$ in U and that

$$H(\sigma(t)) = \exp tY. \tag{2}$$

Replacing the sequence σ_k by $\sigma_k^{-1}(\sigma_k^{-1} \varepsilon U$ if k is sufficiently large), we see that there also exists an element $\sigma(-t) \epsilon U$ such that $H(\sigma(-t)) = \exp(-tY)$.

The element $\sigma(t)$ is now defined for $|t| < a$, and the equality (2) holds for these values. It follows immediately from this equality that the conditions $|t_1| < a$, $|t_2| < a$, $|t_1 + t_2| < a$ imply $\sigma(t_1 + t_2) = \sigma(t_1)\sigma(t_2)$. Moreover, $\sigma(t)$ is a continuous function of t.

Since R is simply connected, there exists a continuous homomorphism Θ of R into $\mathfrak{G}$ such that $\Theta(t) = \sigma(t)$ for $|t|$ sufficiently small. The corresponding element $Y(\Theta) \varepsilon \mathfrak{h}$ is clearly Y. Since we have $H(\Theta t) = \exp tY$ for all t, we have $\Theta(t) = \sigma(t)$ for $|t| < a$, which proves Lemma 1.

Corollary. *Let* Θ *be an element of* $\mathfrak{g}$; *if* $Y(\Theta) = \Sigma v_i Y_i$, *we have* $\Theta(t) \varepsilon U$ *for* $|t| < a(\Sigma v_i^2)^{-\frac{1}{2}}$.

This follows at once by applying the lemma to the sequence $\sigma_k = \Theta(k^{-1}t)$ with $u_i = v_i(\Sigma v_i^2)^{-\frac{1}{2}}$.

We shall now prove that $\mathfrak{h}_1$ is a vector space. First, if $Y = Y(\Theta)\varepsilon\mathfrak{h}_1$ and $a\varepsilon R$, aY belongs to $\mathfrak{h}_1$ because it corresponds to the continuous homomorphism $t \to \Theta(at)$. It remains to prove that, if $Z_1 = \Sigma u_i Y_i$, $Z_2 = \Sigma v_i Y_i$ both belong to $\mathfrak{h}_1$, $Z_1 + Z_2$ also belongs to $\mathfrak{h}_1$. Let Θ_1, Θ_2 be the continuous homomorphisms of R into $\mathfrak{G}$ such that $Z_1 = Y(\Theta_1)$, $Z_2 = Y(\Theta_2)$. We set

$$\sigma_k = \Theta_1(k^{-1})\Theta_2(k^{-1}) \qquad (1 \leqslant k < \infty).$$

For k sufficiently large, $\sigma_k\varepsilon U$. Moreover, we have $y_i(H\Theta_1(k^{-1})) = u_i k^{-1}$, $y_i(H\Theta_2(k^{-1})) = v_i k^{-1}$. By formula (3), §X, p. 120, we get

$$y_i(H\sigma_k) = k^{-1}(u_i + v_i) + k^{-2}A_i(k)$$

where the functions $A_i(k)$ remain bounded when k increases indefinitely.

We may assume without loss of generality that $Z_1 + Z_2 \neq 0$; then we have $\sigma_k \neq 0$ for k sufficiently large, and

$$d(H\sigma_k) = L^{-1}(\Sigma_i(u_i + v_i)^2)^{\frac{1}{2}} + A(k)k^{-2}$$

where $A(k)$ remains bounded. It follows from Lemma 1 that $(\Sigma(u_i + v_i)^2)^{-\frac{1}{2}}(Z_1 + Z_2)\varepsilon\mathfrak{h}_1$ and therefore also $Z_1 + Z_2\varepsilon\mathfrak{h}_1$ which completes the proof of our statement that $\mathfrak{h}_1$ is a vector space.

By making use of the corollary to Lemma 1, we see that any element of the form $\exp \Sigma_1^r u_i Y_i$ is the image under H of an element of U if $\Sigma_1^r u_i^2 < a^2$.

Let U_1 be the set of the elements $\sigma\varepsilon U$ such that $y_{r+1}(H\sigma) = \cdots = y_n(H\sigma) = 0$. We assert that *$U_1$ is a neighbourhood of ϵ in $\mathfrak{G}$.*

If $|u_1|, |u_2|, \cdots, |u_n|$ are sufficiently small, the element

$$(\exp \Sigma_1^r u_i Y_i)(\exp \Sigma_{r+1}^n u_i Y_j)$$

belongs to V and its coordinates may be expressed as analytic functions $\psi_i(u_1, \cdots, u_n)$ of $u_1, \cdots, u_n$. If $u_i = \delta_{ij}u$, we have $\psi_i = \delta_{ij}u$, whence

$$\left(\frac{\partial\psi_i}{\partial u_j}\right)_{u=0} = \delta_{ij} \qquad \text{and} \qquad \left(\frac{D(\psi_1, \cdots, \psi_n)}{D(u_1, \cdots, u_n)}\right)_{u=0} = 1.$$

It follows that there exists a system of coordinates $(z_1, z_2, \cdots, z_n)$ on $\mathfrak{H}$ at η such that the formulas

$$Z_i((\exp \Sigma_1^r u_i Y_i)(\exp \Sigma_{r+1}^n u_i Y_i)) = u_i \qquad (1 \leqslant i \leqslant n)$$

hold whenever $|u_i| < b$ $(1 \leqslant i \leqslant n)$. Let V_2 be a cubic neighbourhood of η with respect to this system; we assume that V_2 is of breadth $< b$ and is contained in V.

We assume for a moment that U_1 is *not* a neighbourhood of ϵ. Then there exists a sequence σ_k of elements of U not belonging to U_1 such that $\lim \sigma_k = \epsilon$, $H\sigma_k \varepsilon V_2$. We have $\lim z_i(\sigma_k) = 0$; for k sufficiently large, the element $\exp \Sigma_1^r z_i(\sigma_k) Y_i$ is the image under H of an element $\sigma'_k \varepsilon U$; moreover, since $\lim \exp \Sigma_1^r z_1(\sigma_k) Y_i = \eta$ and H is a homeomorphism of U into V, we have $\lim \sigma'_k = \epsilon$. We set

$$\sigma''_k = (\sigma'_k)^{-1}\sigma_k.$$

For k sufficiently large, we have $\sigma''_k \epsilon U$. Since $\sigma'_k \epsilon U_1$, we have $\sigma''_k \neq \epsilon$ for k sufficiently large. Moreover, we have

$$H\sigma''_k = \exp \Sigma_{r+1}^n z_i(H\sigma_k) Y_i$$

and hence

$$y_1(H\sigma''_k) = \cdots = y_r(H\sigma''_k) = 0, \qquad y_{r+h}(H\sigma''_k) = z_{r+h}(H\sigma_k)$$
$$(1 \leqq h \leqq n - r).$$

By replacing if necessary the sequence σ_k by a sub-sequence, we may assume without loss of generality that the sequences $\dfrac{y_i(H\sigma_k)}{d(H\sigma_k)}$ have limits u_i $(1 \leqq i \leqq n)$ $\left(\text{because } \left|\dfrac{y_i(H\sigma_k)}{d(H\sigma_k)}\right| \leqslant 1\right)$. We have clearly $u_1 = u_2 = \cdots = u_r = 0$, $\Sigma_{r+1}^n u_i^2 = 1$. By the lemma, it follows that the element $\Sigma_{r+1}^n u_i Y_i$ belongs to $\mathfrak{h}_1$, which is a contradiction with the fact that $\{Y_1, \cdots, Y_r\}$ is a base of $\mathfrak{h}_1$.

Therefore, our statement that U_1 is a neighbourhood of ϵ is proved.

If $\sigma \varepsilon U_1$, we have $H\sigma = \exp \Sigma_1^r y_i(H\sigma) Y_i$. We set $x_i(\sigma) = y_i(H\sigma)$ $(1 \leqq i \leqq r)$. The mapping which assigns to σ the point $\{x_1(\sigma), \cdots, x_r(\sigma)\}$ obviously maps U_1 homeomorphically onto a subset of R^r which contains the cube defined by the inequalities $|x_i| < a$ $(1 \leqq i \leqq r)$.

There exists a number $a' > 0$ such that the conditions $|y_i(\tau_1)| < a'$, $|y_i(\tau_2)| < a'$ imply $\tau_1\tau_2^{-1} \varepsilon V$. If we denote by U'_1 the set of elements $\sigma \varepsilon U_1$ such that $|x_i(\sigma)| < a'$ $(1 \leqq i \leqq r)$, the conditions $\sigma_1, \sigma_2 \varepsilon U'_1$ imply $\sigma_1\sigma_2^{-1} \varepsilon U_1$ and the functions $x_i(\sigma_1\sigma_2^{-1})$ may be expressed as functions of $x_1(\sigma_1) \cdots x_r(\sigma_1), x_1(\sigma_2), \cdots, x_r(\sigma_2)$; these functions are defined and analytic for $|x_i(\sigma_1)| < a'$, $|x_i(\sigma_2)| < a'$.

The fact that $\mathfrak{G}$ is a Lie group will follow if we can prove

Proposition 2. *Let $\mathfrak{G}$ be a topological group. We assume that there exists a neighbourhood U of the neutral element ϵ in $\mathfrak{G}$ and a system of n real valued functions $x_1, x_2, \cdots, x_n$ defined on U, with the following properties:* (1) *the mapping $\sigma \to (x_1(\sigma), \cdots, x_n(\sigma))$ is a homeomorphism of U with the cube Q in R^n defined by the inequalities $|x_i| < a$ $(1 \leqq i \leqq n)$, a being some number > 0;* (2) *there exists a number $a' > 0$ $(a' < a)$ and n functions $f_i(u_1, u_2, \cdots, u_n; v_1, v_2, \cdots, v_n)$, defined and analytic in the domain defined by the inequalities $|u_i| < a'$, $|v_i| < a'$, $(1 \leqq i \leqq n)$ such that the formulas $\sigma\tau^{-1} \varepsilon U$, $x_i(\sigma\tau^{-1}) = f_i(x_1(\sigma), \cdots, x_n(\sigma); x_1(\tau), \cdots, x_n(\tau))$ hold under the conditions $|x_i(\sigma)| < a'$, $|x_i(\tau)| < a'$ $(1 \leqq i \leqq n)$. Then $\mathfrak{G}$ is a Lie group.*

Let $\mathfrak{G}_1$ be the connected component of ϵ in $\mathfrak{G}$. Since U is a neighbourhood of ϵ and is connected, $\mathfrak{G}_1$ is obviously open. It is sufficient to prove that $\mathfrak{G}_1$ is a Lie group. Therefore we may assume without loss of generality that $\mathfrak{G}$ itself is connected.

We denote by φ_σ the left-translation associated with σ in $\mathfrak{G}$, and by $\mathfrak{A}(\sigma)$ the class of real valued functions defined in neighbourhoods of σ in $\mathfrak{G}$ and which depend analytically on the functions $x_1 \circ \varphi\sigma^{-1}$, $x_2 \circ \varphi\sigma^{-1}, \cdots, x_n \circ \varphi\sigma^{-1}$ around σ. We assert that the assignment $\sigma \to \mathfrak{A}(\sigma)$ defines a variety whose underlying space is $\mathfrak{G}$.

The conditions I, II of §1, Chapter III, p. 68, are obviously satisfied. Since φ_σ is a homeomorphism of $\mathfrak{G}$ with itself, the properties (1), (2) of condition III hold in our case. It remains to check the property (3). Let U' be the set of the elements $\zeta \varepsilon U$ such that $|x_i(\zeta)| < a'$; U' is a neighbourhood of ϵ. Hence $\sigma U'^{-1}$ is a neighbourhood of σ. Let τ be any element of this set; we have $\tau^{-1}\sigma = \zeta \varepsilon U'$. We have $(x_i \circ \varphi\sigma^{-1})(\rho) = x_i(\sigma^{-1}\rho) = x_i(\sigma^{-1}\tau\tau^{-1}\rho) = x_i(\zeta^{-1}\tau^{-1}\rho) = f_i(x_i(\zeta^{-1}), \cdots, x_n(\zeta^{-1}); x_1(\rho^{-1}\tau), \cdots, x_n(\rho^{-1}\tau))$, these formulas holding provided $\tau^{-1}\rho \varepsilon U'$, $\rho \varepsilon \tau U'$. This proves that the functions $x_i \circ \varphi_{\sigma-1}$ belong to $\mathfrak{A}(\sigma)$; thus property (4) holds.

The assignment $\sigma \to \mathfrak{A}(\sigma)$ defines therefore a manifold $\mathcal{G}$. It is obvious that the mapping $(\sigma, \tau) \to \sigma\tau^{-1}$ of $\mathcal{G} \times \mathcal{G}$ into $\mathcal{G}$ is analytic at the point $(\epsilon, \epsilon) \varepsilon \mathcal{G} \times \mathcal{G}$. We have yet to prove that it is everywhere analytic.

It follows immediately from the definition that every left-translation φ_σ is an analytic isomorphism of the manifold $\mathcal{G}$ with itself. The same applies to the right-translation ψ_σ associated with σ; in fact, let us first assume that $|x_i(\sigma)| < a'$ $(1 \leqq i \leqq n)$. The mapping $J: \tau \to \tau^{-1}$ is clearly analytic at ϵ, and ψ_σ is obtained by first performing J, and then the mapping $\tau \to \tau^{-1}\sigma$ which is analytic at ϵ provided $|x_i(\sigma)| < a'$ $(1 \leqq i \leqq n)$. The formula $\psi_\sigma\tau = \tau\sigma = \phi_{\tau_0}(\tau_0^{-1}\tau\sigma)$ shows that ψ_σ is analytic at every point τ_0 if $\sigma \varepsilon U'$. But U', being a neighbourhood of ϵ in the connected group $\mathfrak{G}$, is a system of generators of $\mathfrak{G}$. Hence any element $\sigma \varepsilon \mathfrak{G}$ may be written in the form $\sigma_1\sigma_2 \cdots \sigma_k$ with $\sigma_j \varepsilon U'$ $(1 \leqslant j \leqslant k)$. We have $\psi_\sigma = \psi_{\sigma_k} \circ \psi_{\sigma_{k-1}} \circ \cdots \circ \psi_{\sigma_1}$, which shows that ψ_σ is an analytic mapping.

Now, in order to prove that the mapping $(\sigma, \tau) \rightarrow \sigma\tau^{-1}$ is analytic at every point (σ_0, τ_0), we write

$$\sigma\tau^{-1} = \sigma_0(\sigma_0^{-1}\sigma)(\tau_0^{-1}\tau)^{-1}\tau_0^{-1}.$$

It is obvious that the mapping $(\sigma, \tau) \rightarrow (\sigma_0^{-1}\sigma_0, \tau_0^{-1}\tau)$ of $\mathcal{G} \times \mathcal{G}$ into itself is analytic at (σ_0, τ_0). Since φ_{σ_0} and ψ_{τ_0} are analytic mappings, and since the mapping $(\sigma, \tau) \rightarrow \sigma\tau^{-1}$ is analytic at (ϵ, ϵ), it is also analytic at (σ_0, τ_0).

Therefore $\mathcal{G}$ is the underlying manifold of an analytic group, whose underlying topological group is $\mathfrak{G}$, and this completes the proofs of Propositions 1 and 2.

Corollary. *Every closed subgroup of a Lie group is a Lie group.*

In fact, a Lie group being locally compact, the same holds for every closed subgroup of a Lie group.

We observe also that it follows immediately from Proposition 2 that any topological group which is locally isomorphic with a Lie group is itself a Lie group. If $\mathfrak{G}$ is a Lie group, $\mathfrak{G}$ is clearly locally simply connected. Therefore, if $\mathfrak{G}$ is connected, it admits a simply connected covering group $(\tilde{\mathfrak{G}}, f)$, and we see that $\tilde{\mathfrak{G}}$ is a Lie group.

§XV. Groups of Automorphisms

Let H be a subgroup of $GL(n, C)$. Denote by $x_{ij}(\sigma)$ $(1 \leqslant i, j \leqslant n)$ the coefficients of a matrix $\sigma \varepsilon GL(n, C)$. We shall say that H is an *algebraic subgroup* of $GL(n, C)$ if there exists a set of polynomials $P_\alpha(\cdots, x_{ij}, \cdots)$ in n^2 arguments (α running over some set of indices) such that the conditions

$$\sigma \varepsilon H$$

and

$$P_\alpha(\cdots, x_{ij}(\sigma), \cdots) = 0 \qquad \text{(for all } \alpha)$$

are equivalent. For instance, $SL(n, C)$ and $O(n, C)$ are algebraic subgroups of $GL(n, C)$, and $Sp(n, C)$ is an algebraic subgroup of $GL(2n, C)$.

Denote by $x'_{ij}(\sigma)$ and $x''_{ij}(\sigma)$ the real and imaginary parts of $x_{ij}(\sigma)$. If we can find a set of polynomials $Q_\beta(\cdots x'_{ij}, x''_{ij} \cdots)$ in $2n^2$ arguments such that the conditions $\sigma \varepsilon H$ and $Q_\beta(\cdots x'_{ij}(\sigma), x''_{ij}(\sigma), \cdots)$

= 0 (for all β) are equivalent, then we say that H is *a pseudoalgebraic subgroup* of $GL(n, C)$. For instance, $GL(n, R)$, $SL(n, R)$, $O(n)$, $SO(n)$, $U(n)$, $SU(n)$ are pseudo-algebraic subgroups of $GL(n, C)$; $Sp(n)$ is a pseudo-algebraic subgroup of $GL(2n, C)$.

It follows immediately from the Corollary to Proposition 1, §XIV, p. 130 that any algebraic or pseudo-algebraic subgroup of $GL(n, C)$ is a Lie group (when it is considered as a topological subgroup of $GL(n, C)$).

Now, let $\mathfrak{g}$ be any algebra over the field R of real numbers; i.e. $\mathfrak{g}$ is a vector space of finite dimension n over R in which there is defined a bilinear law of composition $(X, Y) \to XY$ (we do not require any other condition than bilinearity for this law of composition). The automorphisms of the algebra $\mathfrak{g}$ clearly form a subgroup $\mathfrak{A}$ of the group of automorphisms of the underlying vector space of $\mathfrak{g}$ (i.e. of $GL(n, R)$). Let $\{X_1, \cdots, X_n\}$ be a base in $\mathfrak{g}$; with respect to this base, every automorphism α of $\mathfrak{g}$ is expressible by a matrix (also denoted by α). Set

$$\alpha(X_i) = \Sigma_{j=1}^{n} a_{ji} X_j \qquad (1 \leqslant i \leqslant n)$$

Then the conditions which express that α is an automorphism (i.e. the conditions $\alpha(X_i X_j) = \alpha(X_i)\alpha(X_j)$ can be expressed by a certain number of algebraic relationships between the coefficients a_{ij} (which, furthermore, must be real). It follows that $\mathfrak{A}$ is a pseudo-algebraic subgroup of $GL(n, R)$ and is therefore a Lie group. We shall determine the Lie algebra of this group. Let $\mathfrak{a}$ denote this Lie algebra. Then $\mathfrak{a}$ is a subalgebra of the Lie algebra of $GL(n, R)$, and the elements of $\mathfrak{a}$ can therefore be considered as matrices of degree n with real coefficients, i.e. also as linear endomorphisms of the underlying vector space of $\mathfrak{g}$. Let A be a matrix belonging to $\mathfrak{a}$. Then, for every real t, exp tA is an automorphism of $\mathfrak{g}$, whence

$$(\exp tA)XY = ((\exp tA)X)((\exp tA)Y)$$

for any $X, Y \varepsilon \mathfrak{g}$.

Let E be the unit matrix; then

$$(E + tA + t^2 A_t)XY = ((E + tA + t^2 A_t)X)((E + tA + t^2 A_t)Y)$$

where A_t is a matrix whose coefficients remain bounded when t approaches 0. It follows easily that

$$(1) \qquad A(XY) = A(X)Y + XA(Y)$$

Any linear endomorphism A of $\mathfrak{g}$ which satisfies (1) is called a *derivation* of the algebra $\mathfrak{g}$. We have seen that every matrix $A \varepsilon \mathfrak{a}$ is a derivation of $\mathfrak{g}$. We shall see that the converse of this assertion holds true.

Let A be a derivation of $\mathfrak{g}$. We have

$$A^p(XY) = \Sigma_{ij} \frac{p!}{i!j!} A^i(X)A^j(Y)$$

where the summation is extended over all pairs (i, j) such that $i \geqslant 0$, $j \geqslant 0$, $i + j = p$ (we set $A^0 = E$). Write

$$\frac{1}{i!} A^i(X). \quad \frac{1}{j!} A^j(Y) = \Sigma_{k=1}^n \lambda_{ijk} X_k$$

We know that the coefficients of A^i are smaller in absolute value than M^i, where M is some constant. It follows easily that the double series $\Sigma_{ij}|\lambda_{ijk}|$ are convergent, whence

$$(\exp tA)XY = \Sigma \frac{1}{i!} A^i(X) \frac{1}{j!} A^j(X) = ((\exp tA)X)((\exp tA)Y)$$

which proves that $\exp tA \varepsilon \mathfrak{A}$, whence $A \varepsilon \mathfrak{a}$. We have proved

Proposition 1. *Let* $\mathfrak{g}$ *be any algebra over the field of real numbers. Then the derivations of* $\mathfrak{g}$ *form a subalgebra of* $\mathfrak{gl}(n, R)$ *(where* n *is the dimension of* $\mathfrak{g}$*), and this algebra is the Lie algebra of the group of automorphisms of* $\mathfrak{g}$.

Assume in particular that $\mathfrak{g}$ is the Lie algebra of a simply connected Lie group $\mathfrak{G}$. Let α be any automorphism of $\mathfrak{g}$; then there corresponds to α a continuous homomorphism θ of $\mathfrak{G}$ into itself such that $\alpha = d\theta$ (cf. Theorem 2, §VI, p. 111). Since α is an automorphism, it has a reciprocal mapping α' which is also an automorphism. Let θ' be the continuous homomorphism of $\mathfrak{G}$ into itself which corresponds to α'; then $d(\theta \circ \theta') = \alpha \circ \alpha'$ is the identity mapping of $\mathfrak{g}$. It follows that $\theta \circ \theta'$ is the identity mapping of $\mathfrak{G}$. In the same way, we see that $\theta' \circ \theta$ is also the identity mapping of $\mathfrak{G}$. It follows immediately that θ is an isomorphism of $\mathfrak{G}$ (considered as a topological group) with itself (i.e. it is an automorphism of the underlying group and a homeomorphism of the underlying space with itself). Such a mapping is called an *automorphism* of the group $\mathfrak{G}$. Conversely, we see easily that to every automorphism θ of $\mathfrak{G}$ there corresponds an automorphism $\alpha = d\theta$ of $\mathfrak{g}$. Therefore we have proved:

Proposition 2. *Let $\mathfrak{G}$ be a simply connected Lie group. Then the group of automorphisms of $\mathfrak{G}$ is isomorphic with a Lie group whose Lie algebra is the algebra of derivations of the Lie algebra of $\mathfrak{G}$.*

Denote by θ_α the automorphism of $\mathfrak{G}$ which corresponds to an automorphism α of $\mathfrak{g}$. If $X \varepsilon \mathfrak{g}$, we have θ_α (exp X) = exp $\alpha(X)$, which shows that θ_α(exp X) depends continuously on α, for X fixed. Since $\mathfrak{G}$ is connected, any $\sigma \varepsilon \mathfrak{G}$ may be written in the form $\sigma_1 \cdots \sigma_h$ where each σ_i is of the form exp Y_i, $Y_i \varepsilon \mathfrak{g}$. It follows immediately that $\theta_\alpha(\sigma)$ depends continuously on α for σ fixed.

Now, let $\mathfrak{G}$ be a connected Lie group which is not simply connected, and let $(\tilde{\mathfrak{G}}, f)$ be a simply connected covering group of $\mathfrak{G}$. Let θ be an automorphism of $\mathfrak{G}$. Then $\theta \circ f$ is a continuous homomorphism of $\tilde{\mathfrak{G}}$ into $\mathfrak{G}$. Denote by ϵ and $\tilde{\epsilon}$ the neutral elements of $\mathfrak{G}$ and $\tilde{\mathfrak{G}}$ respectively. By Proposition 1, chapter II, §VIII, p. 50, there exists a continuous mapping $\tilde{\theta}$ of $\tilde{\mathfrak{G}}$ into itself such that $f \circ \tilde{\theta} = \theta \circ f$ and $\tilde{\theta}(\tilde{\epsilon}) = \tilde{\epsilon}$. The mapping $(\tilde{\sigma}, \tilde{\tau}) \to \tilde{\theta}(\tilde{\sigma}\tilde{\tau})(\tilde{\theta}(\tilde{\sigma}))^{-1}(\tilde{\theta}(\tilde{\tau}))^{-1}$ maps continuously the connected space $\tilde{\mathfrak{G}} \times \tilde{\mathfrak{G}}$ into the kernel F of f, which is discrete. It follows immediately that $\tilde{\theta}(\tilde{\sigma}\tilde{\tau})(\tilde{\theta}(\tilde{\sigma}))^{-1}(\tilde{\theta}(\tilde{\tau}))^{-1} = \tilde{\epsilon}$, which proves that $\tilde{\theta}$ is a homomorphism of $\mathfrak{G}$ into itself. Let θ' be the reciprocal automorphism of θ; then, in the same way, there corresponds to θ' a continuous homomorphism $\tilde{\theta}'$ of $\mathfrak{G}$ into itself. Furthermore, we see immediately that there exists a neighbourhood of $\tilde{\epsilon}$ in $\tilde{\mathfrak{G}}$ on which both $\tilde{\theta} \circ \tilde{\theta}'$ and $\tilde{\theta}' \circ \tilde{\theta}$ coincide with the identity mapping. Since $\tilde{\mathfrak{G}}$ is connected, $\tilde{\theta} \circ \tilde{\theta}'$ and $\tilde{\theta}' \circ \tilde{\theta}$ both coincide with the identity mapping of $\tilde{\mathfrak{G}}$, which shows that $\tilde{\theta}$ is an automorphism of $\mathfrak{G}$. Furthermore, we clearly have $\tilde{\theta}(F) = F$.

Conversely, let $\tilde{\theta}$ be an automorphism of $\tilde{\mathfrak{G}}$ such that $\tilde{\theta}(F) = F$. Let σ be any element of $\mathfrak{G}$, and let $\tilde{\sigma}$ be any element of $\tilde{\mathfrak{G}}$ such that $f(\tilde{\sigma}) = \sigma$. Then the value of $f(\tilde{\theta}(\tilde{\sigma}))$ depends only upon σ, not upon the choice of $\tilde{\sigma}$. If we set $\theta(\sigma) = f(\tilde{\theta}(\tilde{\sigma}))$, we see easily (by arguments of the same type as above) that θ is an automorphism of $\mathfrak{G}$. Therefore, the group of automorphisms of $\mathfrak{G}$ may be identified with the group of those automorphisms of $\mathfrak{G}$ which map F into itself. Since F is closed, it follows from the remark which follows Proposition 2 that this group is a closed subgroup of the group of automorphisms $\mathfrak{A}$ of $\mathfrak{G}$ (in the topology which makes $\mathfrak{A}$ a Lie group). We have therefore proved that *the group of automorphisms of any connected Lie group is isomorphic with a Lie group.*

CHAPTER V

The Differential Calculus of Cartan

Summary. §§I and II are algebraic in character; their object is to construct the Grassmann algebra $\mathfrak{A}$ associated with a given vector space $\mathfrak{M}$. For reasons of convenience we have arranged the construction in such a way that the dual space of $\mathfrak{M}$ (and not $\mathfrak{M}$ itself) is contained in $\mathfrak{A}$; i.e. the elements of $\mathfrak{A}$ are alternating *contravariant* tensors.

In §III, we define the exterior differential forms of Cartan on a manifold and their differentiation. These forms behave in a contravariant way under an analytic mapping (this is why we introduce the notation "δ," which is considered as dual to the "d" of Chapter III). The operation δ is proved to commute with the differentiation (formula (4), §III, p. 146).

In §§IV and V, we apply the differential calculus of Cartan to the theory of Lie groups. The notion of left invariant differential form is defined; the left-invariant differential forms of order 1 are the forms of Maurer-Cartan. Their differentiation is determined in terms of the Lie algebra by formula (2), §IV, p. 152. It is shown how the forms of Maurer-Cartan can be explicitly constructed (in canonical coordinates) if the Lie algebra is known. A very simple example shows how it is possible to arrive by this method at an explicit construction of the law of composition in the group. However, it should be observed that, if one does not insist on using canonical coordinates, there are simpler methods to get the same result which we shall discuss in volume II.

The remainder of the chapter is concerned with the integration of differential forms on a manifold. Only forms of the highest dimension are considered (which means that we do not prove the generalized Stokes's formula). After having defined the orientation of a manifold (§VI), we construct the integral of a function with respect to a differential form on an oriented manifold (§VII); the construction is based on a very useful lemma of Dieudonné. In §VIII, we define invariant integration on a group which will become the main tool in Chapter VI.

§I. MULTILINEAR FUNCTIONS

Let K be a field, and let $\mathfrak{M}_1, \mathfrak{M}_2, \cdots, \mathfrak{M}_r$ be r vector spaces over K, of dimensions $m_1, \cdots, m_r$.

Definition 1. *An r-linear function on* $\mathfrak{M}_1 \times \mathfrak{M}_2 \times \cdots \times \mathfrak{M}_r$ *is a mapping* M *of this set into* K *such that* $\mathsf{M}(\mathbf{e}_1, \cdots, \mathbf{e}_r)$ *is a linear function of any one of its arguments when the* $r - 1$ *others are kept fixed, i.e. we have*

$$\begin{aligned}\mathsf{M}(\mathbf{e}_1, \cdots, \mathbf{e}_{i-1}, a\mathbf{e}_i + a'\mathbf{e}'_i, \mathbf{e}_{i+1}, \cdots, \mathbf{e}_r) &= a\mathsf{M}(\mathbf{e}_1, \cdots, \mathbf{e}_{i-1}, \mathbf{e}_i), \mathbf{e}_{i+1}, \cdots, \mathbf{e}_r) \\ &\quad + a'\mathsf{M}(\mathbf{e}_1, \cdots, \mathbf{e}_{i-1}, \mathbf{e}'_i, \mathbf{e}_{i+1}, \cdots, \mathbf{e}_r)\end{aligned}$$

if $a, a' \varepsilon K$.

Let M_1, M_2 be two such r-linear functions. Then the functions $a_1\mathsf{M}_1 + a_2\mathsf{M}_2$, which maps $(\mathbf{e}_1, \cdots, \mathbf{e}_r)$ into $a_1\mathsf{M}_1(\mathbf{e}_1, \cdots, \mathbf{e}_r) + a_2\mathsf{M}_2(\mathbf{e}_1, \cdots, \mathbf{e}_r)$ is clearly again an r-linear function.

Let $\{\mathbf{a}_{i,1}, \cdots, \mathbf{a}_{i,m_i}\}$ be a base of $\mathfrak{M}_i$. If $\mathbf{e}_i = \Sigma_j x_{ij}\mathbf{a}_{ij}$ $(1 \leqslant j \leqslant m_i;\ x_{ij} \varepsilon K)$, and if M is an r-linear function, we have

$$(1) \qquad \mathsf{M}(\mathbf{e}_1, \cdots, \mathbf{e}_r) = \Sigma x_{1j_1}x_{2j_2} \cdots x_{rj_r}\mathsf{M}(\mathbf{a}_{1j_1}, \cdots, \mathbf{a}_{rj_r}),$$

which shows that M is entirely determined when the quantities $\mathsf{M}(\mathbf{a}_{1j_1}, \cdots, \mathbf{a}_{r,j_r})$ are determined. Conversely, these quantities may obviously be taken arbitrarily in K. It follows that the r-linear functions form a vector space over K of dimension $m_1 \cdots m_r$.

Now let $\mathfrak{N}_1, \mathfrak{N}_2, \cdots, \mathfrak{N}_s$ be a new system of vector spaces over K. Let M be an r-linear function defined on $\mathfrak{M}_1 \times \mathfrak{M}_2 \times \cdots \times \mathfrak{M}_r$, and let N be an s-linear function defined on $\mathfrak{N}_1 \times \mathfrak{N}_2 \times \cdots \times \mathfrak{N}_s$. Then the function Λ defined on $\mathfrak{M}_1 \times \mathfrak{M}_2 \times \cdots \times \mathfrak{M}_r \times \mathfrak{N}_1 \times \mathfrak{N}_2 \times \cdots \times \mathfrak{N}_s$ by the formula

$$(2) \quad \Lambda(\mathbf{e}_1, \cdots, \mathbf{e}_r; \mathbf{f}_1, \cdots, \mathbf{f}_s) = \mathsf{M}(\mathbf{e}_1, \cdots, \mathbf{e}_r)\mathsf{N}(\mathbf{f}_1, \cdots, \mathbf{f}_s),$$

(for $\mathbf{e}_i \varepsilon \mathfrak{M}_i$, $\mathbf{f}_j \varepsilon \mathfrak{N}_j$) is clearly an $(r + s)$-linear function.

Definition 2. *The function* Λ *defined by formula* (2) *is called the Kronecker product of* M *and* N; *it is denoted by* MN.

The following properties of this operation are obvious:

1) It is linear with respect to each argument, i.e.

$$(a_1\mathsf{M}_1 + a_2\mathsf{M}_2)\mathsf{N} = a_1\mathsf{M}_1\mathsf{N} + a_2\mathsf{M}_2\mathsf{N}, \qquad (a_1, a_2 \varepsilon K)$$
$$\mathsf{M}(a_1\mathsf{N}_1 + a_2\mathsf{N}_2) = a_1\mathsf{MN}_1 + a_2\mathsf{MN}_2,$$

2) If $\mathfrak{P}_1, \mathfrak{P}_2, \cdots, \mathfrak{P}_t$ is a third system of vector spaces over K, and if Π is a t-linear function defined on $\mathfrak{P}_1 \times \mathfrak{P}_2 \times \cdots \times \mathfrak{P}_t$ we have

$$(\mathsf{MN})\Pi = \mathsf{M}(\mathsf{N}\Pi).$$

If $r = 1$, the r-linear functions on $\mathfrak{M}_1$ are simply the linear mappings of $\mathfrak{M}_1$ into K.

Definition 3. *The vector space composed of the linear mappings of a vector space* $\mathfrak{M}$ *into* K *is called the dual space of* $\mathfrak{M}$ *and denoted by* $\mathfrak{M}'$.

If $\{\mathbf{a}_1, \cdots, \mathbf{a}_m\}$ is a base of $\mathfrak{M}$, to every i $(1 \leqslant i \leqslant m)$ there corresponds an element $\varphi_i \varepsilon \mathfrak{M}'$ which is defined by the conditions $\varphi_i\mathbf{a}_j = \delta_{ij}$. Moreover, these m elements of $\mathfrak{M}'$ form a base of $\mathfrak{M}'$, which is called the *dual base* of the base $\{\mathbf{a}_1, \cdots, \mathbf{a}_m\}$ of $\mathfrak{M}$.

Let $\mathbf{a}$ be any element of $\mathfrak{M}$. Then, if we consider $\varphi(\mathbf{a})$, for $\varphi \varepsilon \mathfrak{M}'$, as a function of φ, we obtain a linear function $\psi_{\mathbf{a}}$ on $\mathfrak{M}'$: $\psi_{\mathbf{a}} = \varphi(\mathbf{a})$. Moreover, the mapping $\mathbf{a} \rightarrow \psi_{\mathbf{a}}$ is clearly a linear mapping of $\mathfrak{M}$ into

$\mathfrak{M}'' = (\mathfrak{M}')'$. If $\mathbf{a} = \Sigma x_i \mathbf{a}_i$ is $\neq 0$, there exists at least one $\varphi \varepsilon \mathfrak{M}'$ for which $(\mathbf{a}) \neq 0$,—for instance $\varphi = \varphi_i$ if $x_i \neq 0$. Therefore we have $\psi_{\mathbf{a}} \neq 0$, and the linear mapping $\mathbf{a} \to \psi_{\mathbf{a}}$ is univalent. Since $\mathfrak{M}$ and $\mathfrak{M}''$ have the same dimension, this mapping is a linear isomorphism of $\mathfrak{M}$ with $\mathfrak{M}''$. We may call it the *natural isomorphism* of $\mathfrak{M}$ with $\mathfrak{M}''$.

It follows in particular that any base in $\mathfrak{M}'$ is the dual of some base in $\mathfrak{M}$.

Proposition 1. *Let $\mathfrak{M}_1, \mathfrak{M}_2, \cdots, \mathfrak{M}_r$ be r vector spaces over K. Let $\{\varphi_{i1}, \varphi_{i2}, \cdots, \varphi_{im_i}\}$ be a base in the dual space $\mathfrak{M}_i'$ of $\mathfrak{M}_i$. Then the $m_1 \cdots m_r$ elements $\varphi_{1j_1}\varphi_{2j_2} \cdots \varphi_{rj_r}$ $(1 \leqslant j_\alpha \leqslant m_\alpha, 1 \leqslant \alpha \leqslant r)$ form a base of the space of r-linear functions on $\mathfrak{M}_1 \times \mathfrak{M}_2 \times \cdots \times \mathfrak{M}_r$.*

In fact, let $\{\mathbf{a}_{i,1}, \cdots, \mathbf{a}_{i,m_i}\}$ be the base of $\mathfrak{M}$ of which $\{\varphi_{i,1}, \varphi_{i,2}, \cdots, \varphi_{i,m_i}\}$ is the dual base. We have

$$\varphi_{1j_1}\varphi_{2j_2} \cdots \varphi_{rj_r}(\mathbf{a}_{1,k_1}, \mathbf{a}_{2,k_2}, \cdots, \mathbf{a}_{r,k_r}) = \delta_{j_1k_1}\delta_{j_2k_2} \cdots \delta_{j_rk_r}$$

which, by comparison with formula (1), proves Proposition 1.

§II. Alternate Functions

We shall now consider r-linear functions Λ defined on the product $\mathfrak{M}^r$ of r vector spaces identical with a given vector space $\mathfrak{M}$ of dimension m[1] over a basic field K which is assumed to be of characteristic 0. Let $\mathfrak{H}_r$ be the space of these r-linear functions, for $r \geqslant 1$. We shall denote the set K, considered as a vector space of dimension 1 over K, by $\mathfrak{H}_0$. Let us form the set $\Pi_{r=0}^{\infty}\mathfrak{H}_r$. An element of this set is a mapping which assigns to every $r \geqslant 0$ an element $\Lambda_r \varepsilon \mathfrak{H}_r$. We shall consider the subset $\mathfrak{D}$ composed of those elements of $\Pi_0^\infty \mathfrak{H}_r$ of which almost all coordinates Λ_r are equal to 0 (i.e. all except a finite number). If $\tilde{\Lambda} = (\Lambda_0, \Lambda_1, \cdots, \Lambda_r, \cdots)$ and $\tilde{\mathsf{M}} = (\mathsf{M}_0, \mathsf{M}_1, \cdots, \mathsf{M}_r, \cdots)$ are any elements of $\mathfrak{D}$ and if $a, b \varepsilon K$, the element

$$a\tilde{\Lambda} + b\tilde{\mathsf{M}} = (a\Lambda_0 + b\mathsf{M}_0, a\Lambda_1 + b\mathsf{M}_1, \cdots, a\Lambda_r + b\mathsf{M}_r, \cdots)$$

also belongs to $\mathfrak{D}$. It follows immediately that $\mathfrak{D}$ is a vector space (of infinite dimension) over K. An element of $\mathfrak{D}$ is called *homogeneous of order r* if all its coordinates are zero except perhaps the r-coordinate. Such an element may, without trouble, be identified with its r-coordinate. If we make these identifications, the element $\tilde{\Lambda} = (\Lambda_0, \Lambda_1, \cdots, \Lambda_r, \cdots)$ may be written in the form $\Sigma_0^\infty \Lambda_r$. (The symbol Σ_0^∞ stands for Σ_0^R where R is an integer so large that $\Lambda_r = 0$ for $r > R$.)

In §I we have defined the product $\Lambda_r \mathsf{M}_s$ of an element $\Lambda_r \varepsilon \mathfrak{H}_r$ by an element $\mathsf{M}_s \varepsilon \mathfrak{H}_s$, provided $rs > 0$. If, for instance, $r = 0$, Λ_0 is an

[1] These r-linear forms are called "r-times contravariant tensors."

element of K, and $\Lambda_0\mathsf{M}_s$ is also defined since $\mathfrak{H}_s$ is a vector space over K. The same remark applies if $s = 0$. Moreover, the formula $(\Lambda_r\mathsf{M}_s)\mathsf{N}_t = \Lambda_r(\mathsf{M}_s\mathsf{N}_t)(\Lambda_r\varepsilon\mathfrak{H}_r,\ \mathsf{M}_s\varepsilon\mathfrak{H}_s,\ \mathsf{N}_t\varepsilon\mathfrak{H}_t)$ remains valid if one or more of r, s, t is to 0.

We may now define the product $\tilde{\Lambda}\tilde{\mathsf{M}}$ of any two elements $\tilde{\Lambda} = \Sigma\Lambda_r$, $\tilde{\mathsf{M}} = \Sigma\mathsf{M}_r$ of $\mathfrak{D}$ by the formula

$$\tilde{\Lambda}\tilde{\mathsf{M}} = \Sigma_{r,s=0}^{\infty}\Lambda_r\mathsf{M}_s$$

This means that the t-coordinate of $\tilde{\Lambda}\tilde{\mathsf{M}}$ will be

$$(\tilde{\Lambda}\tilde{\mathsf{M}})_t = \Sigma_{r+s=t}\Lambda_r\mathsf{M}_s$$

This obviously vanishes if t is sufficiently large, when $\tilde{\Lambda}\tilde{\mathsf{M}}\epsilon\mathfrak{D}$.

It is a trivial matter to verify that $\mathfrak{D}$ becomes a ring under this multiplication. If E is the element $(1, 0, \cdots, 0, \cdots)$, E is the unit element of this ring.

We shall now define an operation in $\mathfrak{H}_r$, which we call *alternation*. Let Λ_r be an element of $\mathfrak{H}_r$, and let $\bar{\omega}$ be any permutation of the set $\{1, 2, \cdots, r\}$. The function $\Lambda_r^{\bar{\omega}}$ defined by the formula

$$\Lambda_r^{\bar{\omega}}(\mathbf{e}_1, \cdots, \mathbf{e}_r) = \Lambda_r(\mathbf{e}_{\bar{\omega}(1)}, \cdots, \mathbf{e}_{\bar{\omega}(r)})$$

obviously again belongs to $\mathfrak{H}_r$. Moreover the mapping $\Lambda_r \rightarrow \Lambda_r^{\bar{\omega}}$ is a linear mapping of $\mathfrak{H}_r$ into itself. We now define an operation A_r, called *alternation*, which maps any $\Lambda_r\epsilon\mathfrak{H}_r$ on

$$A_r(\Lambda_r) = \frac{1}{r!}\,\Sigma_{\bar{\omega}}\epsilon(\bar{\omega})\Lambda_r^{\bar{\omega}},$$

this sum being extended over all permutations $\bar{\omega}$ of the set $\{1, 2, \cdots, r\}$, with $\epsilon(\bar{\omega}) = 1$ for even permutations $\bar{\omega}$, and $\epsilon(\bar{\omega}) = -1$ for odd permutations $\bar{\omega}$. If $r = 0$ or 1, we set $A_r(\Lambda_r) = \Lambda_r$.

If $\tilde{\Lambda} = \Sigma_0^{\infty}\Lambda_r$ is an element of $\mathfrak{D}$, we set $A(\tilde{\Lambda}) = \Sigma_0^{\infty}A_r(\Lambda_r)$. Then A is a linear mapping of $\mathfrak{D}$ into itself.

We denote by $\mathfrak{J}$ the set of elements $\tilde{\Lambda}\epsilon\mathfrak{D}$ for which $A(\tilde{\Lambda}) = 0$. It is clear that $\mathfrak{J}$ is a vector subspace of $\mathfrak{D}$. We shall now prove the remarkable fact that $\mathfrak{J}$ is actually an *ideal* in $\mathfrak{D}$. In other words, the condition $A(\tilde{\Lambda}) = 0$ implies $A(\tilde{\mathsf{M}}\tilde{\Lambda}) = A(\tilde{\Lambda}\tilde{\mathsf{M}}) = 0$, for any $\tilde{\mathsf{M}}\epsilon\mathfrak{D}$. It is clearly sufficient to prove this for the case where $\tilde{\Lambda} = \Lambda_r\varepsilon\mathfrak{H}_r$, $\tilde{\mathsf{M}} = \mathsf{M}_s\varepsilon\mathfrak{H}_s$. Setting $\mathsf{N}_{r+s} = A_{r+s}(\Lambda_r\mathsf{M}_s)$, we have

$$\begin{aligned}(r + s)!\mathsf{N}_{r+s}(\mathbf{e}_1, &\cdots, \mathbf{e}_{r+s})\\ &= \Sigma_{\bar{\omega}}\epsilon(\bar{\omega})\Lambda_r(\mathbf{e}_{\bar{\omega}(1)}, \cdots, \mathbf{e}_{\bar{\omega}(r)})\mathsf{M}_s(\mathbf{e}_{\bar{\omega}(r+1)}, \cdots, \mathbf{e}_{\bar{\omega}(r+s)})\end{aligned}$$

where the summation extends over all permutations $\bar{\omega}$ of the set $\{1, \cdots, r + s\}$. Let G be the group of these permutations, and H the subgroup composed of those which leave the elements $r + 1, \cdots, r + s$ unchanged. Let us consider the sum

$$\Sigma_{\bar{\omega}\epsilon\bar{\omega}_0 H}\epsilon(\bar{\omega})\Lambda_r(\mathbf{e}_{\bar{\omega}(1)}, \cdots, \mathbf{e}_{\bar{\omega}(r)})\mathsf{M}_s(\mathbf{e}_{\bar{\omega}(r+1)}, \cdots, \mathbf{e}_{\bar{\omega}(r+s)})$$

extended over all permutations $\bar{\omega}$ of a certain coset $\bar{\omega}_0 H$. If we set $\mathbf{e}_{\bar{\omega}_0(i)} = f_i$ $(1 \leqslant i \leqslant r)$, this sum may be written in the form

$$\epsilon(\bar{\omega}_0)\mathsf{M}_s(\mathbf{e}_{\bar{\omega}_0(r+1)}, \cdots, \mathbf{e}_{\bar{\omega}_0(r+s)})(\Sigma_{\bar{\omega}'\epsilon H}\epsilon(\bar{\omega}')\Lambda_r(f_{\bar{\omega}'(1)}, \cdots, f_{\bar{\omega}'(r)})),$$

since $\epsilon(\bar{\omega}_0\bar{\omega}') = \epsilon(\bar{\omega}_0)\epsilon(\bar{\omega}')$. But the operations of H induce the complete set of permutations on the set $\{1, \cdots, r\}$, and hence the second factor is 0. It follows that $A_{r+s}(\Lambda_r\mathsf{M}_s) = 0$, and we can prove in the same way that $A_{r+s}(\mathsf{M}_r\Lambda_s) = 0$.

It follows that the set $\mathfrak{O}/\mathfrak{J}$ of the residue-classes of $\mathfrak{O}$ modulo $\mathfrak{J}$ is again a ring. It is also a vector space over K. We claim that it is of finite dimension. In fact, let us choose a base $\{\varphi_1, \varphi_2, \cdots, \varphi_m\}$ in the dual space $\mathfrak{M}' = \mathfrak{H}_1$ of $\mathfrak{M}$. If φ is any element of $\mathfrak{M}'$, we have $\varphi^2 \varepsilon \mathfrak{J}$. In fact $A(\varphi^2)(\mathbf{e}_1, \mathbf{e}_2) = (\frac{1}{2})(\varphi(\mathbf{e}_1)\varphi(\mathbf{e}_2) - \varphi(\mathbf{e}_2)\varphi(\mathbf{e}_1)) = 0$. It follows that $\varphi_i\varphi_j + \varphi_j\varphi_i = (\frac{1}{2})((\varphi_i + \varphi_j)^2 - \varphi_i^2 - \varphi_j^2)\varepsilon\mathfrak{J}$ $(1 \leqslant i, j \leqslant m)$.

Let $\overset{*}{\varphi_i}$ be the residue class of φ_i modulo $\mathfrak{J}$. We have

$$\overset{*}{\varphi_i}\overset{*}{\varphi_j} = -\overset{*}{\varphi_j}\overset{*}{\varphi_i}, \qquad (\overset{*}{\varphi_i})^2 = 0.$$

It follows that a product $\overset{*}{\varphi}_{i_1}, \overset{*}{\varphi}_{i_2} \cdots \overset{*}{\varphi}_{i_r}$ does not change if we perform an even permutation of the factors, and is changed into its *negative* if we perform an odd permutation. Moreover, it is equal to 0 if any two factors are equal. This certainly happens if $r > m$.

We have seen that the elements $\varphi_{i_1}\varphi_{i_2} \cdots \varphi_{i_r}$ form a base of $\mathfrak{H}_r$ if $r > 0$.[1] It follows that $\mathfrak{H}_r \subset \mathfrak{J}$ if $r > m$, and that every element of $\mathfrak{O}/\mathfrak{J}$ is a linear combination of the elements

$$E^*; \overset{*}{\varphi}_{i_1}\overset{*}{\varphi}_{i_2} \cdots \overset{*}{\varphi}_{i_r} \qquad \text{with } i_1 < \cdots < i_r \leqslant m,$$

where E^* is the residue class of E. There being 2^m such elements, $\mathfrak{O}/\mathfrak{J}$ is a vector space of dimension at most equal to 2^m. We shall see a little later that the dimension of $\mathfrak{O}/\mathfrak{J}$ is exactly 2^m.

For the moment, we shall exhibit a complete system of representatives of the residue classes of $\mathfrak{O}$ modulo $\mathfrak{J}$.

[1] Cf. Proposition 1, §I, p. 139.

Definition 1. *An r-linear function Λ_r is said to be alternate if*

$$\Lambda_r^{\bar{\omega}} = \epsilon(\bar{\omega})\Lambda_r$$

for every permutation $\bar{\omega}$ of the set $\{1, \cdots, r\}$.

(If $r = 0$ or 1, any element of $\mathfrak{H}_r$ is considered to be alternate.)

Similarly, an element $\tilde{\Lambda} = (\Lambda_0, \cdots, \Lambda_r, \cdots)$ of $\mathfrak{D}$ is said to be alternate if every coordinate Λ_r is alternate.

If $\tilde{\Lambda}$ is an arbitrary element of $\mathfrak{D}$, $A(\tilde{\Lambda})$ is alternate. In fact, we have, for $r > 0$, $\Lambda_r \varepsilon \mathfrak{H}_r$,

$$(A_r(\Lambda_r))^{\bar{\omega}_0} = \frac{1}{r!}\Sigma\epsilon(\bar{\omega})\Lambda_r^{\bar{\omega}_0\bar{\omega}} = \frac{\epsilon(\bar{\omega}_0)^{-1}}{r!}\Sigma\epsilon(\bar{\omega}_0\bar{\omega})\Lambda_r^{\bar{\omega}_0\bar{\omega}} = \epsilon(\bar{\omega}_0)A_r(\Lambda_r)$$

where $\bar{\omega}_0$ is any permutation of the set $\{1, \cdots, r\}$.

Moreover, if $\tilde{\Lambda}$ is already alternate, we have $A(\tilde{\Lambda}) = \tilde{\Lambda}$. In fact, if $r > 0$, and if Λ_r is alternate, we have

$$A_r(\Lambda_r) = \frac{1}{r!}\Sigma\epsilon^2(\bar{\omega})\Lambda_r = \Lambda_r$$

It follows that the operation A is idempotent: $AA = A$.

Proposition 1. *Any element $\tilde{\Lambda}\varepsilon\mathfrak{D}$ can be represented in one and only one way as a sum, $\tilde{\mathsf{M}} + \tilde{\mathsf{N}}$, of an alternate element $\tilde{\mathsf{M}}$ and an element $\tilde{\mathsf{N}}\epsilon\mathfrak{J}$.*

In fact, we have $\tilde{\Lambda} = A(\tilde{\Lambda}) + (\tilde{\Lambda} - A(\tilde{\Lambda}))$; $A(\tilde{\Lambda})$ is alternate, and we have $A(\tilde{\Lambda} - A(\tilde{\Lambda})) = A(\tilde{\Lambda}) - AA(\tilde{\Lambda}) = 0$, which shows that $\tilde{\Lambda} - A(\tilde{\Lambda})\varepsilon\mathfrak{J}$. Conversely, suppose that $\tilde{\Lambda} = \tilde{\mathsf{M}} + \tilde{\mathsf{N}}$, with $\tilde{\mathsf{M}}$ alternate and $\tilde{\mathsf{N}}\varepsilon\mathfrak{J}$. We have $A(\tilde{\Lambda}) = A(\tilde{\mathsf{M}}) + A(\tilde{\mathsf{N}}) = A(\tilde{\mathsf{M}}) = \tilde{\mathsf{M}}$, whence $\tilde{\mathsf{N}} = \tilde{\Lambda} - A(\tilde{\Lambda})$.

Let $\mathfrak{A}$ be the set of alternate elements in $\mathfrak{D}$. Then $\mathfrak{A}$ is obviously a vector space over K, and Proposition 1 shows that there exists one and only one element of $\mathfrak{A}$ in any given residue class of $\mathfrak{D}$ modulo $\mathfrak{J}$.

If $i_1, \cdots, i_r$ are indices with $1 \leqslant i_1 < \cdots < i_r \leqslant m$, then the element $A_r(\varphi_{i_1} \cdots \varphi_{i_r})$ is alternate. It is equal to

$$\frac{1}{r!}\Sigma_{\bar{\omega}}\epsilon(\bar{\omega})\varphi_{\bar{\omega}(i_1)}\varphi_{\bar{\omega}(i_2)} \cdots \varphi_{\bar{\omega}(i_r)}$$

where the sum is extended over all permutations $\bar{\omega}$ of the set $I = \{i_1, \cdots, i_r\}$. It follows from this expression that the elements $A_r(\varphi_{i_1} \cdots \varphi_{i_r})$, corresponding to the various subsets I of r elements from the set $\{1, \cdots, m\}$, are linearly independent. Since the residue class of $A_r(\varphi_{i_1} \cdots \varphi_{i_r})$ modulo $\mathfrak{J}$ is the same as the residue class of $\varphi_{i_1} \cdots \varphi_{i_r}$ (i.e. $\overset{*}{\varphi}_{i_1} \cdots \overset{*}{\varphi}_{i_r}$), it follows immediately that the $\binom{m}{r}$ elements $\overset{*}{\varphi}_{i_1} \cdots \overset{*}{\varphi}_{i_r}$ (for $1 \leqslant i_1 < \cdots < i_r = m$) are linearly independent in $\mathfrak{O}/\mathfrak{J}$.

The product $\tilde{\Lambda}\tilde{M}$ of two alternate elements of is in general *not* an alternate element, as can be verified easily from examples. However, there exists one and only one alternate element which belongs to the same residue class as $\tilde{\Lambda}\tilde{M}$ modulo $\mathfrak{J}$, namely the element $A(\tilde{\Lambda}\tilde{M})$. Therefore we may define a law of composition in $\mathfrak{A}$ by the formula

$$\tilde{\Lambda} \square \tilde{M} = A(\tilde{\Lambda}\tilde{M})$$

We shall call this law of composition the *Grassmann multiplication*. It is clear that the vector space $\mathfrak{A}$, equipped with this law of composition, becomes an algebra over K, isomorphic with $\mathfrak{O}/\mathfrak{J}$.

Definition 2. *The algebra composed of the alternate multilinear functions, with the Grassmann multiplication as the law of composition, is called the Grassmann algebra of the space* $\mathfrak{M}$.

It follows from our previous considerations that the Grassmann algebra is of dimension $\mathbf{2}^m$ over K. It contains a unit element E, and contains the dual space $\mathfrak{M}'$ of $\mathfrak{M}$. Moreover, if $\{\varphi_1, \cdots, \varphi_m\}$ is a base of $\mathfrak{M}'$, the elements $\varphi_1, \cdots, \varphi_m$ form a set of generators of the Grassmann algebra. We have

$$\varphi_i \square \varphi_i = 0$$
$$\varphi_i \square \varphi_j + \varphi_j \square \varphi_i = 0, \qquad (1 \leqslant i, j \leqslant m)$$

and the elements $\varphi_{i_1} \square \varphi_{i_2} \square \cdots \square \varphi_{i_r}$ corresponding to the various subsets $\{i_1, \cdots, i_r\}$ of the set $\{1, \cdots, m\}$ (with $i_1 < \cdots < i_r \leqslant m$) are linearly independent; every element of the Grassmann algebra may be written as a linear combination of E and of such elements. An element of the Grassmann algebra which is an alternate r-linear function is said to be *homogeneous of order* r (if $r = 0$, i.e. the element is in K, we adopt the convention of calling it a 0-linear form).

APPLICATION. Let $\psi_1, \psi_2, \cdots, \psi_r$ be any r elements of $\mathfrak{M}'$. If we take r linear combinations

$$\Theta_i = \Sigma_{j=1}^{r} a_{ij}\psi_j \qquad (1 \leqslant i \leqslant r),$$

of these elements, we have

$$\Theta_1 \square \Theta_2 \square \cdots \square \Theta_r = \boxed{a_{ij}}\,\psi_1 \square \psi_2 \square \cdots \square \psi_r$$

as can easily be verified.

If $r = m$, and if $\psi_1, \cdots, \psi_m$ are linearly independent, they form a base of $\mathfrak{M}'$, and we know that $\psi_1 \square \psi_2 \square \cdots \square \psi_m \neq 0$. If now r is arbitrary and if $\psi_1, \cdots, \psi_r$ are linearly independent, we can find $m - r$ elements $\psi_{r+1}, \cdots, \psi_m$ of $\mathfrak{M}'$ such that $\psi_1, \cdots, \psi_r, \psi_{r+1}, \cdots, \psi_m$ are linearly independent. We have $\psi_1 \square \psi_2 \square \cdots \square \psi_r \neq 0$, which proves:

Proposition 2. *If $\psi_1, \cdots, \psi_r$ are r elements of $\mathfrak{M}'$, a necessary and sufficient condition for their linear independence is that $\psi_1 \square \psi_2 \square \cdots \square \psi_r \neq 0$. Moreover, if we replace these elements by linear combinations of them, their product in the Grassmann algebra is multiplied by an element of K.*

If $\mathfrak{N}$ is an $(n - r)$-dimensional subspace of $\mathfrak{M}$, the elements $\mathbf{e}$ of $\mathfrak{N}$ may be characterized as those which satisfy r linear equations,

$$(1) \qquad \psi_1(\mathbf{e}) = \cdots = \psi_r(\mathbf{e}) = 0,$$

where $\psi_1, \cdots, \psi_r$ are linearly independent elements of $\mathfrak{M}'$. Moreover, if the formulas (1) define $\mathfrak{N}$, any other set of r equations of $\mathfrak{N}$ is obtained by replacing $\psi_1, \cdots, \psi_r$ by r linearly independent linear combinations of them. Therefore the subspace $\mathfrak{N}$ may be characterized by the product $\psi_1 \square \psi_2 \square \cdots \square \psi_r$, and this product, conversely, is determined by $\mathfrak{N}$, except for a constant factor.[1]

§III. The Differential Forms of Cartan

Definition 1. *Let $\mathcal{V}$ be a manifold and let p be a point of $\mathcal{V}$. The Grassmann algebra associated with the tangent space to $\mathcal{V}$ at p is called the Cartan differential algebra at p.*

We shall denote this algebra by $\mathfrak{C}_p$.

Definition 2. *If we assign to every point p of a subset A of $\mathcal{V}$ a homogeneous element of order r in $\mathfrak{C}_p$, we obtain what is called a differential form[2] of order r defined on A. A differential form of order 1 is also called a Pfaffian form.*

A differential form of order 0 is therefore simply a real valued function, and we know what it means to say that such a function is analytic at a point p of the domain of definition. We shall now extend this notion to a differential form of any order.

An element of order 1 in $\mathfrak{C}_p$ is an element of the dual space to the tangent space $\mathfrak{L}_p$. But we have already seen that this dual space

[1] It was precisely in order to handle analytically the linear varieties of any dimension that Grassmann developed his "geometrical calculus."

[2] Also called "exterior differential form."

is the space $\mathfrak{D}_p$ of differentials at p.[1] Let $\{x_1, \cdots, x_n\}$ be a coordinate system at p; then, for every point q of a neighbourhood of p, the differentials $(dx_1)_q, \cdots, (dx_n)_q$ form a base of $\mathfrak{D}_q$. Let θ be a differential form of order r, defined in a neighbourhood of p; we may express the value θ_q of θ at a point q in the form

$$(1) \qquad \theta_q = \Sigma u_{i_1} \cdots {}_{i_r}(q)(dx_{i_1})_q \,\square \cdots \square\, (dx_{i_r})_q,$$

this summation being extended over all combinations $\{i_1, \cdots, i_r\}$ such that $1 \leqslant i_1 < \cdots < i_r \leqslant n$.

We shall say that the form θ is *analytic at* p if the functions $u_{i_1 \ldots i_r}$ are all analytic at p. To justify this definition, we have to show that it does not depend on the particular coordinate system used. Let $\{x'_1, \cdots, x'_n\}$ be some other coordinate system at p. We may express $x_1, \cdots, x_n$ in the neighbourhood of p as functions $f_1(x'_1, \cdots, x'_n), \cdots, f_n(x'_1, \cdots, x'_n)$ of the new coordinates x', and these functions are analytic at the point $x'_1 = x'_1(p), \cdots, x'_n = x'_n(p)$. We have, for q sufficiently near p,

$$(dx_i)_q = \Sigma_{j=1}^{n} \left(\frac{\partial f_i}{\partial x'_j}\right)_q (dx'_j)_q,$$

where $\left(\frac{\partial f_i}{\partial x'_j}\right)_q$ is the value of $\frac{\partial f_i}{\partial x'_j}$ for $x'_j = x'_j(q)$ $(1 \leqslant j \leqslant n)$. Hence

$$\theta_q = \Sigma u_{i_1 \ldots i_r}(q) \left(\frac{D(f_{i_1}, f_{i_2}, \cdots, f_{i_r})}{D(x'_{j_1}, x'_{j_2}, \cdots, x'_{j_r})}\right)_q (dx'_{j_1})_q \,\square \cdots \square\, (dx'_{j_r})_q$$

where the summation is extended over all systems $(i_1, \cdots, i_r; j_1, \cdots, j_r)$ such that $i_1 < \cdots < i_r$, $j_1 < \cdots < j_r$. If we set

$$u'_{j_1 j_2 \ldots j_r}(q) = \Sigma_{i_1, \ldots i_r} u_{i_1 \ldots i_r}(q) \left(\frac{D(f_{i_1}, f_{i_2}, \cdots, f_{i_r})}{D(x'_{j_1}, x'_{j_2}, \cdots, x'_{j_r})}\right)_q$$

we have

$$\theta(q) = \Sigma_{j_1 \ldots j_r} u'_{j_1 \ldots j_r}(q)(dx'_{j_1})_q \,\square \cdots \square\, (dx'_{j_r})_q.$$

If the functions $u_{i_1 \ldots i_r}$ are analytic at p, then the same is of course true for the functions $u'_{j_1 \ldots j_r}$, which justifies our definition of analytic differential forms.

Similarly, if the functions $u_{j_1 \ldots i_r}$ are continuous at p, the functions $u'_{j_1 \ldots j_r}$ are also continuous. In this case we shall say that θ is *continuous* at p.

[1] Cf. Chapter III, §IV, p. 76.

We shall now define the operation of *differentiation* on differential forms. Let θ be a differential form of order r, which is analytic at a point $p \varepsilon \mathcal{V}$. If $r = 0$, θ is a function on $\mathcal{V}$, and its differential at p has already been defined. In the general case, we define the *differential*, $(d\theta)_p$, of θ at p to be the element of $\mathfrak{C}_p$ defined by

$$(d\theta)_p = \Sigma(du_{i_1 \ldots i_r})_p \,\square\, (dx_{i_1})_p \,\square\, \cdots \,\square\, (dx_{i_r})_p.$$

Here again we have to show our definition is independent of the coordinate system.

Before doing this, we shall first prove a certain number of properties of the differentiation operation defined by (2) with respect to the special system of coordinates $\{x_1, \cdots x_n\}$.

If θ_1, θ_2 are forms of order r, and a_1, a_2 are real numbers, we have

$$(2) \qquad (d(a_1\theta_1 + a_2\theta_2))_p = a_1(d\theta_1)_p + a_2(d\theta_2)_p.$$

Now, suppose that θ, η are differential forms of orders r, s, both analytic at p. They are both defined in some neighbourhood of p and if we assign to every point q of this neighbourhood the element $\theta_q \,\square\, \eta_q$ we obtain a differential form $\theta \,\square\, \eta$ which is clearly analytic at p. We assert that its differential at p is

$$(3) \qquad (d(\theta \,\square\, \eta))_p = (d\theta)_p \,\square\, \eta_p + (-1)^r\theta_p \,\square\, (d\eta)_p$$

Suppose first that $r > 0$, $s > 0$. Making use of (3) we see that it will be sufficient to prove the formula in the case where θ, η are given by formulas of the type

$$\theta(q) = u(q)(dx_{i_1})_q \,\square\, \cdots \,\square\, (dx_{i_r})_q,$$
$$\eta(q) = v(q)(dx_{j_1})_q \,\square\, \cdots \,\square\, (dx_{j_s})_q,$$

with $i_1 < \cdots < i_r$, $j_1 < \cdots < j_s$, u and v being analytic functions at p.

If the sets $\{i_1, \cdots, i_r\}$, $\{j_1, \cdots, j_s\}$ have an element in common we have $\theta \,\square\, \eta = 0$, $(d\theta)_p \,\square\, \eta_p = 0$, $\theta_p \,\square\, (d\eta)_p = 0$, and formula (3) is proved. If not, let $k_1, \cdots, k_{r+s}$ be the elements of the set $\{i_1, \cdots i_r, j_1, \cdots j_s\}$, arranged in ascending order. We have

$$(\theta \,\square\, \eta)_q = \epsilon u(q)v(q)(dx_{k_1})_q \,\square\, \cdots \,\square\, (dx_{k_{r+s}})_q,$$

where ϵ is $+1$ or -1 according as the permutation

$$\begin{pmatrix} i_1 & \cdots & i_r j_1 & \cdots & j_s \\ k_1 & \cdots & k_r k_{r+1} & \cdots & k_{r+s} \end{pmatrix}$$

is even or odd.   Hence

$$(d(\theta \,\square\, \eta))_p = \epsilon((du)_p v(p) + u(p)(dv)_p) \,\square\, (dx_{k_1})_p \,\square\, \cdots \,\square\, (dx_{k_{r+s}})_p$$
$$= d\theta_p \,\square\, \eta_p + u(p)dv_p \,\square\, (dx_{i_1})_p \,\square\, \cdots \,\square\, (dx_{i_.})_p \,\square\, (dx_{j_1})_p$$
$$\square\, \cdots \,\square\, (dx_{j_.})_p$$
$$= d\theta_p \,\square\, \eta_p + (-1)^r \theta_p \,\square\, d\eta_p$$

since $dv_p \,\square\, (dx_i)_p = -(dx_i)_p \,\square\, dv_p$.

If $s = 0$, $r > 0$, we have $\eta_q = v(q)$, where v is analytic at p, and we may assume that θ is given by the same formula as above.   We have

$$(d(\theta \,\square\, \eta))_p = (du_p v(p) + u(p)dv_p) \,\square\, (dx_{i_1})_p \,\square\, \cdots \,(dx_{i_n})_p$$
$$= d\theta_p \,\square\, \eta_p + (-1)^r \theta_p \,\square\, d\eta_p.$$

A similar argument proves formula (3) if $r = 0$.   Therefore our formula is established in every case.

In particular, if ω_1, ω_2 are Pfaffian forms which are analytic at p, we have

$$4) \qquad (d(\omega_1 \,\square\, \omega_2))_p = (d\omega_1)_p \,\square\, (\omega_2)_p - (\omega_1)_p \,\square\, (d\omega_2)_p$$

It follows easily that if $\omega_1, \cdots, \omega_r$ are Pfaffian forms, all analytic at p, we have

$$(5) \quad (d(\omega_1 \,\square\, \omega_2 \,\square\, \cdots \,\square\, \omega_r))_p$$
$$= \Sigma_1^r (-1)^{i-1} (\omega_1)_p \,\square\, \cdots \,\square\, (\omega_{i-1})_p \,\square\, (d\omega_i)_p \,\square\, (\omega_{i+1})_p \,\square\, \cdots \,\square\, (\omega_r)_p$$

Let f be any function on $\mathcal{U}$, analytic at p.   If we assign to any point q at which f is analytic the element $df_q \epsilon \mathfrak{C}_q$ we obtain a Pfaffian form df, the differential of f.   In the neighbourhood of p we may express f as a function $f^*(x_1, \cdots, x_n)$ of the coordinates x.   If q belongs to this neighbourhood we have

$$(df)_q = \Sigma_1^n \left(\frac{\partial f^*}{\partial x_i}\right)_q (dx_i)_q$$

Since the functions $\dfrac{\partial f^*}{\partial x_i}$ are analytic at the point $x_1 = x_1(q), \cdots,$

$x_n = x_n(q)$, df is analytic at p.   We have

$$(d(df))_p = \Sigma_{ij} \left(\frac{\partial^2 f^*}{\partial x_j \partial x_i}\right)_p (dx_i)_p \,\square\, (dx_j)_p.$$

If we observe that

$$\frac{\partial^2 f^*}{\partial x_j \partial x_i} = \frac{\partial^2 f^*}{\partial x_i \partial x_j} \qquad (dx_j)_p \,\square\, (dx_i)_p = -(dx_i)_p \,\square\, (dx_j)_p$$

we find

$$d(df) = 0. \tag{6}$$

By formula (5) we see that if $f_1, \cdots, f_r$ are functions analytic at p, then

$$d(df_1 \square \cdots \square df_r) = 0. \tag{7}$$

We are now able to prove our differentiation operation is independent of the coordinate system. Let $\{x'_1, \cdots, x'_n\}$ be any other system of coordinates at p, and let us denote by the symbol d' the operation of differentiation defined in terms of this new system. This operation has the same formal properties as d.

If θ is a form of order $r > 0$, expressed by formula (1), we have, by (2)

$$(d'\theta)_p = \Sigma(d'(u_{i_1 \ldots i_r} dx_{i_1} \square \cdots \square dx_{i_r}))_p.$$

By formula (3) we have

$$\begin{aligned}(d'(u_{i_1 i_2 \ldots i_r} dx_{i_1} \square dx_{i_2} \square \cdots \square dx_{i_r}))_p \\ = (d'u_{i_1 i_2 \ldots i_r})_p \square (dx_{i_1} \square dx_{i_2} \square \cdots \square dx_{i_r})_p \\ + u'_{i_1 i_2 \ldots i_r}(d'(dx_{i_1} \square dx_{i_2} \square \cdots \square dx_{i_r}))_p\end{aligned}$$

For any function f, we have $d'f = df$, by definition. Hence $dx_{i_1} \square \cdots \square dx_{i_r} = d'x_{i_1} \square \cdots \square d'x_{i_r}$, and the second term in our last formula is zero, by (7). The first term is equal to $(du_{i_1 \ldots i_r})_p (dx_{i_1})_p \square \cdots \square (dx_{i_r})_p$, which proves that $(d'\theta)_p = (d\theta)_p$.

The differentiation property expressed by (6) may be extended to any differential form: *if θ is any analytic differential form, we have*

$$d(d\theta) = 0.$$

Let θ_q be expressed by formula (1) at any point q of some neighbourhood of a point p where θ is analytic. Then

$$(d\theta)_q = \Sigma(du_{i_1 \ldots i_r})_q \square (dx_{i_1})_q \square \cdots \square (dx_{i_r})_q,$$

whence $d(d\theta) = 0$, by formulas (6) and (7).

The Effect of a Mapping

Now let $\mathcal{W}$ be another manifold and let Φ represent an analytic mapping of $\mathcal{W}$ into $\mathcal{V}$. If $q \varepsilon \mathcal{W}$ and $p = \Phi(q)$, $d\Phi_q$ is a linear mapping of the tangent space $\mathfrak{M}_q$ to $\mathcal{W}$ at q into the tangent space $\mathfrak{L}_p$ to $\mathcal{V}$ at p. Let $\mathfrak{C}_p$ and $\mathfrak{D}_q$ be the Grassmann algebras of $\mathcal{V}$ and $\mathcal{W}$ at the points

p and q respectively. We shall see that there corresponds to $d\Phi_q$ a dual mapping $\delta\Phi_q$ of $\mathfrak{C}_p$ into $\mathfrak{D}_q$.

A homogeneous element θ of order $r > 0$ in $\mathfrak{C}_p$ is an alternate r-linear form $\theta(L_1, \cdots, L_r)$ on $\mathfrak{L}_p$. Let $M_1, \cdots, M_r$ be any r elements of $\mathfrak{M}_q$; we set

$$(1) \qquad \theta_1(M_1, \cdots, M_r) = \theta(d\Phi_q M_1, \cdots, d\Phi_q M_r).$$

It is clear that θ_1 is an r-linear alternate form on $\mathfrak{M}_q$; we set

$$\delta\Phi_q\theta = \delta\Phi_q(\theta) = \theta_1$$

We obtain in this way, for every $r > 0$, a linear mapping of the set of homogeneous elements of order r in $\mathfrak{C}_p$ into the set of homogeneous elements of order r in $\mathfrak{D}_q$. If $r = 0$, a homogeneous element of order 0 in $\mathfrak{C}_p$ is a real number θ, and in this case we simply set $\delta\Phi_q\theta = \theta$. If θ is a non-homogeneous element in $\mathfrak{C}_p$, we represent θ in the form $\theta^{r_1} + \theta^{r_2} + \cdots + \theta^{r_n}$, where θ^{r_i} is homogeneous of order r_i, and we set $\delta\Phi_q\theta = \Sigma_i\delta\Phi_q\theta^{r_i}$.

Hence $\delta\phi_q$ is a linear mapping of $\mathfrak{C}_p$ into $\mathfrak{D}_q$. It is also a ring-homomorphism; i.e., we have

$$(2) \qquad \delta\Phi_q(\theta^r \square \theta^s) = (\delta\Phi_q\theta^r) \square (\delta\Phi_q\theta^s)$$

if θ^r, θ^s are homogeneous elements of orders r, s in $\mathfrak{C}_p$. In fact, we have

$$(\theta^r \square \theta^s)(L_1, L_2, \cdots, L_{r+s})$$
$$= \Sigma \frac{\epsilon(\bar{\omega})}{(r+s)!} \theta^r(L_{\bar{\omega}(1)}, \cdots, L_{\bar{\omega}(r)})\theta^s(L_{\bar{\omega}(r+1)}, \cdots, L_{\bar{\omega}(r+s)})$$

where the summation is extended over all permutations $\bar{\omega}$ of the set $\{1, \cdots, r+s\}$, and where $\epsilon(\bar{\omega})$ is $+1$ or -1 according as $\bar{\omega}$ is even or odd; (2) then follows immediately from the defining formula (1).

Let us take a coordinate system $\{x_1, \cdots, x_n\}$ at p on $\mathfrak{V}$. Then $(dx_i)_p$ $(1 \leqslant i \leqslant m)$ is a homogeneous element of order 1 of $\mathfrak{C}_p$, and we have

$$(dx_i)_p(d\Phi M_q) = (d(x_i \circ \Phi))_q M_q,$$

for every $M_q \epsilon \mathfrak{M}_q$. It follows immediately that

$$(3) \qquad \delta\Phi_q(dx_i)_p = (d(x_i \circ \Phi))_q$$

We can now let q vary on the manifold $\mathfrak{W}$. Let θ be an analytic differential form of order r on $\mathfrak{V}$. Then the assignment $q \to \delta\Phi_q\theta$ defines a differential form on $\mathfrak{W}$, which we may denote by $\delta\Phi\theta$. It follows immediately from (3) and from the analytic character of the functions $x_i \circ \phi$ that $\delta\Phi\theta$ is analytic on $\mathcal{W}$. If

$$\theta_p = \Sigma u_{i_1 \ldots i_r}(p)(dx_{i_1})_p \,\square \cdots \square\, (dx_{i_r})_p$$

is the expression of θ_p we have (simplifying our notation by writing $x_i \circ \Phi = y_i$)

$$(\delta\Phi\theta)_q = \Sigma(u_{i_1 \ldots i_r} \circ \Phi)_q(dy_{i_1})_q \,\square \cdots \square\, (dy_{i_r})_q.$$

Hence

$$(d(\delta\Phi\theta))_q = \Sigma((d(u_{i_1 \ldots i_r} \circ \Phi))_q \,\square\, (dy_{i_1})_q \,\square \cdots \square\, (dy_{i_r})_q.$$

If we observe that

$$(d(u_{i_1 \ldots i_r} \circ \Phi)_q = \delta\Phi_q(du_{i_1, i_2 \ldots i_r})_{\Phi q}$$
$$(dy_i)_q = \delta\Phi_q(dx_i)_{\Phi q}$$

we see that

$$d(\delta\Phi\theta) = \delta\Phi(d\theta). \tag{4}$$

§IV. THE FORMS OF MAURER-CARTAN

Let $\mathcal{G}$ be an analytic group. We denote by Φ_σ the left translation associated with an element $\sigma \varepsilon \mathcal{G}$ If θ is an analytic differential form on $\mathcal{G}$, the same is true of $\delta\Phi_\sigma\theta$.

Definition 1. *The form θ is said to be left-invariant if $\delta\Phi_\sigma\theta = \theta$ for all $\sigma \varepsilon \mathcal{G}$.*

If this is the case, we have $\theta_\sigma = \delta\Phi_{\sigma^{-1}}\theta_\epsilon$, which proves that θ is uniquely determined when θ_ϵ is known (ϵ being the neutral element of $\mathcal{G}$).

The left-invariant differential forms of order 0 are the constants.

Definition 2. *A left-invariant Pfaffian form is called a form of Maurer-Cartan.*

Let ω be a form of Maurer-Cartan and let X be a left-invariant infinitesimal transformation. The value ω_σ of ω at an element σ is a linear function on the tangent space to $\mathcal{G}$ at σ; therefore the symbol $\omega_\sigma(X_\sigma)$ has a meaning. We assert that $\omega_\sigma(X_\sigma)$ does not depend on σ. In fact, we have $\omega_\sigma(X_\sigma) = (\delta\Phi_{\sigma^{-1}}\omega_\epsilon)(X_\sigma) = (\omega_\epsilon)(d\Phi_{\sigma^{-1}}X_\sigma) = \omega_\epsilon(X_\epsilon)$.

Conversely, let ω_ϵ be any linear form on the tangent space at ϵ; if we set $\omega_\sigma = \delta\Phi_{\sigma^{-1}}\omega_\epsilon$, the assignment $\sigma \rightarrow \omega_\sigma$ is a Pfaffian form on $\mathcal{G}$, and we have $\omega_\sigma(X_\sigma) =$ constant, for any left-invariant infinitesimal transformation X. We have $(\delta\Phi_{\sigma_0}\omega_{\sigma_0\sigma})(X_\sigma) = \omega_{\sigma_0\sigma}(d\Phi_{\sigma_0}X_\sigma) = \omega_{\sigma_0\sigma}(X_{\sigma_0\sigma}) = \omega_\epsilon(X_\epsilon) = \omega_\sigma(X_\sigma)$, from which it follows that $\omega_\sigma = \delta\Phi_{\sigma_0}\omega_{\sigma_0\sigma}$ for any $\sigma_0 \epsilon \mathcal{G}$: ω is invariant. We shall prove that ω is also analytic. Let us take a system of coordinates $\{x_1, \cdots, x_n\}$ at an element $\sigma_0 \varepsilon \mathcal{G}$ and a base $\{X_1, \cdots, X_n\}$ for the Lie algebra of $\mathcal{G}$. If σ is sufficiently near to σ_0, we may express ω_σ in the form $\Sigma A_i(\sigma)(dx_i)_\sigma$, and we have

$$(1) \qquad \omega(X_j) = \Sigma_i A_i(\sigma)(X_j x_i)_\sigma \qquad (j = 1, \cdots, n).$$

The left hand sides of these equations are constants. Since $(X_1)_\sigma, \cdots, (X_n)_\sigma$ are linearly independent, the determinant

$$|(X_j x_i)_\sigma|$$

does not vanish, and the linear equations (1) may be solved for $A_1(\sigma), \cdots, A_n(\sigma)$. Since the functions $(X_j x_i)_\sigma$ are analytic at σ_0, the same is true of the functions $A_i(\sigma)$,—which proves the analyticity of ω.

We see that, if n is the dimension of $\mathcal{G}$, there exist exactly n linearly independent forms of Maurer-Cartan, say $\omega_1, \cdots, \omega_n$. It is clear that, if the $a_{i_1 \ldots i_r}$ are any constants, $\Sigma a_{i_1, \ldots, i_r} \omega_{i_1} \square \cdots \square \omega_{i_r}$ is a left-invariant differential form of order r, and that any left-invariant differential form of order $r > 0$ may be written in this form.

Any left-invariant differential form θ of order $r > 0$ may be considered as an r-linear alternate form on the Lie algebra $\mathfrak{g}$, of $\mathcal{G}$, by setting $\theta(Y_1, \cdots, Y_r) = \theta_\epsilon((Y_1)_\epsilon, \cdots, (Y_r)_\epsilon)$. We may therefore identify the left-invariant differential forms with the homogeneous elements of the Grassman algebra associated with $\mathfrak{g}$.

If ω is a form of Maurer-Cartan, we have $\delta\Phi_\sigma\, d\omega = d\delta\Phi_\sigma\omega = d\omega$, and $d\omega$ is also left-invariant. We shall prove that

$$(2) \qquad \boxed{d\omega(X, Y) = (\tfrac{1}{2})\omega([X, Y])},$$

where X and Y are any elements of $\mathfrak{g}$.

Using the above notation, we have $d\omega = \Sigma dA_i \square dx_i$, whence

$$\begin{aligned} d\omega(X, Y) &= (\tfrac{1}{2})\Sigma_i(dA_i(X)dx_i(Y) - dA_i(Y)dx_i(X)) \\ &= (\tfrac{1}{2})\Sigma_i((XA_i)(Yx_i) - (YA_i)(Xx_i)). \end{aligned}$$

On the other hand, we have $\Sigma A_i Y x_i = \text{constant}$, whence

$$0 = X(\Sigma A_i Y x_i) = \Sigma(XA_i)(Yx_i) + \Sigma A_i(XYx_i),$$

and similarly,

$$0 = \Sigma(YA_i)(Xx_i) + \Sigma A_i(YXx_i).$$

We may therefore write

$$d\omega(X, Y) = (\tfrac{1}{2})\Sigma A_i(YXx_i - XYx_i)$$
$$= (\tfrac{1}{2})\Sigma A_i([X, Y]x_i),$$

which proves (2).

Let $\{X_1, \cdots, X_n\}$ be a base of the Lie algebra $\mathfrak{g}$. We can find a dual base $\{\omega_1, \cdots, \omega_n\}$ for the forms of Maurer-Cartan, i.e. a base such that $\omega_i(X_j) = \delta_{ij}$ $(1 \leqslant i, j \leqslant n)$. We have

$$[X_i, X_j] = \Sigma_k c_{ijk} X_k$$

where the c_{ijk} are the constants of structure. It follows from (2) that $d\omega_k(X_i, X_j) = (\tfrac{1}{2})c_{ijk}$. Taking into account the equalities $c_{ijk} + c_{jik} = 0$, it follows that

$$\boxed{d\omega_k = (\tfrac{1}{2})\Sigma_{i,j} c_{ijk}\omega_i \,\Box\, \omega_j}. \tag{3}$$

Let $\{x_1, \cdots, x_n\}$ be a system of coordinates on $\mathcal{G}$ at the neutral element ϵ, and let V be a cubic neighbourhood of ϵ with respect to this system. If $\sigma \varepsilon V$, we can write $(\omega_i)_\sigma$ in the form

$$(\omega_i)_\sigma = \Sigma_{j=1}^n A_{ij}(x_1(\sigma), \cdots, x_n(\sigma))(dx_j)_\sigma \quad . (i = 1, \cdots, n),$$

where the function $A_{ij}(x_1, \cdots, x_n)$ are defined and analytic in the domain defined by the inequalities $|x_i - x_i(\epsilon)| < a$, a being the breadth of V. We set

$$\omega_i(x, dx) = \Sigma_j A_{ij}(x)dx_j.$$

If $\sigma_0 \epsilon V$, the functions $x_i(\sigma_0\sigma) = y_i(\sigma)$ are defined and analytic in a neighbourhood of ϵ. The left-invariance of ω_i gives the relations

$$\Sigma_j A_{ij}(y_1(\sigma), y_2(\sigma), \cdots, y_n(\sigma))(dy_j)_\sigma$$
$$= \Sigma_j A_{ij}(x_1(\sigma), \cdots, x_n(\sigma))(dx_j)_\sigma$$

The functions $y_i(\sigma)$ may be expressed as functions $y_i(x_1(\sigma), \cdots, x_n(\sigma))$ of the x-coordinates of σ (these functions being defined and analytic provided the quantities $|x_i(\sigma) - x_i(\epsilon)|$ are sufficiently small $(1 \leqslant i \leqslant n)$), and the functions $y_1(x_1, \cdots, x_n), \cdots, y_r(x_1, \cdots, x_n)$ satisfy the equations

$$\boxed{\omega_i(y, dy) = \omega_i(x, dx)}, \tag{4}$$

which are called the *equations of Maurer-Cartan.*

The determinant $\boxed{A_{ij}(x_1, \cdots, x_n)}$ does not vanish for $|x_i - x_i(\epsilon)| < a$. Therefore the equations (4) yield expressions

$$(5) \qquad \frac{\partial y_i}{\partial x_j} = F_{ij}(y_1, \cdots, y_n; x_1, \cdots, x_n) \qquad (1 \leqslant i, j \leqslant n),$$

for the partial derivatives $\frac{\partial y_i}{\partial x_j}$ as functions of the y's and x's.

On the other hand, we have

$$(6) \qquad y_i(x_1(\epsilon), \cdots, x_n(\epsilon)) = x_i(\sigma_0) \qquad (i = 1, \cdots, n).$$

Therefore, when the expressions $\omega_i(x, dx)$ of the forms of Maurer-Cartan are known, the problem of determining the functions $x_i(\sigma_0\sigma)$ is reduced to the integration of the equations (5) with the initial conditions (6). The problem depends itself on the integration of systems of ordinary differential equations.

§V. Explicit Construction of the Forms of Maurer-Cartan in Canonical Coordinates

Let $\mathcal{G}$ be an analytic group, $\mathfrak{g}$ its Lie algebra and $\{X_1, \cdots, X_n\}$ a base of $\mathfrak{g}$. There corresponds to this base a canonical coordinate system[1] $\{x_1, \cdots, x_n\}$ at the neutral element ϵ of $\mathcal{G}$. Let $\{\omega_1, \cdots, \omega_n\}$ be the base of the forms of Maurer-Cartan defined by $\omega_i(X_j) = \delta_{ij}$. We want to determine the expressions

$$\omega_i(x, dx) = \Sigma_{j=1}^n A_{ij}(x_1, \cdots, x_n)dx_j$$

of the forms ω_i in terms of the coordinates x.

Let us observe first that the mapping $(x_1, \cdots, x_n) \rightarrow \exp \Sigma x_i X_i$ is an analytic mapping of the whole of R^n into $\mathcal{G}$. We may denote this mapping by the notation "exp." The forms $\Sigma A_{ij}(x)dx_j$ are the forms $(\delta \exp)\omega_i$. It follows that the functions $A_{ij}(x_1, \cdots, x_n)$ are defined and analytic over the whole of R^n.

To every element $X \varepsilon \mathfrak{g}$ there corresponds an analytic homomorphism Θ_X of the additive group R of real numbers into $\mathcal{G}$, and $\delta\Theta_X\omega_i$ is an analytic Pfaffian form on R. We denote by t the coordinate on R, and by L the left-invariant infinitesimal transformation of R defined by $L(t) = 1$. Then $d\Theta_X(L_t) = X_{\Theta_X(t)}$, whence $(\delta\Theta_X\omega_i)L = \omega_i(X) = a_i$ if $X = \Sigma a_i X_i$. It follows that

$$\delta\Theta_X\omega_i = a_i dt$$

Let Θ_X^* be the mapping, $t \rightarrow (a_1 t, \cdots, a_n t)$ of R into R^n. We have $\Theta_X(t) = \exp \Theta_X^*(t)$, whence

$$\delta\Theta_X^*(\omega_i(x, dx)) = a_i dt,$$

[1] Cf. Chapter IV, §VIII, p. 115.

which gives the formula

$$\Sigma_{j=1}^{n} A_{ij}(a_1 t, \cdots, a_n t) a_j dt = a_i dt$$

or

$$(1) \qquad \Sigma_j A_{ij}(x_1 t, \cdots, x_n t) x_j = x_i \qquad (i = 1, \cdots, n).$$

We now introduce the mapping, $(t, x'_1, \cdots, x'_n) \rightarrow (tx'_1, \cdots, tx'_n)$ of R^{n+1} into R^n. Under this mapping, there corresponds to $\omega_i(x, dx)$ an analytic Pfaffian form $\omega'_1(x', t, dx', dt)$ on R^{n+1} whose expression is

$$\begin{aligned} \omega'_i(x', t, dx', dt) &= \Sigma_j A_{ij}(x'_1 t, \cdots, x'_n t)(x'_j dt + t\, dx'_j) \\ &= t \Sigma_j A_{ij}(x' t, \cdots, x' t) dx'_j + x'_i dt, \end{aligned}$$

(making use of formulas (1)).

Since $d\omega_k = (\frac{1}{2}) \Sigma_{i,j} c_{ijk} \omega_i \square \omega_j$, we also have $d\omega'_k = (\frac{1}{2}) \Sigma_{i,j} c_{ijk} \omega'_i \square \omega'_j$. In order to abbreviate, we set $A_{ij}(x'_i t, \cdots, x'_n t) = A_{ij}(x' t)$; then we have

$$d\omega'_k = \Sigma_l A_{kl}(x't) dt \square dx'_l + \Sigma_l t \frac{\partial A_{kl}}{\partial t} (x't) dt \square dx'_l + dx'_k \square dt + \cdots$$

$$\Sigma_{ij} c_{ijk} \omega'_i \square \omega'_j = \Sigma_{ijl} t c_{ijk} A_{il}(x't) x'_j dx'_l \square dt + \Sigma_{ijl} t c_{ijk} A_{jl}(x't) x'_i dt \square dx'_l + \cdots$$

where the terms which are not written do not contain dt. Therefore the identification of the terms which contain dt gives

$$A_{kl}(x't) + t \frac{\partial A_{kl}(x't)}{\partial t} = \frac{t}{2} \Sigma_{ij} c_{ijk} (A_{jl}(x't) x'_i - A_{il}(x't) x'_j) + \delta_{kl}$$

or, since $c_{ijk} = -c_{jik}$,

$$\frac{\partial}{\partial t} (t A_{kl}(x't)) = \delta_{kl} + \Sigma_{ij} c_{ijk} x'_i (t A_{jl}(x't))$$

Let us consider $x'_i, \cdots, x'_n$ as fixed quantities. We denote the matrix $(tA_{kl}(x't))$ by $\mathfrak{A}(t)$, and we denote the matrix whose coefficient in the k-th row and j-th column is $\Sigma_i c_{ijk} x'_i$ by $\mathfrak{X}'$. Then we have

$$\frac{d\mathfrak{A}}{dt} = E + \mathfrak{X}'\mathfrak{A}$$

where E is the unit matrix. Moreover, we have $\mathfrak{A}(0) = 0$.

By the same argument which was used to prove the convergence of the series which represents the exponential of a matrix (cf. Chapter I,

§II, p. 5), we see that the series

$$\Sigma_1^\infty \frac{t^m}{m!} \mathfrak{X}'^{m-1}$$

converges uniformly for t in any bounded interval. If $\mathfrak{A}'(t)$ is its sum, we have, $\mathfrak{A}'(0) = 0$, $\dfrac{d\mathfrak{A}'(t)}{dt} = E + \mathfrak{X}'\mathfrak{A}'$; therefore $\mathfrak{A}'(t) = \mathfrak{A}(t)$.

Putting $t = 1$, we obtain the following result:

Proposition 1. *Let $\mathfrak{G}$ be an analytic group, and let $\{X_1, \cdots, X_n\}$ be a base for its Lie algebra. Let $\{x_1, \cdots, x_n\}$ be the corresponding canonical system of coordinates, and let $\omega_1, \omega_2, \cdots, \omega_n$ be the forms of Maurer-Cartan defined by the formulas $\omega_i(X_j) = \delta_{ij}$. If $\omega_i(x, dx) = \Sigma_{j=1}^n A_{ij}dx_j$ is the expression of ω_i in terms of the coordinates x, the matrix $\mathfrak{A} = (A_{ij})$ is given by the formula*

$$\mathfrak{A} = \Sigma_1^\infty \frac{1}{m!} \mathfrak{X}^{m-1}$$

where $\mathfrak{X}$ is the matrix whose (k, j)-coefficient is $\Sigma_i c_{ijk}x_i$

Remark 1. The series which gives the matrix $\mathfrak{A}$ converges for all real or complex values of the numbers x_i, c_{ijk} and the convergence is uniform for $|x_i|$, $|c_{ijk}|$ restricted to any bounded region.

It follows in particular that the functions $A_{ij}(x_1, \cdots, x_n)$ can be extended to integral monogenic functions of the complex variables $x_1, \cdots, x_n$.

Remark 2. If we set $X = \Sigma x_i X_i$, we have

$$[X_j, X] = \Sigma_i c_{jik} x_i X_k = -\Sigma_i c_{ijk} x_i X_k.$$

The mapping, $Y \to [Y, X]$ is a linear mapping of $\mathfrak{g}$ into itself. Making use of the base $\{X_1, \cdots, X_n\}$, we may represent this mapping by a matrix, and we see that this matrix is $-{}^t\mathfrak{X}$ where ${}^t\mathfrak{X}$ is the transpose of $\mathfrak{X}$.

Example. Let us consider the Lie algebra of order 3 with the law of composition defined by

$$(1) \qquad [X_1\ X_2] = 0, \qquad [X_3, X_2] = X_1, \qquad [X_3, X_1] = 0.$$

Here the matrix $\mathfrak{X}$ is

$$\mathfrak{X} = \begin{pmatrix} 0 & x_3 & -x_2 \\ 0 & 0 & 0 \\ 0 & 0 & 0 \end{pmatrix}$$

whence $\mathfrak{X}^2 = 0$, and

$$\mathfrak{A} = \begin{pmatrix} 1 & (\frac{1}{2})x_3 & -(\frac{1}{2})x_2 \\ 0 & 1 & 0 \\ 0 & 0 & 1 \end{pmatrix}$$

We have $\omega_1 = dx_1 + (\frac{1}{2})x_3dx_2 - (\frac{1}{2})x_2dx_3$, $\omega_2 = dx_2$, $\omega_3 = dx_3$. The equations of Maurer-Cartan are

$$\begin{aligned} dx_1 + (\tfrac{1}{2})x_3dx_2 - (\tfrac{1}{2})x_2dx_3 &= dy_1 + (\tfrac{1}{2})y_2dy_3 - (\tfrac{1}{2})y_3dy_2, \\ dx_2 &= dy_2 \\ dx_3 &= dy_3, \end{aligned}$$

and the law of composition in the group is therefore

$$\begin{aligned} x_1(\sigma\tau) &= x_1(\tau) + (\tfrac{1}{2})(x_2(\sigma)x_3(\tau) - x_3(\sigma)x_2(\tau)), \\ x_2(\sigma\tau) &= x_2(\sigma) + x_2(\tau), \\ x_3(\sigma\tau) &= x_3(\sigma) + x_3(\tau). \end{aligned}$$

It is easy to verify directly that these formulas define a group whose manifold is R^3. This proves the existence of an analytic group whose Lie algebra is the algebra defined by formulas (1).

§VI. ORIENTED MANIFOLDS

Let $\mathfrak{L}$ be a vector space of dimension n over the field R of real numbers. We know that the space $\mathfrak{H}_n$ of alternate n-linear functions on $\mathfrak{L}$ is of dimension 1 over R. If B and B' are two elements of this space, with $B \neq 0$, $B' \neq 0$, we have $B' = aB$, a being a real number $\neq 0$. It follows that the elements $B \neq 0$ in $\mathfrak{H}_n$ fall into two classes, defined in the following ways: B, and $B' = aB$, belong to the same class if $a > 0$, to opposite classes if $a < 0$.

The complex notion formed by giving $\mathfrak{L}$ and one of these two classes is called an *oriented vector space*. The n-linear functions of the class which has been selected will be called the *positive* n-linear functions on the oriented vector space.

Let $(L_1, \cdots, L_n)$ be an element of the product $\mathfrak{L}^n = \mathfrak{L} \times \mathfrak{L} \times \cdots \times \mathfrak{L}$ (i.e., a mapping of the set $\{1, \cdots, n\}$ into $\mathfrak{L}$). If the set $\{L_1, \cdots, L_n\}$ is a base of $\mathfrak{L}$ we shall say that the finite sequence $(L_1, \cdots, L_n)$ is an *ordered base;* every base is thus represented in $n!$ different ways as the set of elements of an ordered base.

If B is an element $\neq 0$ in $\mathfrak{H}_n$, and if $(L_1, \cdots, L_n)$ is an ordered base, we have $B(L_1, \cdots, L_n) \neq 0$. The latter number may be positive or negative; but, if $B' = aB$ $(a > 0)$ is an element of $\mathfrak{H}_n$ belonging to the same class as B, $B'(L_1, \cdots, L_n)$ will have the same sign as $B(L_1, \cdots, L_n)$.

By an *oriented vector space* $\mathfrak{L}$ we mean a pair $\tilde{\mathfrak{L}} = (\mathfrak{L}, \mathfrak{K})$ formed by a vector space $\mathfrak{L}$ over the field of real numbers and by one of the classes, $\mathfrak{K}$, of non vanishing n-linear forms on $\mathfrak{L}$ (where $n = \dim \mathfrak{L}$). The space $\mathfrak{L}$ is called the underlying vector space of $\tilde{\mathfrak{L}}$. The n-linear functions belonging to $\mathfrak{K}$ are called the *positive n-linear functions* on $\tilde{\mathfrak{L}}$. An ordered base $(L_1, \cdots, L_n)$ of $\mathfrak{L}$ is called an *ordered base* of $\tilde{\mathfrak{L}}$ if and only if we have $B(L_1, \cdots, L_n) > 0$ for every $B \varepsilon \mathfrak{K}$.

A given vector space $\mathfrak{L}$ over the field of real numbers is the underlying vector space of exactly two oriented vector spaces $\tilde{\mathfrak{L}}_1$ and $\tilde{\mathfrak{L}}_2$. We shall say that $\tilde{\mathfrak{L}}_1$ and $\tilde{\mathfrak{L}}_2$ are *oppositely oriented.* If $(L_1, \cdots, L_n)$ is an oriented base of $\tilde{\mathfrak{L}}_1$, the same is true of every oriented base of $\mathfrak{L}$ which is deduced from $(L_1, \cdots, L_n)$ by an even permutation of the basic elements; if, on the contrary, we perform an odd permutation on $L_1, \cdots, L_n$, then we obtain an oriented base of $\tilde{\mathfrak{L}}_2$.

Now, let $\mathcal{V}$ be a manifold of dimension n. If p is a point of $\mathcal{V}$ we shall denote by $\mathfrak{L}_p$ the tangent space to $\mathcal{V}$ at p. Suppose that we have given a law which assigns to every point $p \varepsilon \mathcal{V}$ one, say $\tilde{\mathfrak{L}}_p$, of the two oriented vector spaces which admit $\mathfrak{L}_p$ as their underlying vector space. Assume furthermore that the following condition is satisfied: φ being any continuous differential form of order n on $\mathcal{V}$, if φ_p is a positive n-linear function on $\tilde{\mathfrak{L}}_p$, then φ_p is also positive on $\tilde{\mathfrak{L}}_q$ for all points q of some neighbourhood of p. Then we shall say that the pair formed by the manifold $\mathcal{V}$ and by the law $p \rightarrow \tilde{\mathfrak{L}}_p$ is an *oriented manifold* of dimension n. The manifold $\mathcal{V}$ is called the underlying manifold of the oriented manifold. The oriented vector space $\tilde{\mathfrak{L}}_p$ is called the *oriented tangent space* to the oriented manifold at the point p.

Let $\tilde{\mathcal{V}}$ be an oriented manifold, and let $\mathcal{V}$ be the underlying manifold of $\tilde{\mathcal{V}}$. By an ordered system of coordinates at a point p of $\mathcal{V}$ we understand a finite sequence $(x_1, \cdots, x_n)$ of functions such that the set $\{x_1, \cdots, x_n\}$ is a system of coordinates at p. If the n-linear form $dx_1 \square \cdots \square dx_n$ is positive on the oriented tangent space to $\tilde{\mathcal{V}}$ at p, then we say that $(x_1, \cdots, x_n)$ is an ordered system of coordinates at p on $\tilde{\mathcal{V}}$. If this is the case, $(x_1, \cdots, x_n)$ is also an ordered system of coordinates on $\tilde{\mathcal{V}}$ at every point of some neighbourhood of p.

Not every manifold is the underlying manifold of an oriented manifold; for instance, it can be shown that the projective plane is not. A manifold which is the underlying manifold of some orientable manifold is said to be *orientable.* To orient the manifold is to make choice of one of the oriented manifolds of which it is the underlying manifold.

Let $\tilde{\mathcal{V}}$ be an oriented manifold, and denote by $\tilde{\mathfrak{L}}_p$ the oriented tangent space to $\tilde{\mathcal{V}}$ at a point $p \varepsilon \tilde{\mathcal{V}}$. Let also $\tilde{\mathfrak{L}}_p^*$ be the oriented vector space oppositely oriented to $\tilde{\mathfrak{L}}_p$; then it is clear that the pair formed by the underlying manifold $\mathfrak{V}$ of $\tilde{\mathfrak{V}}$ and by the law $p \rightarrow \tilde{\mathfrak{L}}_p^*$ is again an oriented manifold $\tilde{\mathfrak{V}}^*$; we shall say that $\tilde{\mathfrak{V}}$ and $\tilde{\mathfrak{V}}^*$ are *oppositely oriented.* The oriented manifolds $\tilde{\mathfrak{V}}$ and $\tilde{\mathfrak{V}}^*$ are the only ones which admit $\mathfrak{V}$ as underlying manifold. In fact, let $\tilde{\mathfrak{W}}$ be any oriented manifold which admits $\mathfrak{V}$ as its underlying manifold. Denote by E the set of points $q \varepsilon \mathfrak{V}$ such that $\tilde{\mathfrak{L}}_q$ is the oriented tangent space to $\tilde{\mathfrak{W}}$ at q. If $q \varepsilon E$, let $(x_1, \cdot \cdot \cdot, x_n)$ be an ordered system of coordinates on $\tilde{\mathfrak{V}}$ at q; then $(x_1, \cdot \cdot \cdot, x_n)$ is also an ordered system of coordinates on both $\tilde{\mathfrak{V}}$ and $\tilde{\mathfrak{W}}$ at every point of some neighbourhood of q, from which it follows immediately that E is open. Similarly, let E^* be the set of points $r \varepsilon \mathfrak{V}$ such that the oriented tangent space to $\tilde{\mathfrak{W}}$ at r is $\tilde{\mathfrak{L}}_r^*$; then the same argument shows that E^* is open. Since $\mathfrak{V}$ is the union of E and E^* and $E \cap E^* = \phi$, it follows from the connectedness of $\mathfrak{V}$ that one of the sets E, E^* coincides with $\mathfrak{V}$, which proves our assertion.

The underlying manifold of an analytic group $\mathfrak{G}$ is always orientable. In fact, let $\omega_1, \cdot \cdot \cdot, \omega_n$ be n linearly independent forms of Maurer-Cartan on $\mathfrak{G}$ (where $n = \dim \mathfrak{G}$). Then $\omega_1 \square \cdot \cdot \cdot \square \omega_r$ is a differential form of order n on $\mathfrak{G}$ which is continuous and everywhere $\neq 0$. Hence we may orient $\mathfrak{G}$ by the requirement that this form shall be everywhere positive.

Let $\tilde{\mathfrak{L}}$ and $\tilde{\mathfrak{M}}$ be two oriented vector spaces, of dimensions m and n respectively and let $\mathfrak{L}$ and $\mathfrak{M}$ be the underlying vector spaces of $\tilde{\mathfrak{L}}$ and $\tilde{\mathfrak{M}}$ respectively. Let B be a positive m-linear form on $\tilde{\mathfrak{L}}$ and let C be a positive n-linear form on $\tilde{\mathfrak{M}}$. Then BC is an $(m + n)$-linear form on $\mathfrak{L} \times \mathfrak{M}$ and is $\neq 0$; we may orient $\mathfrak{L} \times \mathfrak{M}$ by requiring that BC shall be positive. It is easy to see that the orientation obtained in this way depends only upon $\tilde{\mathfrak{L}}$ and $\tilde{\mathfrak{M}}$, not on the choices of B and C. The oriented space obtained in this manner is called the product of the oriented vector spaces $\tilde{\mathfrak{L}}$ and $\tilde{\mathfrak{M}}$; it is denoted by $\tilde{\mathfrak{L}} \times \tilde{\mathfrak{M}}$.

Now, let $\tilde{\mathfrak{V}}$ and $\tilde{\mathfrak{W}}$ be two oriented manifolds; we denote by $\tilde{\mathfrak{L}}_p$ the oriented tangent space to $\tilde{\mathfrak{V}}$ at a point $p \varepsilon \tilde{\mathfrak{V}}$ and by $\tilde{\mathfrak{M}}_q$ the oriented tangent space to $\tilde{\mathfrak{W}}$ at a point $q \varepsilon \tilde{\mathfrak{W}}$. Let $\mathfrak{V}$ and $\mathfrak{W}$ be the underlying manifolds of $\tilde{\mathfrak{V}}$ and $\tilde{\mathfrak{W}}$; we know that the tangent space at (p, q) to $\mathfrak{V} \times \mathfrak{W}$ may be identified with the product of the tangent spaces to $\mathfrak{V}$ at p and to $\mathfrak{W}$ at q. It is easy to see that the manifold $\mathfrak{V} \times \mathfrak{W}$, together with the law $(p, q) \rightarrow \tilde{\mathfrak{L}}_p \times \tilde{\mathfrak{M}}_q$, gives rise to an oriented manifold. We shall denote this oriented manifold by $\tilde{\mathfrak{V}} \times \tilde{\mathfrak{W}}$, and we shall call it the product of the oriented manifolds $\tilde{\mathfrak{V}}$ and $\tilde{\mathfrak{W}}$. Denote by $\bar{\omega}_1$ and $\bar{\omega}_2$ the projections of $\mathfrak{V} \times \mathfrak{W}$ onto $\mathfrak{V}$ and $\mathfrak{W}$ respectively. Let $(x_1, \cdot \cdot \cdot, x_m)$ be an ordered system of coordinates at p on $\tilde{\mathfrak{V}}$

and let $(y_1, \cdots, y_n)$ be an ordered system of coordinates at q on $\tilde{\mathcal{W}}$; then it is easy to see that $(x_1 \circ \bar{\omega}_1, \cdots, x_m \circ \bar{\omega}_1, y_1 \circ \bar{\omega}_2, \cdots, y_n \circ \bar{\omega}_2)$ is an ordered system of coordinates at (p, q) on $\tilde{\mathcal{V}} \times \tilde{\mathcal{W}}$.

§VII. Integration of Differential Forms

Let $\mathcal{V}$ be an oriented manifold of dimension n, and let φ^n be a differential form of order n on $\mathcal{V}$. We wish to show how φ^n may be used as an element of integral on $\mathcal{V}$.

We shall say that a subset V of $\mathcal{V}$ is a cubic set if it is a cubic neighbourhood of some point p with respect to a coordinate system at p. We shall say that a real valued function f, defined on $\mathcal{V}$, has the property P if f is continuous and if there exists a relatively compact cubic set V outside of which f equals 0.

Let f be such a function. We can find a point p_0, an ordered coordinate system $(x_1, \cdots, x_n)$ at p_0 on $\mathcal{V}$, and a cubic neighbourhood V of p_0 with respect to this system such that f is zero outside V. Let a be the breadth of V, and let Q be the cube in R^n defined by the inequalities $|x_i - x_i(p_0)| < a$. If $p \varepsilon V$, we may write

$$f(p) = f^*(x_1(p), \cdots, x_n(p))$$
$$\varphi_p^n = F(x_1(p), \cdots, x_n(p))(dx_1)_p \square \cdots \square (dx_n)_p,$$

where $f^*(x_1, \cdots, x_n)$, $F(x_1, \cdots, x_n)$ are continuous functions on Q. Moreover, the function f^*F is bounded on Q and approaches 0 when $(x_1, \cdots, x_n)$ approaches the boundary of Q. Hence the integral

$$(1) \qquad I = \int_Q f^*(x_1, \cdots, x_n)F(x_1, \cdots, x_n)dx_1 \cdots dx_n$$

is defined. We shall prove that the value of this integral does not depend on the choice of $p_0, x_1, \cdots, x_n, V$. Let p_0' be another point of $\mathcal{V}$, $(x_1', \cdots, x_n')$ an ordered coordinate system at p_0' on $\mathcal{V}$, and V' a cubic neighbourhood of p_0' with respect to this system, such that f is also zero outside V'. We denote by Q' the cube of R^n defined by the inequalities $|x_i' - x_i'(p_0')| < a'$, where a' is the breadth of V', and we must prove the equality,

$$(2) \quad \int_Q f^*(x_1, \cdots, x_n)F(x_1, \cdots, x_n)dx_1 \cdots dx_n$$
$$= \int_{Q'} f^{*\prime}(x_1', \cdots, x_n')F'(x_1', \cdots, x_n')dx_1' \cdots dx_n',$$

where $f^{*\prime}$ and F' are defined by the formulas

$$f^{*\prime}(x_1'(p) \cdots, x_n'(p)) = f(p)$$
$$F'(x_1'(p), \cdots, x_n'(p))(dx_1')_p \square \cdots \square (dx_n')_p = \varphi_p^n$$

(for $p \varepsilon V'$). The function f is zero outside $V \cap V'$. Let U, U' be the images of $V \cap V'$ under the mappings, $p \to (x_1(p), \cdots, x_n(p))$, and $p \to (x'_1(p), \cdots, x'_n(p))$; U and U' are open subsets of Q, Q' respectively, and the integrals which occur in the formula to be proved do not change if we restrict the domains of integration to U, U' instead of Q, Q'.

If $p \varepsilon V \cap V'$, the x'-coordinates, $x'_1(p), \cdots, x'_n(p)$ of p may be expressed as functions, $g_1(x_1(p), \cdots, x_n(p)), \cdots, g_n(x_1(p), \cdots, x_n(p))$, of the x-coordinates of p; the functions $g_1(x_1, \cdots, x_n), \cdots, g_n(x_1, \cdots, x_n)$ are defined and analytic on U, and the mapping, $(x_1, \cdots, x_n) \to (g_1(x_1, \cdots, x_n), \cdots, g_n(x_1, \cdots, x_n))$ maps U topologically into U'. We set

$$D(x_1, \cdots, x_n) = \frac{D(g_1, \cdots, g_n)}{D(x_1, \cdots, x_n)}$$

whence

$$(dx'_1)_p \square = \cdots \square (dx'_n)_p$$
$$= D(x_1(p), \cdots, x_n(p))(dx_1)_p \square \cdots \square (dx_n)_p.$$

Since $(x_1, \cdots, x_n)$ and $(x'_1, \cdots, x'_n)$ are ordered coordinate systems on the oriented manifold $\mathcal{V}$, we have

$$D(x_1, \cdots, x_n) > 0 \qquad \text{if} \qquad (x_1, \cdots, x_n) \varepsilon U.$$

Moreover, we have

$$F(x_1, \cdots, x_n) = F'(g_1(x), \cdots, g_n(x))D(x_1, \cdots, x_n),$$

and

$$f^*(x_1, \cdots, x_n) = f^{*\prime}(g_1(x), \cdots, g_n(x)).$$

Therefore formula (2) follows at once from the classical formula for changing coordinates in multiple integrals.

It follows that the number I defined in formula (1) depends only on f and φ^n. We shall set

$$\textstyle\int_{\mathcal{V}} f\varphi^n = I,$$

and this formula defines the integration of functions f which have property P.

The following properties are obvious from our definition:

(1) If the continuous functions f_1 and f_2 are zero outside the same cubic set V, we have

$$\textstyle\int_{\mathcal{V}}(a_1 f_1 + a_2 f_2)\varphi^n = a_1 \int_{\mathcal{V}} f_1 \varphi^n + a_2 \int_{\mathcal{V}} f_2 \varphi^n$$

(where a_1 and a_2 are any real numbers).

(2) If f has property P, and g is any continuous function, then the function gf has property P.

Now we shall extend the definition of our integration process to a larger class of functions. A continuous function f is said to be *zero at infinity* if it can be represented as a finite sum of functions having property P. We assert that if $f = f_1 + \cdots + f_h$ and $f = f'_1 + \cdots + f'_{h'}$ are two representations of this kind, we have

$$(3)\quad \int_{\mathcal{V}} f_1\varphi^n + \int_{\mathcal{V}} f_2\varphi^n + \cdots \int_{\mathcal{V}} f_n\varphi^n = \int_{\mathcal{V}} f'_1\varphi^n + \int_{\mathcal{V}} f'_2\varphi^n + \cdots + \int_{\mathcal{V}} f'_{h'}\varphi^n$$

We shall need the following lemma:

Lemma 1.[1] *Let E be a relatively compact subset of $\mathcal{V}$. There exists a continuous function μ, which is zero at infinity and equal to 1 on E.*

Let $\bar{E}$ be the adherence of E in $\mathcal{V}$; then $\bar{E}$ is a compact set. We select at every point $p \varepsilon \bar{E}$ a coordinate system $\{x_{1,p}, \cdots, x_{n,p}\}$ and a cubic neighbourhood V_p of p with respect to this system. We define the function μ_p by the formulas

$$\mu_p(q) = 1 - \max_i \{a_p^{-1}|x_{i,p}(q) - x_{i,p}(p)|\} \qquad \text{if } q \varepsilon V_p,$$
$$\mu_p(q) = 0 \qquad \text{if } q \text{ does not belong to } V_p,$$

where a_p denotes the breadth of V_p. Each function μ_p is continuous. Since $\bar{E}$ is compact, it can be covered by a finite number of the sets V_p, say $V_{p_1}, \cdots, V_{p_k}$. The function $\Sigma_1^k \mu_{p_i}$ is $\neq 0$ everywhere on $\bar{E}$; therefore it has a minimum, $m > 0$, on $\bar{E}$. We set $s(q) = \max \{m, \Sigma_i \mu_{p_i}\}$; the function $s(q)$ is continuous, everywhere $\geqslant m$, and equal to $\Sigma_i \mu_{p_i}$ on $\bar{E}$. The function $\mu = \Sigma_1^k \frac{\mu_{p_i}}{s}$ obviously has the required properties.

Now we can prove formula (3). Let E be the set of points at which one at least of the functions $f_1, \cdots, f_h, f'_1, \cdots, f'_{h'}$ is $\neq 0$; obviously E is relatively compact, and therefore we may apply our lemma to E. Let $g = \Sigma_i^k g_i$ be a continuous function which equals 1 on E, each function g_i having property P. We have $fg_i = f_1g_i + \cdots + f_hg_i = f'_1g_i + \cdots + f'_hg_i$; for fixed i the functions $f_1g_i, \cdots, f_ng_i$ are all equal to zero outside the same relatively compact cubic set. Hence we have

$$(4)\qquad \Sigma_{\alpha=1}^h \int_{\mathcal{V}} f_\alpha g_i\varphi^n = \Sigma_{\alpha=1}^{h'} \int_{\mathcal{V}} f'_\alpha g_i\varphi^n$$

On the other hand, the functions $f_\alpha g_1, \cdots, f_\alpha g_k$, for α fixed, are also all zero outside the same cubic set, and their sum is $f_\alpha g = f_\alpha$.

[1] This lemma is due to Dieudonné.

whence

$$\int_{\mathfrak{V}} f_\alpha \varphi^n = \Sigma_1^k \int_{\mathfrak{V}} f_\alpha g_i \varphi^n,$$

and we have a similar formula for the functions f'. Therefore we obtain formula (3) by adding the k formulas (4).

We may now define the integral $\int_{\mathfrak{V}} f\varphi^n$ of a function f which is zero at infinity as the common value of all expressions $\Sigma_\alpha \int_{\mathfrak{V}} f_\alpha \varphi^n$, for all representations of f as a finite sum of functions with property P.

If f_1, f_2 are functions which are zero at infinity, the functions $a_1 f_1 + a_2 f_2$ is also zero at infinity, and we have

$$(5) \qquad \int_{\mathfrak{V}} (a_1 f_1 + a_2 f_2)\varphi^n = a_1 \int_{\mathfrak{V}} f_1 \varphi^n + a_2 \int_{\mathfrak{V}} f_2 \varphi^n.$$

If the differential form φ^n is everywhere positive on the oriented manifold $\mathfrak{V}$, we may assert that the integral (with respect to φ^n) of a non negative continuous function f, null at infinity, is non negative, and is even positive unless f is identically equal to 0. In fact, it is clearly sufficient to prove these assertions for a function f which has the property P; in this case, our assertions follow immediately from the definition if we observe that the function denoted by F in (1) is positive.

Let (g_p) be a sequence of continuous functions on $\mathfrak{V}$ which converges uniformly to a function g; then we have, for any continuous f null at infinity,

$$\lim_{p \to \infty} \int_{\mathfrak{V}} g_p f \varphi^n = \int_{\mathfrak{V}} g f \varphi^n$$

This is proved by decomposing f into a sum of functions having the property P and observing that our formula follows immediately from the definition if f has the property P.

Remark. *A continuous function is zero at infinity if and only if it is zero outside some compact subset of* $\mathfrak{V}$.

The "only if" is trivial. Conversely, if f is zero outside the compact set E, there exists (by the lemma) a function $g = \Sigma_1^k g_i$, which is equal to 1 on E, the functions g_i having property P. We have $f = \Sigma_i f g_i$, which shows that f is zero at infinity.

In particular, every continuous function on a compact manifold is zero at infinity.

Effect of an analytic isomorphism

Let $\mathfrak{V}$, $\mathfrak{W}$ be oriented manifolds, and let Φ be an analytic isomorphism of the underlying manifold of $\mathfrak{V}$ with the underlying manifold of $\mathfrak{W}$. Let n be the common dimension of $\mathfrak{V}$, $\mathfrak{W}$. If ψ^n is a differential form of order n on $\mathfrak{W}$, $\delta\Phi(\psi^n)$ is a differential form of order n on $\mathfrak{V}$; if p is a point of $\mathfrak{V}$ such that $(\psi^n)_{\Phi p} \neq 0$, we have

$(\delta\Phi(\psi^n))_p \neq 0$. But, if $(\psi^n)_{\Phi p}$ is positive on $\mathcal{W}$, $(\delta\Phi(\psi^n))_p$ may be either positive or negative on $\mathcal{V}$. If for any point p and for any ψ^n such that $(\psi^n)_{\Phi p} \neq 0$, the form $(\delta\Phi(\psi^n))_p$ has the same sign as $(\psi^n)_{\Phi p}$, we shall say that Φ *preserves the orientation.*

An equivalent formulation of this condition is the following: if the functions $(y_1, \cdots, y_n)$ form an ordered coordinate system at $\Phi(p)$ on $\mathcal{W}$, the functions $\{y_1 \circ \Phi, \cdots, y_n \circ \Phi\}$ form an ordered coordinate system at p on $\mathcal{V}$.

Let Φ be an orientation preserving analytic isomorphism of $\mathcal{V}$ with $\mathcal{W}$, ψ^n a continuous differential form of order n on $\mathcal{W}$, and f a continuous function, zero at infinity on $\mathcal{V}$. Since Φ is a homeomorphism, it follows immediately from the remark made above that $f \circ \overset{-1}{\Phi}$ is zero at infinity on $\mathcal{W}$. We assert that we have the formula

$$\int_{\mathcal{V}} f\delta\Phi(\psi^n) = \int_{\mathcal{W}} (f \circ \overset{-1}{\Phi})\psi^n \tag{6}$$

It is obviously sufficient to prove this formula in the case where f is zero outside some relatively compact cubic set, V, of $\mathcal{V}$. In this case we can find a point $p_0 \varepsilon V$ and an ordered coordinate system, $(x_1, \cdots, x_n)$ at p_0 on $\mathcal{V}$ such that V is a cubic neighbourhood of p_0, say of breadth a.

Since Φ preserves orientation, the functions $y_1 = x_1 \circ \overset{-1}{\Phi}, \cdots, y_n = x_n \circ \overset{-1}{\Phi}$ form an ordered coordinate system at $\Phi(p_0)$ on $\mathcal{W}$. Moreover, the set $\Phi(V)$ is the cubic neighbourhood of breadth a of $\Phi(p_0)$ with respect to this system.

We can express ψ^n in $\Phi(V)$ by the formula

$$(\psi^n)_q = F(y_1(q), \cdots, y_n(q))(dy_1)_q \,\square\, \cdots \,\square\, (dy_n)_q.$$

On the other hand, we have, for $p \varepsilon V, f(p) = f^*(x_1(p), \cdots, x_n(p))$.

We have

$$(\delta\Phi(\psi^n))_p = F(x_1(p), \cdots, x_n(p))(dx_1)_p \,\square\, \cdots \,\square\, (dx_n)_p$$

$$(f \circ \overset{-1}{\Phi})_q = f^*(y_1(q), \cdots, y_n(q)).$$

It follows immediately that the ordinary multiple integrals which give (by definition) the values of the two sides of (6) are really the same integral, which proves formula (6).

Integration on the product of two manifolds

Now let $\mathcal{V}$ and $\mathcal{W}$ be oriented manifolds of dimensions m and n respectively. We suppose that there are given an m-linear differential form φ^m on $\mathcal{V}$ and an n-linear differential form ψ^n on $\mathcal{W}$. We form the product $\mathcal{V} \times \mathcal{W}$, and denote by $\bar{\omega}_1$, $\bar{\omega}_2$ the projections of $\mathcal{V} \times \mathcal{W}$ onto $\mathcal{V}$, $\mathcal{W}$ respectively. Then $\delta\bar{\omega}_1(\varphi^m)$ and $\delta\bar{\omega}_2(\psi^n)$ are differential forms on $\mathcal{V} \times \mathcal{W}$; hence $\delta\bar{\omega}_1(\varphi^m) \square \delta\bar{\omega}_2(\psi^n)$ is a differential form of order $m + n$ on $\mathcal{V} \times \mathcal{W}$. We shall denote it simply by $\varphi^m\psi^n$. If φ^m and ψ^n are continuous, so is $\varphi^m\psi^n$.

Let $f = f(p, q)$ be a continuous function which is zero at infinity on $\mathcal{V} \times \mathcal{W}$. Then, for each fixed q, the function $f_q(p) = f(p, q)$, considered as a function on $\mathcal{V}$, is zero at infinity. In fact, if C is a compact subset of $\mathcal{V} \times \mathcal{W}$ such that $f = 0$ outside C, $\bar{\omega}_1(C)$ is compact, and, for every $q \varepsilon \mathcal{W}$, f_q is zero outside $\bar{\omega}_1(C)$. We shall prove moreover that

$$\int_{\mathcal{V}\times\mathcal{W}} f\varphi^m\psi^n = \int_{\mathcal{W}}\left(\int_{\mathcal{V}} f_q\varphi^m\right)\psi^n \tag{7}$$

(we observe that $\int_{\mathcal{V}} f_q\varphi^m = 0$ if q does not belong to $\bar{\omega}_2(C)$).

The sets $V \times W$, where V, W are relatively compact cubic subsets of $\mathcal{V}$, $\mathcal{W}$ are open in $\mathcal{V} \times \mathcal{W}$. The argument used in the proof of the earlier lemma shows immediately that there is a continuous function, g, equal to 1 on C, which may be expressed as a finite sum $\Sigma_1^k g_i$, where each g_i is zero outside one of the sets $V \times W$. We have $f = fg = \Sigma_1^k fg_i$, and therefore it is sufficient to prove (7) under the additional assumption that f is zero outside some set of the form $V \times W$, where V, W are relatively compact cubic subsets of $\mathcal{V}$, $\mathcal{W}$.

By assumption, we can find points $p_0 \varepsilon V$, $q_0 \varepsilon W$ and ordered coordinate systems $(x_1, \cdots, x_n)$, $(y_1, \cdots, y_n)$ at p_0, q_0 on $\mathcal{V}$, $\mathcal{W}$ such that V, W are cubic neighbourhoods of p_0, q_0 with respect to these systems.

If $p \varepsilon V$, $q \varepsilon W$, we have $f(p, q) = f^*(x_1(p), \cdots, x_m(p); y_1(q), \cdots, y_n(q))$, $\varphi_p^m = F(x_1(p), \cdots, x_m(p))(dx_1)_p \square \cdots \square (dx_m)_p$ and $\psi_q^n = G(y_1(q), \cdots, y_n(q))(dy_1)_q \square \cdots \square (dy_n)_q$. If we set $x_1' = x_i \circ \bar{\omega}_1$, $y_j' = y_j \circ \bar{\omega}_2$, we have

$$\begin{aligned}(\varphi^m\psi^n)_{(p,q)} = F(x_1(p), \cdots, x_m(p))G(y_1(q), \cdots, y_n(q))(dx_1')_{(p,q)} \\ \square \cdots \square (dx_m')_{(p,q)} \square (dy_1')_{(p,q)} \square \cdots \square (dy_n')_{(p,q)}\end{aligned}$$

Hence

$$\begin{aligned}&\int_{\mathcal{V}\times\mathcal{W}} f\varphi^m\psi^n \\ &= \int_{Q'\times Q''} f(x_1, \cdots, x_m, y_1, \cdots, y_n)F(x_1, \cdots, x_m)G(y_1, \cdots, y_n) \\ &\qquad dx_1 \cdots dx_m dy_1 \cdots dy_n\end{aligned}$$

where Q' is the subset of R^m defined by the inequalities $|x_i - x_i(p_0)| < a'$, and Q'' the subset of R^n defined by the inequalities $|y_j - y_j(q_0)| < a''$, a' and a'' being the breadths of V, W.

Therefore we have

$$\int_{\mathfrak{V}\times\mathfrak{W}} f\varphi^m\psi^n$$
$$= \int_{Q''}(\int_{Q'} f(x_1, \cdots, x_m, y_1, \cdots, y_n)F(x_1, \cdots, x_m)dx_1dx_2 \cdots dx_m)$$
$$G(y_1, \cdots, y_n)dy_1dy_2 \cdots dy_n$$

and this is exactly the assertion of formula (7).

§VIII. Invariant Integration on a Group

Let $\mathfrak{G}$ be a Lie group, and let $\mathfrak{G}_0$ be the component of the neutral element ϵ in $\mathfrak{G}$. Then $\mathfrak{G}_0$ is the underlying topological group of an analytic group $\mathcal{G}_0$. We have already observed that the underlying manifold $\mathcal{G}_0$ of an analytic group is always orientable. If $\omega_1, \omega_2, \cdots,$ ω_n are n linearly independent forms of Maurer-Cartan, (n being the dimension of $\mathcal{G}_0$), $\omega_1 \square \omega_2 \square \cdots \square \omega_n = \varphi^n$ is a continuous differential form of order r on $\mathcal{G}_0$ which is everywhere $\neq 0$. We may orient $\mathcal{G}_0$ in such a way that φ^n is positive everywhere. This being done, we have an integration process on $\mathcal{G}_0$ for the continuous functions which are zero at infinity.

Let σ_0 be any element of $\mathcal{G}_0$, and let Φ_{σ_0} be the corresponding left translation. Since $\delta\Phi_{\sigma_0}(\varphi^n) = \varphi^n$, Φ_{σ_0} is an orientation preserving analytic isomorphism of $\mathcal{G}_0$ with itself. Therefore we have, by formula (6), §VII, p. 161

$$\int_{\mathcal{G}_0} f\varphi^n = \int_{\mathcal{G}_0}(f \circ \overset{-1}{\Phi_{\sigma_0}})\varphi^n$$

which we may as well write in the form

$$\int_{\mathcal{G}_0} f\varphi^n = \int_{\mathcal{G}_0}(f \circ \Phi_{\sigma_0})\varphi^n$$

In this formula f represents any function which is zero at infinity on $\mathfrak{G}_0$.

It is easy to extend the definition of integration to functions which are zero at infinity on $\mathfrak{G}$ instead of $\mathfrak{G}_0$ (i.e. continuous functions, zero outside some compact subset of $\mathfrak{G}$). Let f be such a function. In every connected component $\mathfrak{G}_\alpha$ of $\mathfrak{G}$ we select a point σ_α; for each α we define a function f_α on $\mathfrak{G}_0$ by the formula

$$f_\alpha(\tau) = f(\sigma_\alpha\tau) \qquad \tau \varepsilon \mathfrak{G}_0$$

Each function f_α is zero at infinity on $\mathfrak{G}_0$. Moreover, only a finite number of these functions can be $\neq 0$. In fact, let C be a compact subset of $\mathfrak{G}$ outside of which f is zero. Since $\mathfrak{G}$ is a Lie group, $\mathfrak{G}_0$ is open in $\mathfrak{G}$ and the topological group $\mathfrak{G}/\mathfrak{G}_0$ is discrete. The image of C under the natural projection of $\mathfrak{G}$ onto $\mathfrak{G}/\mathfrak{G}_0$, being a compact subset of a discrete set, is a finite set, which shows that C meets at most a finite number of components of $\mathfrak{G}$, say $\mathfrak{G}_{\alpha_1}, \mathfrak{G}_{\alpha_2}, \cdots, \mathfrak{G}_{\alpha_k}$. If $\alpha \neq \alpha_1, \alpha_2, \cdots, \alpha_k$, the function f_α is identically zero. The sum

$$\Sigma_\alpha \int_{\mathcal{G}_0} f_\alpha \varphi^n$$

therefore has a meaning. We assert that its value is independent of the choice of the elements σ_α. Let σ'_α be some other element of $\mathfrak{G}_\alpha$, and let f'_α be the function defined by $f'_\alpha(\tau) = f(\sigma'_\alpha\tau)$, $(\tau \varepsilon \mathfrak{G}_0)$. We have $f'_\alpha(\tau) = f_\alpha(\sigma_\alpha^{-1}\sigma'_\alpha\tau)$. We set $\sigma_\alpha^{-1}\sigma'_\alpha = \sigma''_\alpha$; this is an element of $\mathfrak{G}_0$, and we have

$$\int_{\mathcal{G}_0} f'_\alpha \varphi^n = \int_{\mathcal{G}_0} f_\alpha \circ \Phi_{\sigma_\alpha''} \varphi^n = \int_{\mathcal{G}_0} f_\alpha \varphi^n$$

which proves our assertion.

We can therefore define the integral of f over $\mathfrak{G}$ by the formula

$$\int_{\mathfrak{G}} f\varphi^n = \Sigma_\alpha \int_{\mathcal{G}_0} f_\alpha \varphi^n$$

This integration obviously has the properties expressed by formulas (5) and (6) of §VII, p. 161. Moreover, if σ is any element of $\mathfrak{G}$, we have

$$\int_{\mathfrak{G}} f\varphi^n = \int_{\mathfrak{G}} (f \circ \Phi_\sigma)\varphi^n$$

where Φ_σ is the left translation associated with σ. In fact, if we set $g = f \circ \Phi_\sigma$, we have $g(\tau) = f(\sigma\tau)$, $g_\alpha(\tau) = f(\sigma_\alpha\sigma\tau)$. If $\mathfrak{G}_\beta$ is the coset of $\sigma_\alpha\sigma$ modulo $\mathfrak{G}_0$, we have $f(\sigma_\alpha\sigma\tau) = f_\beta(\sigma_\beta^{-1}\sigma_\alpha\sigma\tau)$ and $\sigma_\beta^{-1}\sigma_\alpha\sigma \varepsilon \mathfrak{G}_0$, whence $\int_{\mathcal{G}_0} g_\alpha \varphi^n = \int_{\mathfrak{G}_0} f_\beta \varphi^n$. Since $\mathfrak{G}_\beta = \mathfrak{G}_\alpha\sigma$, $\mathfrak{G}_\beta$ runs through the set of all components when $\mathfrak{G}_\alpha$ does so, which proves our formula.

Finally, we observe that we have oriented $\mathcal{G}_0$ in such a way that φ^n represents an everywhere positive differential form. It follows that if the function f is everywhere $\geqslant 0$, we also have $\int_{\mathfrak{G}} f\varphi^n \geqslant 0$. It is sufficient to prove this for a function f which vanishes outside some relatively compact cubic subset, V, of $\mathfrak{G}_0$. We can find a point $p_0 \varepsilon V$ and an ordered coordinate system $(x_1, \cdots, x_n)$ at p_0 on $\mathfrak{G}_0$ such that V is a cubic neighbourhood of p_0 with respect to this system. If we have, for $p \varepsilon V$,

$$(\varphi^n)_p = F(x_1(p), \cdots, x_n(p))(dx_1)_p \,\square \cdots \square\, (dx_n)_p,$$

the positiveness of φ^n implies that $F(x_1, \cdots, x_n)$ is a positive function. Hence the formula which defines $\int_{\mathfrak{G}_0} f\varphi^n$ in this case shows immediately that the integral is $\geqslant 0$ if f is $\geqslant 0$. Moreover, if f is everywhere $\geqslant 0$ and somewhere $\neq 0$ we have

$$\int_{\mathfrak{G}} f\varphi^n > 0.$$

A convention of notation

When some left-invariant differential form φ^n of order n has been selected, and is thereafter kept fixed, the integral $\int_{\mathfrak{G}} f\varphi^n$ is often denoted by $\int_{\mathfrak{G}} f(\tau)d\tau$, where the symbol τ of the variable of integration may be changed, as usual, provided there is no conflict with the rest of the notation.

With this notation the invariant character of the integration is expressed by the formula

$$\int_{\mathfrak{G}} f(\tau)d\tau = \int_{\mathfrak{G}} f(\sigma\tau)d\tau$$

where σ is any element of $\mathfrak{G}$.

Effect of the right-translations

Let σ_0 be a fixed element in $\mathfrak{G}$. We consider first the mapping $\tau \rightarrow \sigma_0\tau\sigma_0^{-1} = \Theta_{\sigma_0}(\tau)$ of $\mathfrak{G}$ into itself. This mapping induces an analytical isomorphism of the analytic group $\mathcal{G}_0$ with itself. Therefore $\delta\Theta_{\sigma_0}(\varphi^n)$ is again a left-invariant differential form of order n. As such, it can be written in the form $c(\sigma_0)\varphi^n$, where $c(\sigma_0)$ is a constant depending on σ_0. Therefore, if f is any function which is zero at infinity on $\mathfrak{G}$, we have

$$\int_{\mathfrak{G}} (f \circ \overset{-1}{\Theta}_{\sigma_0})\varphi^n = c(\sigma_0)\int_{\mathfrak{G}} f\varphi^n$$

or

$$\int_{\mathfrak{G}} f(\sigma_0^{-1}\tau\sigma_0)d\tau = c(\sigma_0)\int_{\mathfrak{G}} f\varphi^n \tag{1}$$

Since $\Theta_{\sigma_0\sigma_1} = \Theta_{\sigma_0} \circ \Theta_{\sigma_1}$, we have $c(\sigma_0\sigma_1) = c(\sigma_0)c(\sigma_1)$. On the other hand, it is quite easy to see that the function $c(\sigma_0)$ of σ_0 is analytic at the neutral element, and in particular is continuous. Since the mapping $\sigma_0 \rightarrow c(\sigma_0)$ is a homomorphism of $\mathfrak{G}$ into the multiplicative group of real numbers, it is everywhere continuous.

Formula (1), combined with the left-invariance of our integration process, gives

$$\int_{\mathfrak{G}} f(\tau\sigma_0)d\tau = c(\sigma_0)\int_{\mathfrak{G}} f(\tau)d\tau. \tag{2}$$

The case of a compact group

If $\mathfrak{G}$ is compact, the constant 1 can be integrated over $\mathfrak{G}$ with respect to any left-invariant differential form of order n, and $c = \int_{\mathfrak{G}} 1 \cdot \varphi^n$ is a *positive* constant. Replacing φ^n by $c^{-1}\varphi^n$, we see that it is always possible to normalize our integration process in such a way that

$$\int_{\mathfrak{G}} 1 \cdot d\sigma = 1.$$

We shall always assume that this has been done when we deal with integration on a compact group.

If we apply (2) with $f = 1$, we find $c(\sigma_0) = 1$: *the left-invariant integration on a compact group is also right-invariant.*

Let ϕ_ρ^* be the right-translation corresponding to an element $\rho \varepsilon \mathfrak{G}$. Since ϕ_ρ^* commutes with any left-translation, $\delta\phi_\rho^* \varphi^n$ is again a left-invariant form; the integration process defined by φ^n being right-invariant, we must have $\delta\phi_\rho^* \varphi^n = \varphi^n$: φ^n itself is right-invariant.

Now let J be the mapping $\sigma \to \sigma^{-1}$ of $\mathfrak{G}$ into itself; $\delta J \varphi^n$ is a right-invariant differential form of order n, whence $\delta J \varphi^n = k\varphi^n$, with k constant. Since $\int_{\mathfrak{G}} 1 \cdot \delta J \varphi^n = \int_{\mathfrak{G}} 1 \cdot \varphi^n = 1$ we have $k = 1$, $\delta J \varphi^n = \varphi^n$, which gives the formula

$$\int_{\mathfrak{G}} f(\sigma^{-1}) d\sigma = \int_{\mathfrak{G}} f(\sigma) d\sigma.$$

CHAPTER VI

Compact Lie Groups and Their Representations

Summary. The chapter begins with an exposition of the simplest features of the general theory of representations. In order to be able later to apply notions and results in the theory of representations of Lie algebras, we introduce the general idea of an "S-module," where S is any set whatsoever. We interrupt the exposition in §II in order to prove as soon as possible the essential fact that every representation of a compact Lie group is semi-simple.

In §§VII, VIII, IX, we develop the ideas which center around van Kampen's and Tannaka's theorems. The main emphasis is placed on the construction of the complex Lie group which corresponds to a given compact Lie group. It follows from Tannaka's theorem that a compact Lie group $\mathfrak{G}$ may be defined as the group of the "representations" of the set $\mathfrak{R}$ of representations of $\mathfrak{G}$; we obtain the corresponding complex Lie group by dropping one of the conditions which were included by Tannaka in the notion of "representation of $\mathfrak{R}$," viz. the one which refers to imaginary conjugate representations. Our method shows in a natural way the fact that the associated complex group is topologically equivalent to the product of $\mathfrak{G}$ and a cartesian space; this is a particular case of a theorem of Cartan.

In §XI, we give the proof of the famous Peter-Weyl theorem. In §§XII, XIII we are concerned with some simple applications of the Peter-Weyl theorem.

§I. General Notions

Let S be an arbitrary set of elements. By an *S-module* on a field K we mean a pair $(\mathfrak{P}, \mathsf{P})$ formed by a vector space $\mathfrak{P}$ of finite dimension over K and a mapping P which assigns to every element $\sigma \varepsilon S$ a linear endomorphism $\mathsf{P}(\sigma)$ of $\mathfrak{P}$. It follows that an S-module is an additive group with two domains of operators, one being the field K and the other one being the set S.

Two S-modules $(\mathfrak{P}, \mathsf{P})$ and $(\mathfrak{P}', \mathsf{P}')$ are said to be *isomorphic* if there exists a linear isomorphism I of $\mathfrak{P}$ with $\mathfrak{P}'$ such that $I \circ \mathsf{P}(\sigma) = \mathsf{P}'(\sigma) \circ I$ for every $\sigma \varepsilon S$.

In particular, we shall have to consider the case where S is the set of elements of a group G. We denote by ϵ the neutral element of G. An S-module $(\mathfrak{P}, \mathsf{P})$ is called a *representation space* of G if the following conditions are satisfied: 1) $\mathsf{P}(\sigma\tau) = \mathsf{P}(\sigma) \circ \mathsf{P}(\tau)$ for any elements σ, τ of G; 2) $\mathsf{P}(\epsilon)$ is the identity mapping of $\mathfrak{P}$. It follows immediately that $\mathsf{P}(\sigma^{-1})$ is then the reciprocal mapping of $\mathsf{P}(\sigma)$. If

$(\mathfrak{P}, \mathsf{P})$ is a representation space of the group G, the mapping P is called a *representation* of G. The dimension of $\mathfrak{P}$ is called the *degree* of P.

Remark. The correct notation for an S-module or a representation space is the notation $(\mathfrak{P}, \mathsf{P})$. However, we shall often use the symbol $\mathfrak{P}$ alone to indicate the S-module (or representation space); this notation should be considered as an abbreviation of the complete notation, and should be avoided when there is danger of confusion.

Let $(\mathfrak{P}, \mathsf{P})$ be a representation space of a group G. If we select a base $\{\mathbf{e}_1, \cdots, \mathbf{e}_d\}$ in $\mathfrak{P}$, we may represent each linear endomorphism $\mathsf{P}(\sigma)$ (for $\sigma \varepsilon G$) by a matrix $\tilde{\mathsf{P}}(\sigma) = (x_{ij})$, of degree d, whose coefficients are given by the formulas

$$\tilde{\mathsf{P}}(\sigma)\mathbf{e}_i = \Sigma_{j=1}^{d} x_{ji}\mathbf{e}_j$$

We have $\tilde{\mathsf{P}}(\sigma\tau) = \tilde{\mathsf{P}}(\sigma)\tilde{\mathsf{P}}(\tau)$, $\tilde{\mathsf{P}}(\epsilon) = E$ (the unit matrix).

Conversely, any mapping $\tilde{\mathsf{P}}$ of G into the set of matrices of degree d with coefficients in K is called a *representation* (of degree d) of G provided the conditions $\tilde{\mathsf{P}}(\sigma\tau) = \tilde{\mathsf{P}}(\sigma)\tilde{\mathsf{P}}(\tau)$, $\tilde{\mathsf{P}}(\epsilon) = E$ are satisfied. If we wish to distinguish between representations by linear endomorphisms and representations by matrices, we shall speak of *abstract representations* in the first case, of *matricial representations* in the second case. If the matricial representation $\tilde{\mathsf{P}}$ is derived from the abstract representation P by selecting a base in the representation space of P, we shall say that $\tilde{\mathsf{P}}$ is a *matricial form* of P and that P is an *abstract form* of $\tilde{\mathsf{P}}$. Clearly, any matricial representation has at least one abstract form and every abstract representation of degree > 0 has at least one matricial form.

Two abstract representations P_1, P_2 of a group G are said to be *equivalent* if their representation spaces are isomorphic. Two matricial representations $\tilde{\mathsf{P}}_1$, $\tilde{\mathsf{P}}_2$ of G are said to be *equivalent* if there exists a regular matrix γ such that

$$\tilde{\mathsf{P}}_2(\sigma) = \gamma\tilde{\mathsf{P}}_1(\sigma)\gamma^{-1}$$

holds for every $\sigma \varepsilon G$ (this implies that $\tilde{\mathsf{P}}_1$ and $\tilde{\mathsf{P}}_2$ have the same degree). The following statements are easy to prove: if the matricial representations $\tilde{\mathsf{P}}_1$, $\tilde{\mathsf{P}}_2$ are equivalent, they have a common abstract form and every abstract form of $\tilde{\mathsf{P}}_1$ is equivalent to every abstract form of $\tilde{\mathsf{P}}_2$; if P_1 and P_2 are equivalent abstract representations of G of positive degree, then any matricial form of P_1 is equivalent with any matricial form of P_2.

Let $(\mathfrak{P}, \mathsf{P})$ be any S-module. A vector subspace $\mathfrak{Q}$ of $\mathfrak{P}$ is said to be *invariant* (with respect to P) if we have $\mathsf{P}(\sigma)\mathfrak{Q} \subset \mathfrak{Q}$ for every $\sigma \varepsilon S$. This being the case, the contraction $\mathsf{P}_1(\sigma)$ of $\mathsf{P}(\sigma)$ to $\mathfrak{Q}$ is a linear endomorphism of $\mathfrak{Q}$; the pair $(\mathfrak{Q}, \mathsf{P}_1)$ is an S-module, which is called a submodule of $(\mathfrak{P}, \mathsf{P})$. Moreover, if $\mathbf{e}$ is an arbitrary vector in $\mathfrak{P}$, the residue class modulo $\mathfrak{Q}$ of $\mathsf{P}(\sigma)\mathbf{e}$ depends only upon the residue class $\mathbf{e}^*$ of $\mathbf{e}$. If we denote the residue class of $\mathsf{P}(\sigma)\mathbf{e}$ by $\Lambda(\sigma)\mathbf{e}^*$, then $\Lambda(\sigma)$ is a linear endomorphism of $\mathfrak{P}/\mathfrak{Q}$, and the pair $(\mathfrak{P}/\mathfrak{Q}, \Lambda)$ is an S-module. If $\mathfrak{P}$ is a representation space of a group G, and if $\mathfrak{Q}$ is an invariant subspace, $(\mathfrak{P}, \mathsf{P}_1)$ and $(\mathfrak{P}/\mathfrak{Q}, \Lambda)$ are also representation spaces of G.

Let $\mathfrak{Q}$ be an invariant subspace of an S-module $(\mathfrak{P}, \mathsf{P})$ of positive dimension. We may select a base $(\mathbf{e}_1, \cdots, \mathbf{e}_d)$ in $\mathfrak{P}$ such that the elements $\mathbf{e}_1, \cdots, \mathbf{e}_r$ form a base of $\mathfrak{Q}$. If $\sigma \varepsilon S$, the matrix which represents $\mathsf{P}(\sigma)$ with respect to the base $\mathbf{e}_1, \cdots, \mathbf{e}_d$ has the form

$$\tilde{\mathsf{P}}(\sigma) = \begin{pmatrix} \tilde{\mathsf{P}}_1(\sigma) & \mathsf{N}(\sigma) \\ 0 & \tilde{\Lambda}(\sigma) \end{pmatrix}$$

where $\tilde{\mathsf{P}}_1(\sigma)$ and $\tilde{\Lambda}(\sigma)$ are square matrices of degrees r and $d - r$ respectively and where $\mathsf{N}(\sigma)$ is a rectangular $(r, d - r)$-matrix. The matrix $\tilde{\mathsf{P}}_1(\sigma)$ represents the contraction of P_1 to $\mathfrak{Q}$ (with respect to the base $\{\mathbf{e}_1, \cdots, \mathbf{e}_r\}$ in $\mathfrak{Q}$); the matrix $\tilde{\Lambda}(\sigma)$ represents the endomorphism of $\mathfrak{P}/\mathfrak{Q}$ which corresponds to $\mathsf{P}(\sigma)$ (with respect to the base of $\mathfrak{P}/\mathfrak{Q}$ formed by the residue classes of $\mathbf{e}_{r+1}, \cdots, \mathbf{e}_d$).

Now, let $\mathfrak{P}$, $\mathfrak{Q}$ and $\mathfrak{R}$ be three invariant subspaces of some S-module. We have the following "homomorphism theorems," due to Fr. Noether:

I. *If* $\mathfrak{R} \subset \mathfrak{Q} \subset \mathfrak{P}$, $\mathfrak{Q}/\mathfrak{R}$ *is an invariant subspace of* $\mathfrak{P}/\mathfrak{R}$, *and* $\mathfrak{P}/\mathfrak{Q}$ *is isomorphic to* $(\mathfrak{P}/\mathfrak{R})/(\mathfrak{Q}/\mathfrak{R})$.

II. *The spaces* $\mathfrak{P} + \mathfrak{Q}$ *and* $\mathfrak{P} \cap \mathfrak{Q}$ *are invariant, and* $\mathfrak{P} + \mathfrak{Q}/\mathfrak{Q}$ *is isomorphic to* $\mathfrak{P}/\mathfrak{P} \cap \mathfrak{Q}$.

Moreover, the isomorphisms whose existence is asserted are "natural isomorphisms"; i.e. they can be defined without reference either to the set S or to the mapping of S in the set of endomorphisms of the S-module under consideration.

For the proof of these facts, we refer the reader to *van der Waerden*, Moderne Algebra, I, Chap. VI, p. 148.

Definition 1. *An S-module* $\mathfrak{P}$ *is said to be simple if it is of dimension* > 0 *and if the only invariant subspaces of* $\mathfrak{P}$ *are* $\{0\}$ *and* $\mathfrak{P}$.

This definition includes in particular the definition of a simple representation space of a group G. Such a representation space is also very often called *irreducible.* The corresponding representation of G is also said to be simple or irreducible. A matricial representation of a group G is said to be simple or irreducible if an abstract form of it is simple.

Definition 2. *An S-module is called semi-simple if it can be represented as a sum of simple sub-modules.*

Remark 1. The sum of a collection $\{\mathfrak{Q}_\alpha\}$ of subspaces of a vector space is the set of vectors of the form $\Sigma_\alpha \mathbf{e}_\alpha$, with $\mathbf{e}_\alpha \varepsilon \mathfrak{Q}_\alpha$, only a finite number of the vectors $\mathbf{e}_\alpha$ being $\neq 0$. It is clear that a sum of invariant subspaces of an S-module is an invariant subspace.

Remark 2. An S-module of dimension 0 will be considered as semi-simple; such a module can be considered as the sum of an empty collection of sub-modules.

Proposition 1. *A semi-simple S-module $\mathfrak{P}$ can be represented as the direct sum $\mathfrak{P} = \Sigma_{i=1}^h \mathfrak{Q}_i$ of a finite collection $\Phi = \{\mathfrak{Q}_i\}$ of simple sub-modules. Moreover, if we have a representation of this kind, and if $\mathfrak{Q}$ is any invariant subspace of $\mathfrak{P}$, then there exists a sub-collection Φ_0 of Φ such that $\mathfrak{P}$ is the direct sum of $\mathfrak{Q}$ and of the sum of the sub-modules belonging to Φ_0.*

By assumption, $\mathfrak{P}$ is the sum of some (finite or infinite) collection Ψ of simple sub-modules. We take a base of $\mathfrak{P}$, and we represent every element of this base as a sum of vectors belonging to the sub-modules of the collection Ψ; in this way, we see that $\mathfrak{P}$ can certainly be represented as the sum of some finite collection Ψ_1 of simple sub-modules. Among all finite subcollections of Ψ which have the property that the sum of their terms is $\mathfrak{P}$, we select one, say Φ, with the smallest possible number of elements. Let $\mathfrak{Q}_1, \cdots, \mathfrak{Q}_h$ be the distinct elements of Φ. We have $\mathfrak{P} = \Sigma_{i=1}^h \mathfrak{Q}_i$. We assert that this sum is direct. In fact, assume that we have a relation of the form $\mathbf{f}_1 + \cdots + \mathbf{f}_h = 0$, with $\mathbf{f}_i \varepsilon \mathfrak{Q}_i$ $(1 \leqslant i \leqslant h)$. We have $\mathbf{f}_1 \varepsilon \mathfrak{Q}_1 \cap (\mathfrak{Q}_2 + \cdots + \mathfrak{Q}_h)$. Now, $\mathfrak{Q}_1 \cap (\mathfrak{Q}_2 + \cdots + \mathfrak{Q}_h)$ is an invariant subspace of $\mathfrak{Q}_1$; if we had $\mathfrak{Q}_1 \cap (\mathfrak{Q}_2 + \cdots + \mathfrak{Q}_h) = \mathfrak{Q}_1$, $\mathfrak{Q}_1$ would be contained in $\mathfrak{Q}_2 + \cdots + \mathfrak{Q}_h$, and $\mathfrak{Q}_2 + \cdots + \mathfrak{Q}_h$ would be equal to $\mathfrak{P}$, in contradiction with our choice of Φ. Since $\mathfrak{Q}_1$ is simple, we have $\mathfrak{Q}_1 \cap (\mathfrak{Q}_2 + \cdots + \mathfrak{Q}_h) = \{0\}$, $\mathbf{f}_1 = 0$. In the same way, we see that $\mathbf{f}_i = 0$ $(1 \leqslant i \leqslant h)$. Our assertion is thereby proved.

Now, let $\mathfrak{Q}$ be any invariant subspace of $\mathfrak{P}$. We consider those subsets Φ' of Φ which have the property that $\mathfrak{P}$ is the sum of $\mathfrak{Q}$ and of the sum of the modules of Φ' (for instance $\Phi' = \Phi$). Among these subsets, we select one, say Φ_0, with the smallest possible number of elements. If $\mathfrak{Q}_{i_1}, \cdots, \mathfrak{Q}_{i_p}$ are the elements of Φ_0, an argument entirely similar to the one used above shows that $\mathfrak{P}$ is the direct sum of $\mathfrak{Q}, \mathfrak{Q}_{i_1}, \cdots, \mathfrak{Q}_{i_p}$.

Proposition 2. *Let $\mathfrak{P}$ be an S-module which has the following property: if $\mathfrak{Q}$ is any invariant subspace of $\mathfrak{P}$, there exists an invariant subspace $\mathfrak{Q}'$ such that $\mathfrak{P}$ is the direct sum of $\mathfrak{Q}$ and $\mathfrak{Q}'$. Then $\mathfrak{P}$ is semi-simple.*

Let $\mathfrak{P}_1$ be the sum of all simple sub-modules of $\mathfrak{P}$. By assumption, $\mathfrak{P}$ is the direct sum of $\mathfrak{P}_1$ and of an other invariant subspace $\mathfrak{R}$. If we had dim $\mathfrak{R} > 0$, $\mathfrak{R}$ would contain some simple sub-module (for instance, an invariant subspace of $\mathfrak{R}$ of smallest positive dimension would give a simple sub-module contained in $\mathfrak{R}$). But any simple sub-module is contained in $\mathfrak{P}_1$ and $\mathfrak{P}_1 \cap \mathfrak{R} = \{0\}$: the assumption dim $\mathfrak{R} > 0$ leads to a contradiction. It follows that $\mathfrak{R} = \{0\}$, $\mathfrak{P} = \mathfrak{P}_1$.

Proposition 3. *Let $\mathfrak{P} = \mathfrak{Q}_1 + \cdots \mathfrak{Q}_h = \mathfrak{Q}'_1 + \cdots + \mathfrak{Q}'_{h'}$ be two representations of a semi-simple S-module $\mathfrak{P}$ as a direct sum of simple S-modules. Then we have $h = h'$ and there exists a permutation $\bar{\omega}$ of the set $\{1, \cdots, h\}$ such that $\mathfrak{Q}_i$ is isomorphic to $\mathfrak{Q}'_{\bar{\omega}(i)} (1 \leqslant i \leqslant h)$.*

We shall construct the permutation $\bar{\omega}$. Suppose that $\bar{\omega}(i)$ is already defined for $i < k$ (where $k \leqslant h$) and has the following properties: *a*) $\bar{\omega}(i) \neq \bar{\omega}(j)$ for $i < j < k$; *b*) $\mathfrak{Q}'_{\bar{\omega}(i)}$ is isomorphic to $\mathfrak{Q}_i$ (for $i < k$); *c*) we have

$$\mathfrak{P} = \Sigma_{i<k}\mathfrak{Q}'_{\bar{\omega}(i)} + \Sigma_{i\geqslant k}\mathfrak{Q}_i$$

We consider the invariant subspace

$$\mathfrak{Q} = \Sigma_{i<k}\mathfrak{Q}'_{\bar{\omega}(i)} + \Sigma_{i>k}\mathfrak{Q}_i$$

By Proposition 2, there exists an invariant subspace $\mathfrak{Q}'$ which is the direct sum of a certain number of the spaces $\mathfrak{Q}'_j$ and which is such that $\mathfrak{P}$ is the direct sum of $\mathfrak{Q}$ and $\mathfrak{Q}'$. Therefore $\mathfrak{Q}'$ is isomorphic to $\mathfrak{P}/\mathfrak{Q}$, i.e. to $\mathfrak{Q}_k$. It follows that $\mathfrak{Q}'$ is simple and therefore that $\mathfrak{Q}'$ is one of the modules $\mathfrak{Q}'_j$, say $\mathfrak{Q}' = \mathfrak{Q}'_{j_0}$. Since $\mathfrak{Q}'_{\bar{\omega}(i)} \subset \mathfrak{Q}$ for $i < k$, we have $j_0 \neq \bar{\omega}(i)$ for $i < k$. We define $\bar{\omega}(k)$ to be the number j_0. It is clear that the function $\bar{\omega}(i)$, now defined for $i < k + 1$, satisfies the conditions *a*), *b*), *c*) above (with k replaced by $k + 1$).

Because we can define the univalent function $\bar{\omega}$ on the set $\{1, \cdots, h\}$, we must have $h' \geqslant h$. Since the two decompositions play symmetric roles, we have also $h' \leqslant h$, whence $h = h'$. Proposition 4 is thereby proved.

Now, let $\mathfrak{P}$ be the representation space of a semi-simple representation P of a group G, and let Λ be any irreducible representation of G.

If we decompose $\mathfrak{P}$ into a direct sum of simple subspaces, we may count the number of these subspaces which yield representations which are equivalent to the given representation $\mathsf{\Lambda}$. It follows from Proposition 3 that this number is independent of the particular decomposition of $\mathfrak{P}$ which is used. The number in question is called the *number of times that the representation* $\mathsf{\Lambda}$ *is contained in the representation* P.

Proposition 4. *Let* $(\mathfrak{P}, \mathsf{P})$ *be a simple S-module over an algebraically closed field K. Assume that all the endomorphisms* $\mathsf{P}(\sigma)$ $(\sigma \varepsilon S)$ *mutually commute. Then* $\mathfrak{P}$ *is of dimension* 1.

In fact, let σ be any element of S. Since K is algebraically closed, there exists an element $u \varepsilon K$ and a vector $\mathbf{e} \neq 0$ in $\mathfrak{P}$ such that $\mathsf{P}(\sigma)\mathbf{e} = u\mathbf{e}$. Let $\mathfrak{Q}$ be the set of vectors $\mathbf{e}$ satisfying this condition. It is clear that $\mathfrak{Q}$ is a vector subspace of $\mathfrak{P}$. Furthermore, if $\sigma \varepsilon S$, $\mathbf{e} \varepsilon \mathfrak{Q}$, we have $\mathsf{P}(\sigma)\mathsf{P}(\tau)\mathbf{e} = \mathsf{P}(\tau)\mathsf{P}(\sigma)\mathbf{e} = u\mathsf{P}(\tau)\mathbf{e}$, whence $\mathsf{P}(\tau)\mathbf{e} \varepsilon \mathfrak{Q}$, which means that $\mathfrak{Q}$ is invariant. Since $\mathfrak{P}$ is simple, it follows that $\mathfrak{Q} = \mathfrak{P}$. In other words, for every $\sigma \varepsilon S$, there exists an element $u(\sigma) \varepsilon K$ such that $\mathsf{P}(\sigma)\mathbf{e} = u(\sigma)\mathbf{e}$ for every $\mathbf{e} \varepsilon \mathfrak{P}$. It follows immediately that any vector subspace of $\mathfrak{P}$ is invariant. Since $\mathfrak{P}$ is simple, it coincides with the subspace generated by any vector $\mathbf{e} \neq 0$ in $\mathfrak{P}$, whence $\dim \mathfrak{P} = 1$.

§II. REPRESENTATIONS OF COMPACT LIE GROUPS

Let $\mathfrak{G}$ be a topological group. By a (matricial) representation of $\mathfrak{G}$ we mean a continuous homomorphism of $\mathfrak{G}$ into either the group $GL(n; C)$ ("complex representation")—or the group $GL(n; R)$ ("real representation").

Theorem 1. *Any real representation of a compact Lie group is equivalent to a representation by orthogonal matrices. Any complex representation is equivalent to a representation by unitary matrices.*

Let us consider the case of a complex representation P of a compact Lie group $\mathfrak{G}$. We set

$$\alpha_1(\sigma) = {}^t\bar{\mathsf{P}}(\sigma) \cdot \mathsf{P}(\sigma)$$

The matrix $\alpha_1(\sigma)$ is always hermitian positive definite (cf. the proof of Proposition 1, §V, Chapter 1, p. 14). The coefficients of $\alpha_1(\sigma)$ are continuous functions of σ on $\mathfrak{G}$.

We shall now make use of the invariant process of integration on $\mathfrak{G}$, normalized as usual by the condition $\int_{\mathfrak{G}} 1 \cdot d\sigma = 1$ (cf. §VIII, Chapter 5, p. 167). We set

$$\alpha_1 = \int_{\mathfrak{G}} \alpha_1(\sigma) d\sigma$$

(i.e. the coefficients of α_1 are the integrals over $\mathfrak{G}$ of the coefficients of $\alpha_1(\sigma)$). Since we have ${}^t\alpha_1(\sigma) = \bar{\alpha}_1(\sigma)$ for every σ, we have also ${}^t\alpha_1 = \bar{\alpha}_1 : \alpha_1$ is hermitian. If $\mathbf{a}$ is any vector in C^n (where n is the degree of our representation) we have

$$\mathbf{a} \cdot \alpha_1 \mathbf{a} = \int_{\mathfrak{G}} \mathbf{a} \cdot \alpha_1(\sigma)\mathbf{a}\, d\sigma$$

Since $\alpha_1(\sigma)$ is positive for every σ, we have $\mathbf{a} \cdot \alpha_1(\sigma)\mathbf{a} \geqslant 0$, whence $\mathbf{a} \cdot \alpha_1\mathbf{a} \geqslant 0$; this proves that α_1 is positive. Since $\alpha_1(\sigma)$ is definite, we have $\mathbf{a} \cdot \alpha_1(\sigma)\mathbf{a} > 0$ provided $\mathbf{a} \neq 0$, whence $\mathbf{a} \cdot \alpha_1\mathbf{a} > 0$; we see that α_1 is a hermitian positive definite matrix.

If τ is any fixed element in $\mathfrak{G}$, we have

$$\begin{aligned} {}^t\bar{\mathsf{P}}(\tau)\alpha_1\mathsf{P}(\tau) &= \int_{\mathfrak{G}} {}^t\bar{\mathsf{P}}(\tau){}^t\bar{\mathsf{P}}(\sigma)\mathsf{P}(\sigma)\mathsf{P}(\tau)d\sigma = \int_{\mathfrak{G}} {}^t\bar{\mathsf{P}}(\sigma\tau)\mathsf{P}(\sigma\tau)d\sigma \\ &= \int_{\mathfrak{G}} {}^t\bar{\mathsf{P}}(\sigma)\mathsf{P}(\sigma)\, d\sigma = \alpha_1 \end{aligned}$$

in virtue of the invariant character of our integration.

We have seen in the course of the proof of Proposition 1, §V, Chapter 1, p. 14 that a positive definite hermitian matrix α_1 may be written in the form α^2, where α is also hermitian positive definite. We set $\mathsf{P}'(\tau) = \alpha\mathsf{P}(\tau)\alpha^{-1}$. Since ${}^t\bar{\alpha} \cdot \alpha = \alpha^2 = \alpha_1$ and ${}^t\bar{\mathsf{P}}(\tau)\alpha_1\mathsf{P}(\tau) = \alpha_1$, it follows from an easy computation that the matrix ${}^t(\bar{\alpha}\bar{\mathsf{P}}(\tau)\bar{\alpha}^{-1})(\alpha\mathsf{P}(\tau)\alpha^{-1})$ is the unit matrix. This means that the representation $\alpha\mathsf{P}\alpha^{-1}$ is unitary.

If now P is a real representation, then α_1 is a real matrix, and we may assume α to be real. It follows that the matrices $\alpha\mathsf{P}(\tau)\alpha^{-1}$ are real and unitary, i.e. orthogonal, which completes the proof of Theorem 1.

Corollary. *Every representation of a compact Lie group is semi-simple.*

By Theorem 1, we may limit ourselves to the consideration of a matricial representation P of our compact Lie group $\mathfrak{G}$ by unitary or orthogonal matrices.

In the complex case, we may consider that the representation space is C^n. Let $\mathfrak{P}$ be any invariant subspace, and let $\mathfrak{P}'$ be the vector subspace of C^n composed of the vectors $\mathbf{f}$ such that $\mathbf{e} \cdot \mathbf{f} = 0$ for all $\mathbf{e} \varepsilon \mathfrak{P}$. If $\mathbf{f} \varepsilon \mathfrak{P}'$, we have

$$\mathbf{e} \cdot \mathsf{P}(\sigma)\mathbf{f} = {}^t\bar{\mathsf{P}}(\sigma)\mathbf{e} \cdot \mathbf{f} = \mathsf{P}(\sigma^{-1})\mathbf{e} \cdot \mathbf{f} = 0$$

for all $\mathbf{e} \varepsilon \mathfrak{P}$ and $\sigma \varepsilon \mathfrak{G}$. It follows that $\mathfrak{P}'$ is an invariant subspace of C^n.

We know that $\mathbf{e} \cdot \mathbf{e} = 0$ implies $\mathbf{e} = 0$; it follows that $\mathfrak{P} \cap \mathfrak{P}' = \{0\}$.

Let $\{\mathbf{e}_1, \cdots, \mathbf{e}_d\}$ be a base $\mathfrak{P}$. The vectors $\mathbf{f}\varepsilon\mathfrak{P}'$ are those whose components satisfy the d linear homogeneous equations $\mathbf{f} \cdot \mathbf{e}_i = 0$ $(1 \leqslant i \leqslant d)$; it follows that the dimension of $\mathfrak{P}'$ is at least $n - d$. Since $\mathfrak{P} \cap \mathfrak{P}' = \{0\}$, the dimension of $\mathfrak{P} + \mathfrak{P}'$ is at least $d + (n - d) = n$. Therefore, we have $\mathfrak{P} + \mathfrak{P}' = C^n$. Corollary 1 then follows from Proposition 2, §I, p. 171.

The argument would be entirely similar in the real case.

§III. OPERATIONS ON REPRESENTATIONS

1. The star representation

Let φ be an endomorphism of a vector space $\mathfrak{P}$ over a field K, and let $\mathfrak{P}'$ be the dual space of $\mathfrak{P}$ (i.e. the space of linear functions on $\mathfrak{P}$ with values in K). If λ is any element of $\mathfrak{P}'$, we define ${}^t\varphi(\lambda)$ to be the element of $\mathfrak{P}'$ which is defined by $({}^t\varphi(\lambda))(\mathbf{e}) = \lambda(\varphi\mathbf{e})$ for every $\mathbf{e}\varepsilon\mathfrak{P}$. It is clear that ${}^t\varphi$ is an endomorphism of $\mathfrak{P}'$. Moreover, if φ_1 and φ_2 are endomorphisms of $\mathfrak{P}$, we have

$$ {}^t(\varphi_1 \circ \varphi_2) = {}^t\varphi_2 \circ {}^t\varphi_1 $$

Let $\{\mathbf{e}_1, \cdots, \mathbf{e}_d\}$ be a base of $\mathfrak{P}$. There corresponds to this base a dual base $\{\lambda_1, \cdots, \lambda_d\}$ of $\mathfrak{P}'$, such that $\lambda_i(\mathbf{e}_j) = \delta_{ij}$ $(1 \leqslant i, j \leqslant d)$. If α is the matrix which represents the endomorphism φ with respect to the base $\{\mathbf{e}_1, \cdots, \mathbf{e}_d\}$, then the matrix which represents ${}^t\varphi$ with respect to the base $\{\lambda_1, \cdots, \lambda_d\}$ is the transpose ${}^t\alpha$ of α.

Now, let $(\mathfrak{P}, \mathsf{P})$ be a representation space of a group G. The formula ${}^t\mathsf{P}(\sigma\tau) = {}^t\mathsf{P}(\tau) \circ {}^t\mathsf{P}(\sigma)$ shows that ${}^t\mathsf{P}$ is not in general a representation of $\mathfrak{G}$, but that the mapping $\sigma \rightarrow {}^t\mathsf{P}(\sigma^{-1})$ is a representation. If $\tilde{\mathsf{P}}$ is a matricial form of P, then the mapping $\sigma \rightarrow (\tilde{\mathsf{P}}(\sigma))^*$ is a matricial form of the representation $\sigma \rightarrow {}^t\mathsf{P}(\sigma^{-1})$.

Definition 1. *If* P *is an abstract representation of a group* G, *the mapping* $\sigma \rightarrow {}^t\mathsf{P}(\sigma^{-1})$ *is called the star of the representation* P *and is denoted by* P^*. *If* P *is a matricial representation of* G, *the mapping* $\sigma \rightarrow (\mathsf{P}(\sigma))^*$ *is called the star representation of* P *and is denoted by* P^*.

Proposition 1. *Let* P *be a unitary matricial representation of a group* G (i.e. *a representation which assigns to every* $\sigma\varepsilon G$ *a unitary matrix*). *Then* P^* *coincides with the imaginary conjugate representation* $\bar{\mathsf{P}}$ *of* P (i.e. $\bar{\mathsf{P}}(\sigma) = \overline{\mathsf{P}(\sigma)}$). *If* P *is an orthogonal representation of* G, *then* $\mathsf{P}^* = \mathsf{P}$.

This follows immediately from the definitions.

On the other hand, we observe that $(\mathsf{P}^*)^* = \mathsf{P}$ for any matricial representation P.

2. The addition of representations

Let $(\mathfrak{P}_1, \mathsf{P}_1)$ and $(\mathfrak{P}_2, \mathsf{P}_2)$ be representation spaces of a group G. We construct the product $\mathfrak{P}_1 \times \mathfrak{P}_2$ of $\mathfrak{P}_1$ and $\mathfrak{P}_2$, and we assign to every $\sigma \varepsilon G$ the linear endomorphism $\mathsf{P}(\sigma)$ of $\mathfrak{P}_1 \times \mathfrak{P}_2$ defined by

$$\mathsf{P}(\sigma)(\mathbf{e}_1, \mathbf{e}_2) = (\mathsf{P}_1(\sigma)\mathbf{e}_1, \mathsf{P}_2(\sigma)\mathbf{e}_2) \qquad (\mathbf{e}_i \varepsilon \mathfrak{P}_i, \cdots i = 1, 2)$$

It is clear that P is a representation of G. We say that P is the *sum* of the representations P_1 and P_2, and we write $\mathsf{P} = \mathsf{P}_1 \dotplus \mathsf{P}_2$.

Let $\{\mathbf{e}_{m1}, \cdots, \mathbf{e}_{md_m}\}$ be a base in $\mathfrak{P}_m$ $(m = 1, 2)$. Let (a_{ij}) and (b_{kl}) be the matrices which represent $\mathsf{P}_1(\sigma)$ and $\mathsf{P}_2(\sigma)$ with respect to these bases. Then $\mathbf{f}_1 = (\mathbf{e}_{11}, 0), \cdots, \mathbf{f}_{d_1} = (\mathbf{e}_{1d_1}, 0), \mathbf{f}_{d_1+1} = (0, \mathbf{e}_{21}), \cdots, \mathbf{f}_{d_1+d_2} = (0, \mathbf{e}_{2d_2})$ form a base in $\mathfrak{P}_1 \times \mathfrak{P}_2$, and the matrix which represents $\mathsf{P}(\sigma)$ with respect to this base is

$$\begin{pmatrix} (a_{ij}) & (0) \\ (0) & (b_{kl}) \end{pmatrix}$$

This leads to the following definition: if α and β are square matrices of degrees d_1 and d_2 respectively, we shall denote by $\alpha \dotplus \beta$ the matrix

$$\begin{pmatrix} \alpha & 0 \\ 0 & \beta \end{pmatrix}$$

of degree $d_1 + d_2$.

Now, if P_1 and P_2 are matricial representations of a group G, we shall of course denote by $\mathsf{P}_1 \dotplus \mathsf{P}_2$ the matricial representation which assigns to every $\sigma \varepsilon G$ the matrix $\mathsf{P}_1(\sigma) \dotplus \mathsf{P}_2(\sigma)$. It follows immediately that, if P_1, P_2 and P_3 are three matricial representations of G then $(\mathsf{P}_1 \dotplus \mathsf{P}_2) \dotplus \mathsf{P}_3 = \mathsf{P}_1 \dotplus (\mathsf{P}_2 \dotplus \mathsf{P}_3)$ and $(\mathsf{P}_1 \dotplus \mathsf{P}_2)^* = \mathsf{P}_1^* \dotplus \mathsf{P}_2^*$. If P_1, P_2 and P_3 are abstract representations, the two sides of the preceding formulas are not equal, but equivalent.

If P_1 and P_2 are either abstract or matricial representations, then $\mathsf{P}_2 \dotplus \mathsf{P}_1$ is equivalent to $\mathsf{P}_1 \dotplus \mathsf{P}_2$. In fact, assuming P_1 and P_2 to be abstract representations, let $(\mathfrak{P}_1, \mathsf{P}_1)$ and $(\mathfrak{P}_2, \mathsf{P}_2)$ be their representation spaces. The linear isomorphism $(\mathbf{e}_1, \mathbf{e}_2) \to (\mathbf{e}_2, \mathbf{e}_1)$ of $\mathfrak{P}_1 \times \mathfrak{P}_2$ with $\mathfrak{P}_2 \times \mathfrak{P}_1$ is clearly an isomorphism of $(\mathfrak{P}_1 \times \mathfrak{P}_2, \mathsf{P}_1 \dotplus \mathsf{P}_2)$ with $(\mathfrak{P}_2 \times \mathfrak{P}_1, \mathsf{P}_2 \dotplus \mathsf{P}_1)$.

3. The Kronecker product

Let $(\mathfrak{P}_1, \mathsf{P}_1)$ and $(\mathfrak{P}_2, \mathsf{P}_2)$ be representation spaces of a group G. Denote by $\mathfrak{B}$ the space of bilinear functions on $\mathfrak{P}_1 \times \mathfrak{P}_2$ with values in K. Let φ_i $(i = 1, 2)$ be a linear endomorphism of $\mathfrak{P}_i$. If we assign to every bilinear form $B \varepsilon \mathfrak{B}$ the bilinear form $\psi(B)$ defined by

$$\psi(B)(\mathbf{e}_1, \mathbf{e}_2) = B(\varphi_1 \mathbf{e}_1, \varphi_2 \mathbf{e}_2) \qquad (\mathbf{e}_i \varepsilon \mathfrak{P}_i,\ i = 1, 2)$$

we clearly obtain a linear endomorphism ψ of $\mathfrak{B}$. We shall say that ψ is the linear endomorphism of $\mathfrak{B}$ which corresponds to the pair (φ_1, φ_2). Let θ_i be some other endomorphism of $\mathfrak{P}_i$ $(i = 1, 2)$, and let π be the linear endomorphism of $\mathfrak{B}$ which corresponds to the pair (θ_1, θ_2). Then we see easily that the linear endomorphism which corresponds to the pair $(\varphi_1 \circ \theta_1, \varphi_2 \circ \theta_2)$ is $\pi \circ \psi$. We may compensate for this inversion of order by going over to the corresponding endomorphisms in the dual space $\mathfrak{B}'$ of $\mathfrak{B}$, because ${}^t(\pi \circ \psi) = {}^t\psi \circ {}^t\pi$.

Definition 2. *Let $\mathfrak{P}_1$ and $\mathfrak{P}_2$ be vector spaces over a field K. The dual space of the space of bilinear functions on $\mathfrak{P}_1 \times \mathfrak{P}_2$ is called the Kronecker product of $\mathfrak{P}_1$ and $\mathfrak{P}_2$ and is denoted by $\mathfrak{P}_1 \mathbf{x} \mathfrak{P}_2$.*[1]

Let $\mathbf{e}_i$ be an element of $\mathfrak{P}_i$ $(i = 1, 2)$. To the pair $(\mathbf{e}_1, \mathbf{e}_2)$ there corresponds a linear function on $\mathfrak{B}$ which assigns to every $B \varepsilon \mathfrak{B}$ the value $B(\mathbf{e}_1, \mathbf{e}_2)$. But a linear function on $\mathfrak{B}$ is an element of $\mathfrak{P}_1 \mathbf{x} \mathfrak{P}_2$. Hence we have a mapping of $\mathfrak{P}_1 \times \mathfrak{P}_2$ into $\mathfrak{P}_1 \mathbf{x} \mathfrak{P}_2$. We shall denote by $\mathbf{e}_1 \mathbf{x} \mathbf{e}_2$ the element of $\mathfrak{P}_1 \mathbf{x} \mathfrak{P}_2$ which corresponds to $(\mathbf{e}_1, \mathbf{e}_2)$.

The mapping $(\mathbf{e}_1, \mathbf{e}_2) \to \mathbf{e}_1 \mathbf{x} \mathbf{e}_2$ is *not* a linear mapping of $\mathfrak{P}_1 \times \mathfrak{P}_2$ into $\mathfrak{P}_1 \mathbf{x} \mathfrak{P}_2$, but it is bilinear; i.e. we have

$$\begin{aligned} (a\mathbf{e}_1 + a'\mathbf{e}_1') \mathbf{x} \mathbf{e}_2 &= a\mathbf{e}_1 \mathbf{x} \mathbf{e}_2 + a'\mathbf{e}_1' \mathbf{x} \mathbf{e}_2 \\ \mathbf{e}_1 \mathbf{x} (a\mathbf{e}_2 + a'\mathbf{e}_2') &= a\mathbf{e}_1 \mathbf{x} \mathbf{e}_2 + a'\mathbf{e}_1 \mathbf{x} \mathbf{e}_2' \end{aligned} \qquad \begin{pmatrix} \mathbf{e}_i, \mathbf{e}_i' \varepsilon \mathfrak{P}_i \\ a, a' \varepsilon K \end{pmatrix}$$

If φ_i is a linear endomorphism of $\mathfrak{P}_i$ $(i = 1, 2)$, the linear endomorphism ${}^t\psi$ of $\mathfrak{P}_1 \mathbf{x} \mathfrak{P}_2$ transforms $\mathbf{e}_1 \mathbf{x} \mathbf{e}_2$ according to the formula

$${}^t\psi(\mathbf{e}_1 \mathbf{x} \mathbf{e}_2) = \varphi_1 \mathbf{e}_1 \mathbf{x} \varphi_2 \mathbf{e}_2.$$

We shall denote ${}^t\psi$ by $\varphi_1 \mathbf{x} \varphi_2$.

Now, let $(\mathfrak{P}_1, \mathsf{P}_1)$ and $(\mathfrak{P}_2, \mathsf{P}_2)$ be representation spaces of a group G. It follows from what we have said that the mapping $\sigma \to \mathsf{P}_1(\sigma) \mathbf{x} \mathsf{P}_2(\sigma)$ is again a representation of G. This representation is called the *Kronecker product* of the representations P_1 and P_2, and is represented by $\mathsf{P}_1 \mathbf{x} \mathsf{P}_2$.

Let $\{\mathbf{e}_{m1}, \cdots, \mathbf{e}_{md_m}\}$ be a base in $\mathfrak{P}_m$ $(m = 1, 2)$. Then the $d_1 d_2$ elements $\mathbf{e}_{1i} \mathbf{x} \mathbf{e}_{2j}$ $(1 \leqslant i \leqslant d_1,\ 1 \leqslant j \leqslant d_2)$ form a base in $\mathfrak{P}_1 \mathbf{x} \mathfrak{P}_2$. In fact, since $\mathfrak{P}_1 \mathbf{x} \mathfrak{P}_2$ has the same dimension as $\mathfrak{B}$, i.e. $d_1 d_2$, it will be sufficient to prove that the elements $\mathbf{e}_{1i} \mathbf{x} \mathbf{e}_{2j}$ are linearly independent. Assume that $\Sigma_{ij} a_{ij} \mathbf{e}_{1i} \mathbf{x} \mathbf{e}_{2j} = 0$, $a_{ij} \varepsilon K$; then $\Sigma_{ij} a_{ij} B(\mathbf{e}_{1i}, \mathbf{e}_{2j}) = 0$ for every $B \varepsilon \mathfrak{B}$. For every pair (i, j) there exists a bilinear function B_{ij}

[1] If $\mathfrak{P}_1 = \mathfrak{P}_2 = \mathfrak{P}$, the elements of $\mathfrak{P} \mathbf{x} \mathfrak{P}$ are the covariant tensors of order 2.

such that $B_{ij}(\mathbf{e}_{1k}, \mathbf{e}_{2l}) = \delta_{ik}\delta_{jl}$. If we set $B = B_{ij}$, we obtain $a_{ij} = 0$, which proves our assertion.

If φ_i is a linear endomorphism of $\mathfrak{P}_i$ $(i = 1, 2)$, we set

$$\varphi_1 \mathbf{e}_{1i} = \Sigma_{k=1}^{d_1} a_{ki}\mathbf{e}_{1k} \qquad \varphi_2 \mathbf{e}_{2j} = \Sigma_{l=1}^{d_2} b_{lj}\mathbf{e}_{2l}$$

and we have

$$(\varphi_1 \times \varphi_2)(\mathbf{e}_{1i} \times \mathbf{e}_{2j}) = \Sigma_{kl} a_{ki} b_{lj} \mathbf{e}_{1k} \times \mathbf{e}_{2l}$$

If we set $\mathbf{f}_{i+d_1(j-1)} = \mathbf{e}_{1i} \times \mathbf{e}_{2j}$, we have

$$(\varphi_1 \times \varphi_2)\mathbf{f}_r = \Sigma_{s=1}^{d_1 d_2} c_{sr}\mathbf{f}_s$$

with

(1) $$c_{i+d_1(j-1), k+d_1(l-1)} = a_{ik} b_{jl}$$

This leads to the following definition:

Definition 3. *Let $\alpha = (a_{ik})$ and $\beta = (b_{jl})$ be matrices of degrees d_1 and d_2 respectively. We denote by $\alpha \times \beta$ (and call Kronecker product of α and β) the matrix (c_{rs}) of degree $d_1 d_2$ whose coefficients are given by formula* (1). *If P_1 and P_2 are matricial representations of a group G, we call the Kronecker product of P_1 and P_2 the representation $\mathsf{P}_1 \times \mathsf{P}_2$ which assigns to every $\sigma \in G$ the matrix $\mathsf{P}_1(\sigma) \times \mathsf{P}_2(\sigma)$.*

It follows immediately from our previous considerations that, if α and β are matrices of degree d_1 and α' and β' matrices of degree d_2, then

$$(\alpha\beta) \times (\alpha'\beta') = (\alpha \times \alpha')(\beta \times \beta')$$

On the other hand, we see immediately that ${}^t(\alpha \times \beta) = {}^t\alpha \times {}^t\beta$. Therefore, if α and β are regular matrices, we have $(\alpha \times \beta)^* = \alpha^* \times \beta^*$. Since $(\alpha^*)^* = \alpha$, we also have $(\alpha \times \beta^*)^* = \alpha^* \times \beta$.

Furthermore, we see easily that $\alpha \times (\beta_1 + \beta_2) = \alpha \times \beta_1 + \alpha \times \beta_2$.

It follows that, if P_1, P_2 and P_3 are abstract representations of a group G, then $(\mathsf{P}_1 \times \mathsf{P}_2)^*$ is equivalent to $\mathsf{P}_1^* \times \mathsf{P}_2^*$, and $\mathsf{P}_1 \times (\mathsf{P}_2 \dotplus \mathsf{P}_3)$ is equivalent to $\mathsf{P}_1 \times \mathsf{P}_2 \dotplus \mathsf{P}_1 \times \mathsf{P}_3$. Furthermore, $\mathsf{P}_1 \times \mathsf{P}_2$ is changed into an equivalent representation if P_1 and P_2 are changed into equivalent representations.

Although we do not have $\alpha \times \beta = \beta \times \alpha$, the representation $\mathsf{P}_1 \times \mathsf{P}_2$ is nevertheless equivalent to $\mathsf{P}_2 \times \mathsf{P}_1$. In fact, the representation spaces of these two representations are isomorphic under an isomorphism which maps every element of the form $\mathbf{e}_1 \times \mathbf{e}_2$ (with $\mathbf{e}_i \in \mathfrak{P}_i$, $i = 1, 2$) onto $\mathbf{e}_2 \times \mathbf{e}_1$. It follows that $(\mathsf{P}_1 \dotplus \mathsf{P}_2) \times \mathsf{P}_3$ is equivalent to $\mathsf{P}_1 \times \mathsf{P}_3 \dotplus \mathsf{P}_2 \times \mathsf{P}_3$.

Another simple argument of the same kind shows that $(\mathsf{P}_1 \times \mathsf{P}_2) \times \mathsf{P}_3$ is equivalent to $\mathsf{P}_1 \times (\mathsf{P}_2 \times \mathsf{P}_3)$.

4. A remark on the representation $\mathsf{P}_1 \times \mathsf{P}_2^*$

Let P_1 and P_2 be matricial representations, of degrees d_1 and d_2 respectively, of a group G. Let $\mathfrak{Q}$ be the set of all rectangular matrices with d_1 lines and d_2 columns; $\mathfrak{Q}$ is a vector space of dimension d_1d_2. Let us assign to every $\sigma \varepsilon G$ the endomorphism Λ_σ of $\mathfrak{Q}$ which maps any matrix $\alpha \varepsilon \mathfrak{Q}$ on $\Lambda_\sigma \alpha = \mathsf{P}_1(\sigma)\alpha\mathsf{P}_2(\sigma^{-1})$. It is easy to verify that $\Lambda_{\sigma\tau} = \Lambda_\sigma \circ \Lambda_\tau$ $(\sigma,\tau \varepsilon G)$ and that Λ_ϵ is the identity mapping of $\mathfrak{Q}$ if ϵ is the neutral element of G. Therefore, the mapping $\sigma \to \Lambda_\sigma$ is an abstract representation Λ of G. We assert that Λ is an abstract form of the representation $\mathsf{P}_1 \times \mathsf{P}_2^*$.

Let us denote by $\alpha_{i+d_1(j-1)}$ the matrix in $\mathfrak{Q}$ which contains a 1 at the intersection of the i-th line and the j-th column and has zeros elsewhere. The d_1d_2 elements $\alpha_{i+d_1(j-1)}$ $(1 \leqslant i \leqslant d_1,\ 1 \leqslant j \leqslant d_2)$ form a base of $\mathfrak{Q}$. A simple computation gives

$$\mathsf{P}_1(\sigma)\alpha_{i+d_1(j-1)}\mathsf{P}_2(\sigma^{-1}) = \Sigma_{kl}a_{ki}(\sigma)b_{jl}(\sigma^{-1})\alpha_{k+d_1(l-1)}$$

where $(a_{ki}(\sigma))$ and $(b_{jl}(\sigma))$ are the matrices $\mathsf{P}_1(\sigma)$ and $\mathsf{P}_2(\sigma)$. If we set $\mathsf{P}_2^*(\sigma) = (b_{jl}^*(\sigma))$, we have $b_{jl}^*(\sigma) = b_{lj}(\sigma^{-1})$, and we see that the matricial form of Λ corresponding to our choice of a base in $\mathfrak{Q}$ is $\mathsf{P}_1 \times \mathsf{P}_2^*$, which proves our assertion.

Let $\tilde{\mathsf{P}}_i$ be an abstract form of P_i $(i = 1, 2)$, and let $(\mathfrak{P}_i, \tilde{\mathsf{P}}_i)$ be the representation space of $\tilde{\mathsf{P}}_i$. The space $\mathfrak{Q}$ may be interpreted as the space of linear mappings of $\mathfrak{P}_2$ into $\mathfrak{P}_1$. From this point of view, Λ_σ may be defined to be the endomorphism of $\mathfrak{Q}$ which assigns to every $\alpha \varepsilon \mathfrak{Q}$ the mapping $\Lambda_\sigma(\alpha)$ defined by

$$\Lambda_\sigma(\alpha)(\mathbf{e}_2) = (\mathsf{P}_1(\sigma) \circ \alpha \circ \mathsf{P}_2(\sigma^{-1}))\mathbf{e}_2 \qquad (\mathbf{e}_2 \varepsilon \mathfrak{P}_2).$$

It is easy to extract from this fact a new proof of the equivalence of Λ with $\mathsf{P}_1 \times \mathsf{P}_2^*$.

§IV. SCHUR'S LEMMA

Proposition 1 (Schur's Lemma). *Let P_1 and P_2 be two irreducible matricial representations of a group G in a field K, of degrees d_1 and d_2 respectively. A necessary and sufficient condition for the equivalence of P_1 and P_2 is that there should exist a rectangular matrix $\alpha \neq 0$ with coefficients in K, with d_1 rows and d_2 columns, such that $\mathsf{P}_1(\sigma)\alpha = \alpha\mathsf{P}_2(\sigma)$ for every $\sigma \varepsilon G$.*

If P_1 is equivalent to P_2, we have $d_1 = d_2$ and there exists a regular matrix γ such that $\mathsf{P}_2(\sigma) = \gamma^{-1}\mathsf{P}_1(\sigma)\gamma$, whence $\mathsf{P}_1(\sigma)\gamma = \gamma\mathsf{P}_2(\sigma)$.

Conversely, let us assume that there exists a matrix $\alpha \neq 0$ such that $\mathsf{P}_1(\sigma)\alpha = \alpha\mathsf{P}_2(\sigma)$ for every $\sigma \varepsilon G$. We construct representation spaces $\mathfrak{P}_1$ and $\mathfrak{P}_2$ for P_1 and P_2, and we interpret α as a linear mapping of $\mathfrak{P}_2$ into $\mathfrak{P}_1$. Let $\mathfrak{Q}_1$ be the image of $\mathfrak{P}_2$ under this mapping; since $\alpha \neq 0$, we have $\mathfrak{Q}_1 \neq \{0\}$. The formula $\mathsf{P}_1(\sigma)\alpha = \alpha\mathsf{P}_2(\sigma)$ shows at once that $\mathfrak{Q}_1$ is invariant. Since $\mathfrak{P}_1$ is irreducible, we have $\mathfrak{Q}_1 = \mathfrak{P}_1$. Let $\mathfrak{Q}_2$ be the set of vectors in $\mathfrak{P}_2$ which are mapped on 0 by α. The same formula as above shows that $\mathfrak{Q}_2$ is an invariant subspace of $\mathfrak{P}_2$. Since $\mathfrak{Q}_2 \neq \mathfrak{P}_2$, we have $\mathfrak{Q}_2 = \{0\}$. It follows that α is a univalent linear mapping of $\mathfrak{P}_2$ onto $\mathfrak{P}_1$, which proves that $d_1 = d_2$ and that α is a square matrix of determinant $\neq 0$. Hence we may write $\mathsf{P}_2(\sigma) = \alpha\mathsf{P}_1(\sigma)\alpha^{-1}$, which shows that P_1 and P_2 are equivalent. Schur's lemma is thereby proved.

Let P be an irreducible matricial representation of the group G. The matrices α such that $\mathsf{P}(\sigma)\alpha = \alpha\mathsf{P}(\sigma)$ for all $\sigma \varepsilon G$ obviously form an algebra $\mathfrak{o}$ (i.e., if α_1 and α_2 are such matrices, then $\alpha_1 + \alpha_2$, $\alpha_1\alpha_2$ and $a\alpha_1$ also have the property in question, where a is any element of K). It follows from the proof of Schur's lemma (applied to the case where $\mathsf{P}_1 = \mathsf{P}_2 = \mathsf{P}$) that every matrix $\alpha \neq 0$ in $\mathfrak{o}$ has an inverse α^{-1}; it is clear that α^{-1} also belongs to $\mathfrak{o}$. We express this fact by saying that $\mathfrak{o}$ is a *division algebra*.

Let γ be any element $\neq 0$ in $\mathfrak{o}$, and let Z be the set of elements of $\mathfrak{o}$ which may be expressed in the form $R(\gamma)$, R being a rational function with coefficients in K. It is clear that Z is a field which contains K and is of finite degree over K (because Z is contained in the ring of all matrices of a certain degree d with coefficients in K). It follows that, if K is algebraically closed, we have $Z = K$, which means that γ is of the form aE, $a \varepsilon K$, E being a unit matrix. We have proved

Proposition 2. *Let* P *be an irreducible representation of a group* G *in an algebraically closed field* K. *The only matrices which commute simultaneously with all matrices* $\mathsf{P}(\sigma)$, $\sigma \varepsilon G$, *are the scalar multiples of the unit matrix.*

Proposition 3. *Let* Λ *and* P *be two representations of a group* G *in an algebraically closed field* K, *and assume that* P *is irreducible. If* Λ *and* $\Lambda \times \mathsf{P}^*$ *are semi-simple, the number of times that* P *is contained in* Λ *is equal to the number of times that the unit representation is contained in* $\Lambda \times \mathsf{P}^*$.

By the unit representation of a group G we mean of course the representation which assigns to every $\sigma \varepsilon G$ the number 1 (considered as a matrix of degree 1). An abstract form of this representation has a representation space $\mathfrak{E}$ of dimension 1 and assigns to every $\sigma \varepsilon G$ the identity mapping of $\mathfrak{E}$ onto itself.

Let M be any semi-simple representation of G in K, and let $\mathfrak{M} = \mathfrak{M}_1 + \cdots + \mathfrak{M}_h$ be a decomposition of its representation space into a direct sum of irreducible subspaces. We may assume that $\mathfrak{M}_1, \cdots, \mathfrak{M}_n$ are all the terms (if any) which are isomorphic to the space $\mathfrak{E}$ of the unit representation (n is an integer such that $0 \leqslant n \leqslant h$). Then each $\mathfrak{M}_i$ with $1 \leqslant i \leqslant n$ is spanned by a vector $\mathbf{e}_i \neq 0$ such that $\mathsf{M}(\sigma)\mathbf{e}_i = \mathbf{e}_i$ for every $\sigma \varepsilon G$. Conversely, let $\mathbf{f}$ be any vector such that $\mathsf{M}(\sigma)\mathbf{f} = \mathbf{f}$ for every σ. We may write $\mathbf{f} = \Sigma_1^h \mathbf{f}_i$, $\mathbf{f}_i \varepsilon \mathfrak{M}_i$. We have by assumption $\Sigma \mathbf{f}_i = \Sigma \mathsf{M}(\sigma)\mathbf{f}_i$, $\mathsf{M}(\sigma)\mathbf{f}_i \varepsilon \mathfrak{M}_i$; the sum of the spaces $\mathfrak{M}_i$ being direct, we have $\mathsf{M}(\sigma)\mathbf{f}_i = \mathbf{f}_i$ $(1 \leqslant i \leqslant h)$. If $i > n$, $\mathfrak{M}_i$ cannot contain any vector $\mathbf{f}_i \neq 0$ with this property. It follows that $\mathbf{f}$ is a linear combination of $\mathbf{e}_1, \cdots, \mathbf{e}_n$. We conclude that the number of times that the unit representation is contained in M is equal to the maximal number of linearly independent vectors $\mathbf{e}$ in $\mathfrak{M}$ such that $\mathsf{M}(\sigma)\mathbf{e} = \mathbf{e}$ for all $\sigma \varepsilon G$.

We come now to the proof of Proposition 3. The representation Λ is equivalent to a sum $\Lambda_1 \dotplus \cdots \dotplus \Lambda_k$ of irreducible representations, and $\Lambda \times \mathsf{P}^*$ is equivalent to $\Lambda_1 \times \mathsf{P}^* \dotplus \cdots \dotplus \Lambda_k \times \mathsf{P}^*$. This shows that it is sufficient to prove Proposition 3 in the case where Λ is itself irreducible.

We may take as representation space of $\Lambda \times \mathsf{P}^*$ the space of all linear mappings A of the representation space $\mathfrak{P}$ of P into the representation space $\mathfrak{L}$ of Λ (cf. §III, p. 182). The representation $\Lambda \times \mathsf{P}^*$ then assigns to every $\sigma \varepsilon G$ the mapping $A \rightarrow A^\sigma = (\Lambda(\sigma))A(\mathsf{P}(\sigma))^{-1}$. If $A^\sigma = A$ for all σ, we have $\Lambda(\sigma)A = A\mathsf{P}(\sigma)$, and Schur's lemma says that this can happen with an $A \neq 0$ only in the case where Λ and P are equivalent. Assuming that this is the case, Proposition 2 shows that all elements A such that $A^\sigma = A$ for all $\sigma \varepsilon G$ are the scalar multiples of one of them. This shows that the unit representation is not contained in $\Lambda \times \mathsf{P}^*$ if Λ is not equivalent to P and is contained exactly once in $\Lambda \times \mathsf{P}^*$ if Λ is equivalent to P. Proposition 3 is thereby proved.

§V. ORTHOGONALITY RELATIONS

Let $\mathfrak{G}$ be a compact Lie group, and let M be a matricial representation of $\mathfrak{G}$ in the field of complex numbers. We shall show how it is possible to compute the number of times the unit representation E is contained in M.

We introduce the invariant integration process on $\mathfrak{G}$, normalized as usual by the condition that $\int_{\mathfrak{G}} 1 \cdot d\sigma = 1$. We denote by $\mathsf{M}_0 = \int_{\mathfrak{G}} \mathsf{M}(\sigma)\, d\sigma$ the matrix whose coefficients are the integrals over $\mathfrak{G}$ of the coefficients of $\mathsf{M}(\sigma)$.

We may consider M as a matricial form of an abstract representation, which we also denote by M. Let $\mathfrak{M}$ be the representation space, and let $\{\mathbf{e}_1, \cdots, \mathbf{e}_d\}$ be the base in $\mathfrak{M}$ by which the matricial form M is derived. If $\mathbf{e}$ is any vector in $\mathfrak{M}$, we assert that $\mathsf{M}(\sigma)\mathsf{M}_0\mathbf{e} = \mathsf{M}_0\mathbf{e}$ for every $\sigma \varepsilon G$. In fact, we set $\mathsf{M}(\tau)\mathbf{e} = \Sigma_{i=1}^d u_i(\tau)\mathbf{e}_i$. We have

$$\begin{aligned} \mathsf{M}_0\mathbf{e} &= \Sigma_i(\textstyle\int_{\mathfrak{G}} u_i(\tau)\, d\tau)\mathbf{e}_i \\ \mathsf{M}(\sigma)\mathsf{M}_0\mathbf{e} &= \Sigma_i(\textstyle\int_{\mathfrak{G}} u_i(\tau)\, d\tau)\mathsf{M}(\sigma)\mathbf{e}_i \\ &= \Sigma_{ij}\textstyle\int_{\mathfrak{G}}(m_{ji}(\sigma)u_i(\tau)\, d\tau)\mathbf{e}_j \end{aligned}$$

if $\mathsf{M}(\sigma) = (m_{ij}(\sigma))$. But we have

$$\begin{aligned} \Sigma_{ij}m_{ji}(\sigma)u_i(\tau)\mathbf{e}_j &= \mathsf{M}(\sigma)\mathsf{M}(\tau)\mathbf{e} = \mathsf{M}(\sigma\tau)\mathbf{e} \\ &= \Sigma_i u_i(\sigma\tau)\mathbf{e}_i \end{aligned}$$

and our assertion follows from the formulas

$$\textstyle\int_{\mathfrak{G}} u_i(\sigma\tau)\, d\tau = \int_{\mathfrak{G}} u_i(\tau)\, d\tau$$

Conversely, if $\mathbf{f}$ is any vector such that $\mathsf{M}(\sigma)\mathbf{f} = \mathbf{f}$ for every $\sigma\varepsilon G$, we have clearly $\mathsf{M}_0\mathbf{f} = \mathbf{f}$. It follows that the vectors $\mathbf{f}\varepsilon\mathfrak{M}$ such that $\mathsf{M}(\sigma)\mathbf{f} = \mathbf{f}$ for all $\sigma\varepsilon G$ are exactly those vectors which are of the form $\mathsf{M}_0\mathbf{e}$, $\mathbf{e}\varepsilon\mathfrak{M}$. In other words, *the number of times that the unit representation is contained in* M *is equal to the rank of the matrix* M_0.

Now, let Λ and P be two irreducible representations of $\mathfrak{G}$ in the field of complex numbers. We set

$$\begin{aligned} \mathsf{M}(\sigma) &= \Lambda(\sigma) \times \mathsf{P}^*(\sigma) \\ \mathsf{M}_0 &= \textstyle\int_{\mathfrak{G}} \mathsf{M}(\sigma)\, d\sigma \end{aligned}$$

We know that the unit representation E is not contained in M if Λ is not equivalent to P and that E is contained exactly once in M if Λ is equivalent to P. Hence, M_0 is the null matrix in the first case, and is a matrix of rank 1 in the second case.

The coefficients of $\mathsf{M}(\sigma)$ are the products of the coefficients of $\Lambda(\sigma)$ by the coefficients of $\mathsf{P}(\sigma^{-1})$. It follows that, if Λ and P are not equivalent, we have

$$\textstyle\int_{\mathfrak{G}} a(\sigma)b(\sigma^{-1})\, d\sigma = 0$$

where $a(\sigma)$ and $b(\sigma)$ are arbitrary coefficients of $\Lambda(\sigma)$ and $\mathsf{P}(\sigma)$ respectively.

In order to investigate the case where Λ and P are equivalent, we shall assume that $\Lambda = \mathsf{P}$ and that Λ is a representation by unitary matrices (we know that, in any case, Λ is equivalent to a representation by unitary matrices). Let $\mathfrak{L}$ be a representation space for Λ, and let

$\{\mathbf{e}_1, \cdots, \mathbf{e}_d\}$ be the base in $\mathfrak{L}$ which gives rise to the matricial representation Λ. We know that we may consider the space $\mathfrak{M}$ of linear mappings of $\mathfrak{L}$ into itself as a representation space for $\Lambda \times \Lambda^* = \mathsf{M}$. Let Θ_{ij} be the element of $\mathfrak{M}$ which maps $\mathbf{e}_j$ upon $\mathbf{e}_i$ and $\mathbf{e}_{j'}$ upon 0 for $j' \neq j$. Then we have

$$\mathsf{M}(\sigma)\Theta_{ij} = \Lambda(\sigma)\Theta_{ij}\Lambda(\sigma^{-1}) = \Sigma_{kl}a_{ki}(\sigma)a_{jl}(\sigma^{-1})\Theta_{kl}$$

if $\Lambda(\sigma) = (a_{ik}(\sigma))$.

Let Θ_1 be the identity mapping of $\mathfrak{L}$ into itself (i.e. $\Theta_1 = \Sigma_i\Theta_{ii}$). We know that the scalar multiples of Θ_1 are the only mappings of $\mathfrak{L}$ into itself which commute with every $\Lambda(\sigma)$, $\sigma \varepsilon G$. It follows that $\mathsf{M}_0\Theta$ is a scalar multiple of Θ_1 for every $\Theta \varepsilon \mathfrak{M}$. We conclude that

$$\int_{\mathfrak{G}} a_{ki}(\sigma)a_{jl}(\sigma^{-1})\, d\sigma = \delta_{kl}c_{ij}$$

where c_{ij} is a number which does not depend on k. We know that, $\int_{\mathfrak{G}} f(\sigma)\, d\sigma = \int_{\mathfrak{G}} f(\sigma^{-1})\, d\sigma$ for any continuous function f on $\mathfrak{G}$. Therefore we have

$$\int_{\mathfrak{G}} a_{jl}(\sigma)a_{ki}(\sigma^{-1})\, d\sigma = \delta_{kl}c_{ij}$$

Comparing our two formulas, we conclude easily that $c_{ij} = \delta_{ij}c$ with some constant c. The value of c can be determined; in fact, we have clearly $\mathsf{M}(\sigma)\Theta_1 = \Theta_1$ for every σ, whence $\mathsf{M}_0\Theta_1 = \Theta_1$, and it follows immediately that $c = d^{-1}$.

Let us now observe that, if Λ is a representation by unitary matrices, we have $a_{jl}(\sigma^{-1}) = \bar{a}_{lj}(\sigma)$.

Definition 1. *Let $\mathfrak{G}$ be a compact Lie group. Any function on $\mathfrak{G}$ which appears as coefficient of some irreducible representation of $\mathfrak{G}$ by unitary matrices is called a simple representative function on $\mathfrak{G}$. Any linear combination of simple representative functions is called a representative function.*

We have proved

Theorem 2. *Let f and g be two simple representative functions on a compact Lie group $\mathfrak{G}$. If f and g appear as coefficients in two inequivalent irreducible representations, we have $\int_{\mathfrak{G}} f(\sigma)g(\sigma^{-1})\, d\sigma = 0$. If f and g are coefficients of the same irreducible representation of degree d, the integral $\int_{\mathfrak{G}} f(\sigma)\bar{g}(\sigma)\, d\sigma$ is equal to d^{-1} if $f = g$ and to 0 if $f \neq g$.*

§VI. THE CHARACTERS

Definition 1. *Let P be a matricial representation of a group G. The trace of the matrix $\mathsf{P}(\sigma)$, considered as a function of the element $\sigma \varepsilon G$, is called the character of the representation P.*

Proposition 1. *Two equivalent representations have the same character. If σ and τ are conjugate elements of G and if χ is the character of any representation of G, we have $\chi(\sigma) = \chi(\tau)$.*

Both assertions follow immediately from the formula $Sp\alpha\beta\alpha^{-1} = Sp\beta$, where α and β are matrices, α having an inverse.

It is clear that the relationship of equivalence between representations defines a division of the set of all representations into classes of mutually equivalent representations.

Definition 2. *A class of representations of a group G is a set of representations which is composed of all representations equivalent to one of them.*

It follows from Proposition 1 that to every class of representations of the group G there is associated a function defined on G, the character of any representation of the class. This function is called the character of the class.

Let $\mathfrak{R}_1$ and $\mathfrak{R}_2$ be two classes of representations of a group G in a field K. From what has been said in §III, it follows that:

1) the star representation P^* of a representation $\mathsf{P} \varepsilon \mathfrak{R}_1$ belongs to a class $\mathfrak{R}_1^*$ which depends only on $\mathfrak{R}_1$;

2) the sum $\mathsf{P}_1 \dot{+} \mathsf{P}_2$ of a representation $\mathsf{P}_1 \varepsilon \mathfrak{R}_1$ and a representation $\mathsf{P}_2 \varepsilon \mathfrak{R}_2$ belongs to a class $\mathfrak{R}_1 \dot{+} \mathfrak{R}_2$ which depends only on $\mathfrak{R}_1$ and $\mathfrak{R}_2$; moreover $\mathfrak{R}_2 \dot{+} \mathfrak{R}_1 = \mathfrak{R}_1 \dot{+} \mathfrak{R}_2$;

3) the Kronecker product $\mathsf{P}_1 \times \mathsf{P}_2$ belongs to a class $\mathfrak{R}_1 \times \mathfrak{R}_2$ which depends only on $\mathfrak{R}_1$ and $\mathfrak{R}_2$; moreover, $\mathfrak{R}_2 \times \mathfrak{R}_1 = \mathfrak{R}_1 \times \mathfrak{R}_2$.

Furthermore, the operations of addition and Kronecker multiplication are associative in the domain of the classes of representations, and the Kronecker multiplication is distributive with respect to the addition. However, the classes of representations do not form a ring, because subtraction is generally impossible.

Let us denote by $\chi_{\mathfrak{R}}$ the character of a class of representations $\mathfrak{R}$.

Proposition 1. *If $\mathfrak{R}_1$ and $\mathfrak{R}_2$ are two classes of representations, we have*

$$\chi_{\mathfrak{R}_1 \dot{+} \mathfrak{R}_2} = \chi_{\mathfrak{R}_1} + \chi_{\mathfrak{R}_2}$$
$$\chi_{\mathfrak{R}_1 \times \mathfrak{R}_2} = \chi_{\mathfrak{R}_1}\chi_{\mathfrak{R}_2}$$

In fact, if α and β are matrices, we see easily that $Sp(\alpha \dot{+} \beta) = Sp\alpha + Sp\beta$, $Sp\alpha \times \beta = (Sp\alpha)(Sp\beta)$.

If a class of representations contains an irreducible (a semi-simple) representation, every representation of the class is irreducible (semi-simple). We then say that the class itself is irreducible (semi-simple).

Every semi-simple class $\mathfrak{R}$ can be represented in the form $\Sigma_i x_i \mathfrak{R}_i$, the x_i's being non negative integers and the $\mathfrak{K}_i$'s being irreducible classes. The number x_i is the number of times that a representation of the class $\mathfrak{K}_i$ is contained in a representation of the class $\mathfrak{K}$. This number depends only upon $\mathfrak{K}$ and $\mathfrak{K}_i$; it is called the number of times that $\mathfrak{K}_i$ is contained in $\mathfrak{K}$.

Proposition 2. *Let $\mathfrak{K}_1$ and $\mathfrak{K}_2$ be two irreducible classes of representations of a compact Lie group $\mathfrak{G}$ in the field of complex numbers. The integral*

$$\int_{\mathfrak{G}} \chi_{\mathfrak{K}_1}(\sigma)\bar{\chi}_{\mathfrak{K}_2}(\sigma)\, d\sigma$$

is equal to 0 if $\mathfrak{K}_1 \neq \mathfrak{K}_2$ and to 1 if $\mathfrak{K}_1 = \mathfrak{K}_2$.

In fact, if P_i is a representation of the class $\mathfrak{K}_i$ ($i = 1, 2$), $\chi_{\mathfrak{K}_i}(\sigma)$ is the sum of the elements of the main diagonal of $\mathsf{P}_i(\sigma)$ and our statement follows immediately from Theorem 2, §V, p. 186.

Corollary 1. *The number of times that an irreducible class $\mathfrak{K}_1$ of representations of a compact Lie group is contained in a class $\mathfrak{K}$ is equal to*

$$\int_{\mathfrak{G}} \chi_{\mathfrak{K}}(\sigma)\bar{\chi}_{\mathfrak{K}_1}(\sigma)\, d\sigma$$

In fact, if we write $\mathfrak{K} = \Sigma_i x_i \mathfrak{K}_i$, we have $\chi_{\mathfrak{K}} = \Sigma_i x_i \chi_{\mathfrak{K}_1}$ and our statement follows immediately from Proposition 2.

Corollary 2. *Two classes of representations of a compact Lie group coincide if and only if they have the same character.*

In fact, Corollary 1 shows that, if two classes $\mathfrak{K}$ and $\mathfrak{K}'$ have the same character, every irreducible class is contained the same number of times in $\mathfrak{K}$ and in $\mathfrak{K}'$.

§VII. THE REPRESENTATIVE RING

Definition 1. *The representative ring of a compact Lie group $\mathfrak{G}$ is the ring generated over the field of complex numbers by the coefficients of all representations of $\mathfrak{G}$.*

In other words, the elements of the representative ring are the complex valued functions on $\mathfrak{G}$ which may be expressed as polynomials in the coefficients of the representations of $\mathfrak{G}$.

More generally, let $\mathcal{E}$ be any set of representations of $\mathfrak{G}$. We shall denote by $\mathfrak{o}(\mathcal{E})$ the ring generated by the coefficients of the representations belonging to $\mathcal{E}$.

If $\mathcal{E}_1$, $\mathcal{E}_2$ are two sets of representations, we shall now find under what condition it is true that $\mathfrak{o}(\mathcal{E}_1) = \mathfrak{o}(\mathcal{E}_2)$.

We shall say that the set $\mathcal{E}$ is *closed* if the following conditions are satisfied:

1) If $\mathsf{P}_1 \varepsilon \mathcal{E}$, $\mathsf{P}_2 \varepsilon \mathcal{E}$, then also $\mathsf{P}_1 \dotplus \mathsf{P}_2 \varepsilon \mathcal{E}$, $\mathsf{P}_1 \times \mathsf{P}_2 \varepsilon \mathcal{E}$

2) If an irreducible representation P is contained in a representation belonging to $\mathcal{E}$, then $\mathsf{P} \varepsilon \mathcal{E}$.

3) A representation which is equivalent to a representation belonging to $\mathcal{E}$ belongs to $\mathcal{E}$.

Let $\mathcal{E}_1$ be any set of representations of $\mathfrak{G}$. Let us consider the set $\mathfrak{F}$ of all irreducible representations which are contained in representations of the form $\Lambda_1 \times \cdots \times \Lambda_h$, with $\Lambda_i \varepsilon \mathcal{E}_1$ $(1 \leqslant i \leqslant h)$. If P_1 and P_2 belong to $\mathfrak{F}$, then every irreducible representation which is contained in $\mathsf{P}_1 \times \mathsf{P}_2$ also belongs to $\mathfrak{F}$. Let $\mathcal{E}$ be the set of representations which are equivalent to representations of the form $\mathsf{P}_1 \dotplus \cdots \dotplus \mathsf{P}_k$ with $\mathsf{P}_j \varepsilon \mathfrak{F}$ $(1 \leqslant j \leqslant k)$. It is clear that $\mathcal{E}$ is closed and is the smallest closed set of representations which contains $\mathcal{E}_1$.

Proposition 1. *Let $\mathcal{E}_1$ be a set of representations of $\mathfrak{G}$ and let $\mathcal{E}$ be the smallest closed set of representations containing $\mathcal{E}_1$. The ring $\mathfrak{o}(\mathcal{E}_1)$ coincides with the set A of all linear combinations with complex coefficients of coefficients of irreducible representations belonging to $\mathcal{E}$.*

Let P be any irreducible representation belonging to $\mathcal{E}$. Then there exist representations $\Lambda_1, \cdots, \Lambda_h$ in $\mathcal{E}_1$ such that P is contained in $\Lambda_1 \times \cdots \times \Lambda_h$, i.e. there exists a regular matrix γ such that

$$\gamma(\Lambda_1 \times \cdots \times \Lambda_h)\gamma^{-1} = \mathsf{P} \dotplus \mathsf{N}$$

where N is some representation. Every coefficient of the representation $\Lambda_1 \times \cdots \times \Lambda_h$ is a product of coefficients of the representations $\Lambda_1, \cdots, \Lambda_h$ and hence belongs to $\mathfrak{o}(\mathcal{E}_1)$. It follows that the coefficients of $\mathsf{P} \dotplus \mathsf{N}$ (and in particular those of P) belong to $\mathfrak{o}(\mathcal{E}_1)$, whence $A \subset \mathfrak{o}(\mathcal{E}_1)$.

If Λ is any representation belonging to $\mathcal{E}$, the coefficients of Λ belong to A. In fact, we know that $\Lambda = \delta(\mathsf{P}_1 \dotplus \cdots \dotplus \mathsf{P}_k)\delta^{-1}$ where δ is a regular matrix and where $\mathsf{P}_1, \cdots, \mathsf{P}_k$ are irreducible and belong to $\mathcal{E}$.

If P and P' are two irreducible representations belonging to $\mathcal{E}$, we have $\mathsf{P} \times \mathsf{P}' \varepsilon \mathcal{E}$. It follows that the product of any coefficient of P by any coefficient of P' belongs to A, from which we deduce immediately that A is a ring. Since $A \subset \mathfrak{o}(\mathcal{E}_1)$ and every coefficient of a representation of the set $\mathcal{E}_1$ belongs to A, it follows that $A = \mathfrak{o}(\mathcal{E}_1)$.

Proposition 2. *Let $\mathcal{E}_1$ and $\mathcal{E}_2$ be two sets of representations of $\mathfrak{G}$. A necessary and sufficient condition for the equality $\mathfrak{o}(\mathcal{E}_1) = \mathfrak{o}(\mathcal{E}_2)$ to hold is for the smallest closed sets of representations containing $\mathcal{E}_1$ and $\mathcal{E}_2$ respectively to be equal.*

Proposition 1 shows that the condition is sufficient. To prove the converse, let us assume that there exists an irreducible representation P which belongs to the smallest closed set containing $\mathcal{E}_2$ but not to the smallest closed set containing $\mathcal{E}_1$. Let f be any coefficient $\neq 0$ of P. If g is any coefficient of an irreducible representation belonging to the smallest closed set containing $\mathcal{E}_1$, we have, by the orthogonality relations,

$$\int_{\mathfrak{G}} g(\sigma)\bar{f}(\sigma)\, d\sigma = 0 \tag{1}$$

It follows that the same formula holds for any $g \varepsilon \mathfrak{o}(\mathcal{E}_1)$. Since

$$\int_{\mathfrak{G}} \bar{f}(\sigma)f(\sigma)\, d\sigma > 0,$$

the function f does not belong to $\mathfrak{o}(\mathcal{E}_1)$, whence $\mathfrak{o}(\mathcal{E}_1) \neq \mathfrak{o}(\mathcal{E}_2)$.

Definition 2. *We say that a set $\mathcal{E}$ of representations of $\mathfrak{G}$ contains sufficiently many representations if the smallest closed set containing $\mathcal{E}$ is the set of all representations of $\mathfrak{G}$.*

It follows from Proposition 2 that this will happen if and only if $\mathfrak{o}(\mathcal{E})$ is the whole representative ring. Moreover, it follows from the proof of Proposition 2 that, if $\mathcal{E}$ does not contain sufficiently many representations, there exists an irreducible representation P of $\mathfrak{G}$ such that (1) holds for every $g\varepsilon\mathfrak{o}(\mathcal{E})$ and for every coefficient f of P.

Proposition 3. *If $\mathfrak{G}$ admits a faithful representation P_0, the set $\{\mathsf{P}_0, \bar{\mathsf{P}}_0\}$ contains sufficiently many representations ($\bar{\mathsf{P}}_0$ denotes the the imaginary conjugate representation of P_0).*

We denote by d_0 the degree of P_0 and we set $\mathsf{P}_0(\sigma) = (x_{ij}(\sigma))$ $(1 \leqslant i, j \leqslant d_0)$.

Lemma 1. *Let f be any continuous function on $\mathfrak{G}$ and let a be a number $\neq 0$. The ring generated by the $2d_0^2$ functions $x_{ij}(\sigma)$, $\bar{x}_{ij}(\sigma)$ contains a function f_1 such that $|f(\sigma) - f_1(\sigma)| \leqslant a$ for all $\sigma\varepsilon\mathfrak{G}$.*

We denote by $y_{i+d_0(j-1)}(\zeta)$, $y_{i+d_0(j-1)+d_0^2}(\zeta)$ the real and imaginary parts of the coefficients of a matrix ζ of degree d_0 $(1 \leqslant i, j \leqslant d_0)$.

The representation P_0 maps $\mathfrak{G}$ in a continuous univalent way onto a subgroup $\mathfrak{G}_1$ of $GL(d_0, C)$. Since $\mathfrak{G}$ is compact, P_0 is a homeomorphism and $\mathfrak{G}_1$ is compact. To every $\sigma\varepsilon\mathfrak{G}$ let us assign the point $\varphi(\sigma)\varepsilon R^{2d_0^2}$ whose coordinates are the numbers $y_1(\mathsf{P}_0(\sigma)), \cdots, y_{2d_0^2}(\mathsf{P}_0(\sigma))$. We clearly obtain a homeomorphism φ of $\mathfrak{G}$ with a compact subset K of $R^{2d_0^2}$. The function $f_2 = f \circ \varphi^{-1}$ is a continuous function defined on K. From a well known theorem in topology[1] it follows that f_2 may be extended to a continuous function defined on the whole of $R^{2d_0^2}$; we

[1] Cf. Tietze's extension theorem, Lefschetz, Algebraic Topology, 34.2, p. 28.

still denote by f_2 this extended function. Since K is compact, it is bounded; let M be an upper bound for the coordinates of the points of K. By the Weierstrass approximation theorem, we know that there exists a polynomial $Q(y_1, \cdots, y_{2d_0^2}) = Q(y)$ such that the inequality

$$|f_2(y) - Q(y)| \leqslant a$$

holds for all points $y = (y_1, \cdots, y_{2d_0^2})$ such that $|y_k| \leqslant M$ $(1 \leqslant k \leqslant 2d_0^2)$. We define the function f_1 by the formula

$$f_1(\sigma) = Q(y_1(\mathsf{P}_0(\sigma)), \cdots, y_{2d_0^2}(\mathsf{P}_0(\sigma)))$$

Then we have $|f(\sigma) - f_1(\sigma)| \leqslant a$ for all $\sigma \varepsilon \mathfrak{G}$. Since

$$y_{i+d_0(j-1)}(\zeta) = \tfrac{1}{2}(x_{ij}(\zeta) + \bar{x}_{ij}(\zeta))$$
$$y_{i+d_0(j-1)+d_0^2}(\zeta) = -\tfrac{1}{2}\sqrt{-1}\,(x_{ij}(\zeta) - \bar{x}_{ij}(\zeta))$$

the function f_1 may be expressed as a polynomial in the functions $x_{ij}(\sigma)$, $\bar{x}_{ij}(\sigma)$. Lemma 1 is proved.

Now, we can prove Proposition 3. Assume for a moment that the set $\mathcal{E}$ does not contain sufficiently many representations. From the remark which follows Proposition 2, it follows that there exists a continuous function $f \not\equiv 0$ on $\mathfrak{G}$ such that (1) holds for every $g \varepsilon \mathfrak{o}(\{\mathsf{P}_0, \bar{\mathsf{P}}_0\})$. Let m be an upper bound for the absolute value of f. Since $\int_{\mathfrak{G}} f(\sigma)\bar{f}(\sigma)\, d\sigma > 0$, we can find a number $a > 0$ such that

$$am < \int_{\mathfrak{G}} f(\sigma)\bar{f}(\sigma)\, d\sigma$$

By Lemma 1, there exists a function $f_1 \varepsilon \mathfrak{o}(\{\mathsf{P}_0, \bar{\mathsf{P}}_0\})$ such that $|f(\sigma) - f_1(\sigma)| \leqslant a$ for all $\sigma \varepsilon \mathfrak{G}$. It follows that

$$\left|\int_{\mathfrak{G}} f(\sigma)\bar{f}(\sigma)\, d\sigma\right| = \left|\int_{\mathfrak{G}} (f(\sigma) - f_1(\sigma))\bar{f}(\sigma)\, d\sigma\right| \leqslant am$$

which gives a contradiction. Proposition 3 is thereby proved.

Remark. It follows from Lemma 1 that, if $\mathfrak{G}$ is a compact Lie group which admits at least one faithful representation, then every continuous function on $\mathfrak{G}$ may be approximated as closely as we wish by a function belonging to the representative ring of $\mathfrak{G}$. Later on, we shall prove that this result holds independently of any assumption on the existence of representations and we shall derive from it the existence of a faithful representation.

Now, let $\mathfrak{H}$ be a closed subgroup of our compact Lie group $\mathfrak{G}$. If P is a representation of $\mathfrak{G}$, the contraction to $\mathfrak{H}$ of the mapping $\sigma \to \mathsf{P}(\sigma)$ is a representation of $\mathfrak{H}$, which we shall call the contraction of P to $\mathfrak{H}$.

Proposition 4. *Assume that a compact Lie group $\mathfrak{G}$ admits at least one faithful representation, and let $\mathfrak{H}$ be a closed subgroup of $\mathfrak{G}$. Then* *every irreducible representation of* $\mathfrak{H}$ *is contained in the contraction to* $\mathfrak{H}$ *of some representation of* $\mathfrak{G}$.

Let P_0 be a faithful representation of $\mathfrak{G}$ and let Λ_0 be the contraction of P_0 to $\mathfrak{H}$. Then Λ_0 is a faithful representation of $\mathfrak{H}$, and therefore the set $\{\Lambda_0, \bar{\Lambda}_0\}$ contains sufficiently many representations of $\mathfrak{H}$. It follows that every irreducible representation of $\mathfrak{H}$ is contained in some representation of the form $\Lambda_0 \times \cdots \times \Lambda_0 \times \bar{\Lambda}_0 \times \cdots \times \bar{\Lambda}_0$. But any such representation is clearly the contraction to $\mathfrak{H}$ of some representation of $\mathfrak{G}$.

Proposition 5. *Let* $\mathfrak{G}$ *be a compact Lie group which admits at least one faithful representation and let* $\mathfrak{H}$ *be a closed subgroup of* $\mathfrak{G}$. *If* $\mathfrak{H} \neq \mathfrak{G}$, *there exists at least one irreducible representation of* $\mathfrak{G}$, *distinct from the unit representation, whose contraction to* $\mathfrak{H}$ *contains the unit representation of* $\mathfrak{H}$.

Let us select a representation $\mathsf{P}_{\mathfrak{K}}$ in each class $\mathfrak{K}$ of equivalent representations of $\mathfrak{G}$. Denote by $f(i, j; \mathsf{P}_{\mathfrak{K}})$ the coefficients of $\mathsf{P}_{\mathfrak{K}}$. Every function f of the representative ring $\mathfrak{o}$ of $\mathfrak{G}$ may be written in the form

$$f = \Sigma_{i,j,\mathfrak{K}} a(i, j; \mathsf{P}_{\mathfrak{K}}) f(i, j; \mathsf{P}_{\mathfrak{K}})$$

where $\mathfrak{K}$ runs over all classes of irreducible representations and the $a(i, j; \mathsf{P}_{\mathfrak{K}})$ are constants of which only a finite number are $\neq 0$. Making use of the orthogonality relations, we obtain

$$\int_{\mathfrak{G}} f(\sigma)\, d\sigma = a(1, 1; \mathsf{E}) \tag{2}$$

where E is the unit representation.

Let us assume for a moment that the contractions to $\mathfrak{H}$ of the representations $\mathsf{P}_{\mathfrak{K}} \neq \mathsf{E}$ never contain the unit representation of $\mathfrak{H}$. Then the contraction of each $f(i, j; \mathsf{P}_{\mathfrak{K}})$ to $\mathfrak{H}$ will be a linear combination of coefficients of irreducible representations of $\mathfrak{H}$, and, if $\mathsf{P}_{\mathfrak{K}} \neq \mathsf{E}$, this expression will involve only coefficients of irreducible representations of $\mathfrak{H}$ distinct from the unit representation.

By the same argument which was used in proving (2), we see that we shall have

$$I(f; \mathfrak{H}) = a(1, 1; \mathsf{E}) = \int_{\mathfrak{G}} f(\sigma)\, d\sigma$$

where $I(*; \mathfrak{H})$ is the invariant integral over the compact group $\mathfrak{H}$, normalized as usual in such a way that $I(1; \mathfrak{H}) = 1$.

The equality $I(f; \mathfrak{H}) = \int_{\mathfrak{G}} f(\sigma)\, d\sigma$, which holds for all functions f of the representative ring $\mathfrak{o}$, holds also for every continuous function on $\mathfrak{G}$, since such a function can be approximated as closely as we wish by a function of $\mathfrak{o}$.

Since $\mathfrak{H} \neq \mathfrak{G}$, there exists a continuous function f on $\mathfrak{G}$ which satisfies the following conditions: 1) f vanishes everywhere on $\mathfrak{H}$; 2) f does not vanish identically on $\mathfrak{G}$.[1] Then we have $I(f\bar{f}; \mathfrak{H}) = 0$, $\int_{\mathfrak{G}} f(\sigma)\bar{f}(\sigma)\,d\sigma \neq 0$, which brings a contradiction.

Let us now return to the study of an arbitrary compact Lie group $\mathfrak{G}$, without making the assumption that $\mathfrak{G}$ admits a faithful representation. Let $\mathfrak{N}$ be the set of those elements $\sigma \varepsilon \mathfrak{G}$ which are represented by unit matrices in all representations of $\mathfrak{G}$. It is clear that $\mathfrak{N}$ is a closed distinguished subgroup of $\mathfrak{G}$. We denote by $\mathfrak{G}_1$ the group $\mathfrak{G}/\mathfrak{N}$. Every representation of $\mathfrak{G}$ maps $\mathfrak{N}$ upon the unit matrix, and hence defines a representation of $\mathfrak{G}_1$. Conversely, every representation of $\mathfrak{G}_1$ will correspond to a representation of $\mathfrak{G}$. It follows that the representative rings of $\mathfrak{G}$ and $\mathfrak{G}_1$ are isomorphic.

Furthermore, if σ is any element of $\mathfrak{G}_1$ other than the neutral element, there exists a representation P of $\mathfrak{G}_1$ such that $\mathsf{P}(\sigma)$ is not the unit matrix. We shall deduce from this fact that $\mathfrak{G}_1$ admits a faithful representation.

Since $\mathfrak{G}_1$ is a Lie group, there exists an open neighbourhood V_1 of the neutral element ϵ_1 in $\mathfrak{G}_1$ which does not contain any subgroup of $\mathfrak{G}_1$;[2] we denote by F the complement of V_1 in $\mathfrak{G}_1$. If P is any representation of $\mathfrak{G}_1$, we denote by $\mathfrak{N}(\mathsf{P})$ the kernel of the representation P. It is clear that $\mathfrak{N}(\mathsf{P})$ is a closed subgroup of $\mathfrak{G}_1$ and that the intersection of the groups $\mathfrak{N}(\mathsf{P})$ for all representations P is the set $\{\epsilon_1\}$. Hence we have

$$\bigcap_{\mathsf{P}} (\mathfrak{N}(\mathsf{P}) \cap F) = \phi$$

Since F is compact, there exists a finite set $\{\mathsf{P}_1, \cdots, \mathsf{P}_k\}$ of representations of $\mathfrak{G}_1$ such that $\bigcap_{i=1}^{k} (\mathfrak{N}(\mathsf{P}_i) \cap F) = \phi$. It follows that $\bigcap_{i=1}^{k} \mathfrak{N}(\mathsf{P}_i) \subset V_1$; since the left side of this inclusion is a group, this group must be $\{\epsilon_1\}$. It follows immediately that $\mathsf{P}_1 \dotplus \cdots \dotplus \mathsf{P}_k$ is a faithful representation of $\mathfrak{G}_1$.

From this, we deduce easily.

Proposition 6. *Let $\mathfrak{G}$ be a compact Lie group. There exists a representation P of $\mathfrak{G}$ with the following property: every element of the kernel of P is also mapped upon a unit matrix by any other representation*

[1] Let σ be an element of $\mathfrak{G}$ not contained in $\mathfrak{H}$. The function defined on $\mathfrak{H} \cup \{\sigma\}$ as being equal to 0 on $\mathfrak{H}$ and to 1 at σ may be extended to a continuous function on $\mathfrak{G}$ (cf. l.c. note 1, p. 190).

[2] This follows easily from Lemma 1, §XIII, Chapter IV, p. 127.

of $\mathfrak{G}$. The coefficients of the representations P *and* $\bar{\mathsf{P}}$ *form a system of generators of the representative ring of* $\mathfrak{G}$.

The last statement follows from Proposition 3, applied to the group $\mathfrak{G}_1 = \mathfrak{G}/\mathfrak{N}$, where $\mathfrak{N}$ is the kernel of P.

§VIII. THE ALGEBRAIC STRUCTURE OF THE REPRESENTATIVE RING

We shall denote by $\mathfrak{G}$ a compact Lie group and by $\mathfrak{o}$ the representative ring of $\mathfrak{G}$.

If P is any representation of $\mathfrak{G}$, we shall denote by $d(\mathsf{P})$ the degree of P and by $f(i, j; \mathsf{P})$ the (i, j)-coefficient of P $(1 \leqslant i, j \leqslant d(\mathsf{P}))$; $\mathfrak{o}$ is therefore the ring generated by all functions of the form $f(i, j; \mathsf{P})$. We know that there exists a finite set of representations of $\mathfrak{G}$ which contains sufficiently many representations; if $\{\mathsf{P}_1, \cdots, \mathsf{P}_h\}$ is such a set, the functions $f(i, j; \mathsf{P}_k)$ $(1 \leqslant i, j \leqslant d(\mathsf{P}_k), 1 \leqslant k \leqslant h)$ form a set of generators of $\mathfrak{o}$. We wish to find the algebraic relations which hold among these generators.

It will be convenient to introduce new independent variables $u(i, j; \mathsf{P})$ $(1 \leqslant i, j \leqslant d(\mathsf{P})$, P running over all representations of $\mathfrak{G})$. Let $\mathfrak{u}$ be the ring of polynomials in the variables $u(i, j; \mathsf{P})$ with coefficients in C (there are infinitely many variables, but each polynomial contains only a finite number of them). There exists a homomorphism of $\mathfrak{u}$ onto $\mathfrak{o}$ which maps each $u(i, j; \mathsf{P})$ upon the corresponding $f(i, j; \mathsf{P})$. Let $\mathfrak{a}$ be the kernel of this homomorphism. We propose to determine $\mathfrak{a}$ by exhibiting a set of generators of this ideal.

To every representation P let us assign the matrix $U(\mathsf{P})$ of degree $d(\mathsf{P})$ whose coefficients are the $u(i, j; \mathsf{P})$. Among the polynomials belonging to $\mathfrak{a}$, we find in particular the following ones:

1) The coefficients of the matrices $U(\mathsf{P}_1 \dotplus \mathsf{P}_2) - (U(\mathsf{P}_1) \dotplus U(\mathsf{P}_2))$, $U(\mathsf{P}_1 \times \mathsf{P}_2) - U(\mathsf{P}_1) \times U(\mathsf{P}_2)$ for all possible choices of the representations $\mathsf{P}_1, \mathsf{P}_2$;

2) the coefficients of the matrices $U(\gamma\mathsf{P}\gamma^{-1}) - \gamma U(\mathsf{P})\gamma^{-1}$, where γ is any regular matrix of degree $d(\mathsf{P})$;

3) the polynomial $u(1, 1; \mathsf{E}) - 1$, where E stands for the unit representation of $\mathfrak{G}$.

Proposition 1. *The polynomials listed under the headings* 1), 2), 3) *form a set of generators of the ideal* $\mathfrak{a}$.

Let $\mathfrak{a}_1$ be the smallest ideal containing the polynomials listed in 1), 2), 3). We have to prove that $\mathfrak{a}_1 = \mathfrak{a}$.

Let us select a representation in each class of equivalent irreducible representations; let $\{\mathsf{P}_\alpha\}_{\alpha \in A}$ be the set of representations obtained in this way (A being some set of indices). We assert that every polynomial in $\mathfrak{u}$ is congruent modulo $\mathfrak{a}_1$ to some finite linear combination of the variables $u(i, j; \mathsf{P}_\alpha)$ $(1 \leqslant i, j \leqslant d(\mathsf{P}_\alpha); \alpha \varepsilon A)$. It is sufficient to prove this for the constant 1, for each variable $u(i, j; \mathsf{P})$ and for the products of any two of these variables. We have $1 \equiv u(1, 1; \mathsf{E})$ $(\text{mod } \mathfrak{a}_1)$, and E is certainly one of the representations P_α. If P is any representation, there exists a matrix γ such that $\mathsf{P} = \gamma(\mathsf{P}_{\alpha_1} \dot{+} \cdots \dot{+} \mathsf{P}_{\alpha_h})\gamma^{-1}$ $(\alpha_1, \cdots, \alpha_h \varepsilon A)$; it follows that the coefficients of $U(\mathsf{P})$ are congruent modulo $\mathfrak{a}_1$ to the corresponding coefficients of $\gamma(U(\mathsf{P}_{\alpha_1}) \dot{+} \cdots \dot{+} U(\mathsf{P}_{\alpha_h}))\gamma^{-1}$, which are themselves linear combinations of the variables $u(i, j; \mathsf{P}_\alpha)$. Finally, we observe that $u(i, j; \mathsf{P})u(i', j'; \mathsf{P}')$ is a coefficient of $U(\mathsf{P}) \times U(\mathsf{P}')$ and is therefore congruent modulo $\mathfrak{a}_1$ to a coefficient of $U(\mathsf{P} \times \mathsf{P}')$, i.e. also to a linear combination of the variables $u(i, j; \mathsf{P}_\alpha)$. Our assertion is thereby proved.

Let P be any polynomial in the ideal $\mathfrak{a}$. We have

$$P \equiv \Sigma a(i, j; \alpha)u(i, j; \mathsf{P}_\alpha) \qquad (\text{mod } \mathfrak{a}_1)$$

where the $a(i, j; \alpha)$'s are constants. It follows that $\Sigma a(i, j; \alpha)u(i, j; \mathsf{P}_\alpha) \varepsilon \mathfrak{a}$, whence $\Sigma a(i, j; \alpha)f(i, j; \mathsf{P}_\alpha) = 0$. Multiplying by $\bar{f}(k, l; \mathsf{P}_\alpha)$ and integrating over the group, we obtain (by the orthogonality relations) $a(k, l; \alpha) = 0$ for all combinations $(k, l; \alpha)$. It follows that $P \varepsilon \mathfrak{a}_1$; Proposition 1 is proved.

If we know any system of generators $\{z_1, \cdots, z_m\}$ of the representative ring $\mathfrak{o}$, we may obtain the algebraic relations among these generators in the following way: we express each $f(i, j; \mathsf{P})$ as a polynomial in the quantities z, and we substitute these expressions in the relations among functions f which result from 1), 2), 3).

We shall now study the homomorphisms $\bar{\omega}$ of the representative ring $\mathfrak{o}$ into the field of complex numbers. If $z_1, \cdots, z_m$ form a system of generators of $\mathfrak{o}$, a homomorphism $\bar{\omega}$ is uniquely determined when the numbers $\bar{\omega}(z_1) = a_1, \cdots, \bar{\omega}(z_m) = a_m$ are given. These numbers cannot be taken arbitrarily but must satisfy the relations $P(a_1, \cdots, a_m) = 0$, where the $P(z_1, \cdots, z_m) = 0$ are the relations which hold among $z_1, \cdots, z_m$ in $\mathfrak{o}$. It follows that the homomorphisms of $\mathfrak{a}$ into C may be identified with the points of the algebraic variety defined by the equations $P = 0$.

Definition 1. *Let $\mathfrak{o}$ be the representative ring of a compact Lie group $\mathfrak{G}$. The set of homomorphisms of $\mathfrak{o}$ into the field C of complex numbers is called the algebraic variety associated with $\mathfrak{G}$. This variety will be denoted by $\mathfrak{M}(\mathfrak{G})$.*

If a system of generators $\{z_1, \cdots, z_m\}$ is given, the set of points

$(\bar{\omega}(z_1), \cdots, \bar{\omega}(z_m))(\bar{\omega}\varepsilon\mathfrak{M}(\mathfrak{G}))$ is called the *model* of $\mathfrak{M}(\mathfrak{G})$ corresponding to the generators $z_1, \cdots, z_m$.

Definition 2. *Let $\mathfrak{R}$ be the set of all representations of a compact Lie group $\mathfrak{G}$. By a representation of $\mathfrak{R}$ we understand a mapping ζ which assigns to every $\mathsf{P}\varepsilon\mathfrak{R}$ a regular matrix $\zeta(\mathsf{P})$ of degree equal to the degree $d(\mathsf{P})$ of P, in such a way that the equalities*

$$(1)\quad \zeta(\mathsf{P}_1 \dot{+} \mathsf{P}_2) = \zeta(\mathsf{P}_1) \dot{+} \zeta(\mathsf{P}_2); \qquad \zeta(\mathsf{P}_1 \times \mathsf{P}_2) = \zeta(\mathsf{P}_1) \times \zeta(\mathsf{P}_2); \quad \zeta(\gamma\mathsf{P}\gamma^{-1}) = \gamma\zeta(\mathsf{P})\gamma^{-1}$$

hold for any representations P_1, P_2, P of $\mathfrak{G}$ and any regular matrix γ of degree $d(\mathsf{P})$.

Let $\bar{\omega}$ be any homomorphism of the representative ring $\mathfrak{o}$ into C. Let us assign to any representation P the matrix $\zeta_{\bar{\omega}}(\mathsf{P})$ whose coefficients are the numbers $\bar{\omega}(f(i, j; \mathsf{P}))$ $(1 \leqslant i, j \leqslant d(\mathsf{P}))$; we obtain a mapping $\zeta_{\bar{\omega}}$ which assigns a matrix to every element of $\mathfrak{R}$.

Proposition 2. *If $\bar{\omega}\varepsilon\mathfrak{M}(\mathfrak{G})$, the mapping $\zeta_{\bar{\omega}}$ is a representation of $\mathfrak{R}$ and the mapping $\bar{\omega} \to \zeta_{\bar{\omega}}$ is a one-to-one mapping of $\mathfrak{M}(\mathfrak{G})$ onto the set of representations of $\mathfrak{R}$.*

The conditions (1) are obviously satisfied for $\zeta_{\bar{\omega}}$. In order to prove that $\zeta_{\bar{\omega}}$ is a representation of $\mathfrak{R}$, it will therefore be sufficient to prove that $\zeta_{\bar{\omega}}(\mathsf{P})$ is a regular matrix. We have $\mathsf{P}^*(\sigma) \cdot {}^t\mathsf{P}(\sigma) = 1$ (the unit matrix) for every $\sigma\varepsilon\mathfrak{G}$. It follows that

$$\Sigma_j f(i, j; \mathsf{P}^*)f(k, j; \mathsf{P}) = \delta_{ik} \qquad (1 \leqslant i, k \leqslant d(\mathsf{P}))$$

Since $\bar{\omega}$ is a homomorphism, we have

$$\Sigma_j\bar{\omega}(f(i, j; \mathsf{P}^*))\bar{\omega}(f(k, j; \mathsf{P})) = \delta_{ik}$$

whence $\zeta_{\bar{\omega}}(\mathsf{P}^*) \cdot {}^t\zeta_{\bar{\omega}}(\mathsf{P}) = 1$, which proves that $\zeta_{\bar{\omega}}(\mathsf{P})$ is regular and that

$$(2)\qquad \zeta_{\bar{\omega}}(\mathsf{P}^*) = (\zeta_{\bar{\omega}}(\mathsf{P}))^*$$

If $\bar{\omega}_1$, $\bar{\omega}_2$ are distinct elements of $\mathfrak{M}(\mathfrak{G})$, there exists a representative function $f(i, j; \mathsf{P})$ such that $\bar{\omega}_1(f(i, j; \mathsf{P})) \neq \bar{\omega}_2(f(i, j; \mathsf{P}))$; it follows that $\zeta_{\bar{\omega}_1}(\mathsf{P}) \neq \zeta_{\bar{\omega}_2}(\mathsf{P})$.

Finally, let ζ be any representation of $\mathfrak{R}$. To each variable $u(i, j; \mathsf{P})$ let us assign the number $\bar{\bar{\omega}}(u(i, j; \mathsf{P}))$ which stands at the intersection of the i-th row and the j-th column of $\zeta(\mathsf{P})$. Since the variables $u(i, j; \mathsf{P})$ are algebraically independent, $\bar{\bar{\omega}}$ may be extended to a homomorphism (also denoted by $\bar{\bar{\omega}}$) of the ring of polynomials in the variables u. Since ζ is a representation of $\mathfrak{R}$, the polynomials in the categories 1), 2) above are obviously mapped upon 0. Moreover, let E be the unit representation; then $\zeta(\mathsf{E})$ is a number $\neq 0$ which is equal to its own square in virtue of the equality $\mathsf{E} \times \mathsf{E} = \mathsf{E}$. It follows that $\zeta(\mathsf{E}) = 1$, whence $\bar{\bar{\omega}}(u(1, 1; \mathsf{E}) - 1) = 0$.

By Proposition 1, the polynomials of the ideal $\mathfrak{a}$ are all mapped on 0 by the homomorphism $\bar{\bar{\omega}}$. Since $\mathfrak{o}$ may be identified with the factor ring of $\mathfrak{a}$ in the polynomial ring, we see that $\bar{\bar{\omega}}$ defines in a natural way a homomorphism $\bar{\omega}$ of $\mathfrak{o}$ into C. It is clear that $\zeta = \zeta_{\bar{\omega}}$. Proposition 2 is thereby proved.

Remark. We have incidentally proved that, if ζ is any representation of $\mathfrak{R}$, then we have

$$\zeta(\mathsf{P}^*) = (\zeta(\mathsf{P}))^*$$

for any representation P of $\mathfrak{G}$.

Let ζ_1, ζ_2 be any two representations of $\mathfrak{R}$. We see immediately that the mapping $\mathsf{P} \rightarrow \zeta_1(\mathsf{P})\zeta_2^{-1}(\mathsf{P})$ is again a representation of $\mathfrak{R}$. It follows immediately that the representations of $\mathfrak{R}$ form a group.

Definition 3. *Making use of the correspondence established by Proposition* 2 *between the elements of* $\mathfrak{M}(\mathfrak{G})$ *and the representations of* $\mathfrak{R}$, *a group structure is defined in* $\mathfrak{M}(\mathfrak{G})$. *The group obtained in this way is called the algebraic group associated with* $\mathfrak{G}$.

We shall now introduce a topology in $\mathfrak{M}(\mathfrak{G})$. To every set of generators $\{z_1, \cdots, z_m\}$ of $\mathfrak{o}$ there corresponds a model M_z of $\mathfrak{M}(\mathfrak{G})$, and the elements of $\mathfrak{M}(\mathfrak{G})$ are in a one-to-one correspondence with the points of this model. Since $M_z \subset C^m$, M_z carries a natural topology (induced by the topology in C^m); the one-to-one correspondence between M_z and $\mathfrak{M}(\mathfrak{G})$ therefore defines a topology in $\mathfrak{M}(\mathfrak{G})$. We shall see that this topology does not depend upon the choice of the model M_z. Let $M_{z'}$ be an other model, defined by a set of generators $\{z'_1, \cdots, z'_{m'}\}$. Then each z_i can be expressed as a polynomial in $z'_1, \cdots, z'_{m'}$, and each $z'_{i'}$ can be expressed as a polynomial in $z_1, \cdots, z_m$. It follows immediately that the correspondence between points of M_z and $M_{z'}$ which correspond to the same element of $\mathfrak{M}(\mathfrak{G})$ is a homeomorphism, which proves our assertion.

Let $\{\mathsf{P}_1, \cdots, \mathsf{P}_h\}$ be a system of sufficiently many representations of $\mathfrak{G}$. We set $\mathsf{P}_0 = \mathsf{P}_1 \dotplus \cdots \dotplus \mathsf{P}_h + \bar{\mathsf{P}}_1 \dotplus \cdots \dotplus \bar{\mathsf{P}}_h$ and we denote by d_0 the degree of P_0. We know that the d_0^2 functions $f(i, j; \mathsf{P}_0)$ $(1 \leqslant i, j \leqslant d)$ form a set of generators of $\mathfrak{o}$. Let us assign to every $\bar{\omega} \in \mathfrak{M}(\mathfrak{G})$ the matrix $\zeta_{\bar{\omega}}(\mathsf{P}_0)$, which we also denote by $\bar{\omega}(\mathsf{P}_0)$. It is clear that we obtain a representation of the group $\mathfrak{M}(\mathfrak{G})$ by matrices of degree d_0. Since the functions $f(i, j; \mathsf{P}_0)$ give rise to a model of

$\mathfrak{M}(\mathfrak{G})$, this representation is faithful and is a homeomorphism. It follows immediately that the topological structure and the group structure on $\mathfrak{M}(\mathfrak{G})$ combine to define $\mathfrak{M}(\mathfrak{G})$ as a topological group.

More generally, if P is any representation of $\mathfrak{G}$, the mapping $\bar{\omega} \to \zeta_{\bar{\omega}}(\mathsf{P})$ is a representation of $\mathfrak{M}(\mathfrak{G})$. We shall denote this representation by $\tilde{\mathsf{P}}$.

Proposition 3. *Let* P *be a representation of* $\mathfrak{G}$ *whose coefficients form a system of generators of the representative ring of* $\mathfrak{G}$. *Then* $\tilde{\mathsf{P}}$ *maps* $\mathfrak{M}(\mathfrak{G})$ *onto the set of all matrices* (x_{ij}) *which have the following property: if* P *is any polynomial such that* $P(\cdots f(i, j; \mathsf{P}) \cdots) = 0$ *identically, then* $P(\cdots x_{ij} \cdots) = 0$.

In fact, if $\bar{\omega} \varepsilon \mathfrak{M}(\mathfrak{G})$, the coefficients of $\tilde{\mathsf{P}}(\bar{\omega}) = \bar{\omega}(\mathsf{P})$ are the numbers $\bar{\omega}(f(i, j; \mathsf{P}))$. Since $\bar{\omega}$ is a homomorphism, the equality $P(\cdots f(i, j; \mathsf{P}) \cdots) = 0$ implies $P(\cdots \bar{\omega}(f(i, j; \mathsf{P})) \cdots) = 0$.

Conversely, let (x_{ij}) be any matrix which satisfies our condition. Since the elements $f(i, j; \mathsf{P})$ form a system of generators of the representative ring $\mathfrak{o}$, our condition implies the existence of a homomorphism $\bar{\omega}$ of $\mathfrak{o}$ into C such that $\bar{\omega}(f(i, j; \mathsf{P})) = x_{ij}$ $(1 \leqslant i, j \leqslant d(\mathsf{P}))$. It follows that the matrix (x_{ij}) is equal to $\bar{\omega}(\mathsf{P})$, which completes the proof of Proposition 3.

Remark. It can be proved that, if P is any representation whatsoever of $\mathfrak{G}$, then $\tilde{\mathsf{P}}$ maps $\mathfrak{M}(\mathfrak{G})$ onto the set of those matrices which are regular and satisfy the condition stated in Proposition 3. We shall omit this proof, which is a little more difficult.

Corollary. *The algebraic group associated with* $\mathfrak{G}$ *is a Lie group.*

In fact, this group is isomorphic (as a topological group) with a subgroup of $GL(d; C)$ which is defined by algebraic relations between coefficients and is therefore closed in $GL(d; C)$. At the same time, it now becomes clear why we have called this group the *algebraic* group associated with $\mathfrak{G}$.

§IX. TOPOLOGICAL STRUCTURE OF THE ASSOCIATED GROUP

Let $\mathfrak{G}$ be a compact Lie group, and let $\mathfrak{M}(\mathfrak{G})$ be its associated algebraic group. If f is any function belonging to the representative ring $\mathfrak{o}$ of $\mathfrak{G}$, then the imaginary conjugate $\bar{f}$ of f also belongs to $\mathfrak{o}$. The mapping $f \to \bar{f}$ is an automorphism of $\mathfrak{o}$ which changes every complex constant into its imaginary conjugate. Now, let $\bar{\omega}$ be any homomorphism of $\mathfrak{o}$ into C; it is quite easy to verify that the mapping $f \to \overline{\bar{\omega}(\bar{f})}$ is again a homomorphism of $\mathfrak{o}$ into C. We shall denote this new homomorphism by $\bar{\omega}^{\iota}$.

Let P be any representation of $\mathfrak{G}$. We have

$$\tilde{\mathsf{P}}(\bar{\omega}^{\iota}) = \bar{\omega}^{\iota}(\mathsf{P}) = \overline{\bar{\omega}(\bar{\mathsf{P}})} \tag{1}$$

where $\bar{\mathsf{P}}$ is the imaginary conjugate representation of P. It follows that $(\bar{\omega}_1\bar{\omega}_2)^{\iota} = \bar{\omega}_1^{\iota}\bar{\omega}_2^{\iota}$. Since $(\bar{\omega}^{\iota})^{\iota} = \bar{\omega}$, the operation $^{\iota}$ appears as an automorphism of order 2 of the group $\mathfrak{M}(\mathfrak{G})$. This automorphism is obviously also a homeomorphism of $\mathfrak{M}(\mathfrak{G})$ with itself.

It follows that the set of elements $\bar{\omega}$ for which $\bar{\omega} = \bar{\omega}^{\iota}$ is a closed subgroup $\mathfrak{G}_1$ of $\mathfrak{M}(\mathfrak{G})$.

Let σ be any element in $\mathfrak{G}$. The mapping $f \rightarrow f(\sigma)(f \varepsilon \mathfrak{o})$ is clearly a homomorphism $\bar{\omega}_\sigma$ of $\mathfrak{o}$ into C. If P is any representation of $\mathfrak{G}$, we have $\tilde{\mathsf{P}}(\bar{\omega}_\sigma) = \bar{\omega}_\sigma(\mathsf{P}) = \mathsf{P}(\sigma)$; it follows that the mapping $\sigma \rightarrow \bar{\omega}_\sigma$ is a continuous homomorphism of $\mathfrak{G}$ into $\mathfrak{M}(\mathfrak{G})$. This homomorphism maps $\mathfrak{G}$ onto some compact subgroup $\mathfrak{G}_2$ of $\mathfrak{M}(\mathfrak{G})$. Since $\bar{f}(\sigma) = \overline{f(\sigma)}$, we have $\bar{\omega}_\sigma^{\iota} = \bar{\omega}_\sigma$, whence $\mathfrak{G}_2 \subset \mathfrak{G}_1$. We shall prove that $\mathfrak{G}_1 = \mathfrak{G}_2$.

Proposition 1. *Let $\bar{\omega}$ be an element of $\mathfrak{M}(\mathfrak{G})$ such that $\bar{\omega} = \bar{\omega}^{\iota}$, being the automorphism defined by* (1). *Then there exists an element $\sigma \varepsilon \mathfrak{G}$ such that $\bar{\omega}(f) = f(\sigma)$ for every $f \varepsilon \mathfrak{o}$.*

We first prove

Lemma 1. *The group $\mathfrak{G}_1$ is compact.*

Let P_0 be a unitary representation of $\mathfrak{G}$ whose coefficients form a system of generators of $\mathfrak{o}$. We have, for any $\bar{\omega} \varepsilon \mathfrak{M}(\mathfrak{G})$,

$$\overline{\bar{\omega}^{\iota}(\mathsf{P}_0)} = \bar{\omega}(\bar{\mathsf{P}}_0) = \bar{\omega}(\mathsf{P}_0^*) = (\bar{\omega}(\mathsf{P}_0))^*$$

(cf. Remark after Proposition 2, §VIII, p. 197). Therefore, if $\bar{\omega} = \bar{\omega}^{\iota}$, the matrix $\bar{\omega}(\mathsf{P}_0) = \tilde{\mathsf{P}}_0(\bar{\omega})$ is unitary. On the other hand, $\tilde{\mathsf{P}}_0$ is a faithful representation of $\mathfrak{M}(\mathfrak{G})$ and therefore maps $\mathfrak{G}_1$ topologically onto a closed subgroup of $GL(d(\mathsf{P}_0), C)$. Since this subgroup is contained in $U(d(\mathsf{P}_0))$, it is compact, which proves Lemma 1.

Since $\mathfrak{G}_1$ and $\mathfrak{G}_2$ are compact Lie groups, in order to prove that $\mathfrak{G}_1 = \mathfrak{G}_2$, we may apply the criterion of Proposition 5, §VII, p. 192. Let Λ be any representation of $\mathfrak{G}_1$; the contraction of Λ to $\mathfrak{G}_2$ is a representation P of $\mathfrak{G}_2$. Assuming that P contains the unit representation of $\mathfrak{G}_2$, we want to prove that Λ contains the unit representation of $\mathfrak{G}_1$. There exists by assumption a regular matrix γ such that

$$\gamma\Lambda(\bar{\omega}_\sigma)\gamma^{-1} = \begin{pmatrix} 1 & 0 \ldots 0 \\ 0 & \\ . & \mathsf{M}(\sigma) \\ . & \\ 0 & \end{pmatrix}$$

for all $\sigma \varepsilon \mathfrak{G}$. Let $f_{kl}(\bar{\omega})$ be the coefficients of the representation $\gamma\Lambda\gamma^{-1}$ of $\mathfrak{G}_1$ $(1 \leqslant k, l \leqslant d(\Lambda))$. Making use of the representation P_0 introduced in the proof of Lemma 1, we denote by $x_{ij}(\bar{\omega})$ the coefficients of the representation $\tilde{\mathsf{P}}_0$ of $\mathfrak{M}(\mathfrak{G})$. The contraction of $\tilde{\mathsf{P}}_0$ to $\mathfrak{G}_1$ being a faithful representation of $\mathfrak{G}_1$, we know that the functions $x_{ij}(\bar{\omega})$, $\bar{x}_{ij}(\bar{\omega})$ form a system of generators of the representative ring of $\mathfrak{G}_1$. It follows that we have, for $\bar{\omega} \varepsilon \mathfrak{G}_1$,

$$f_{kl}(\bar{\omega}) = F_{kl}(\cdot \cdot \cdot, x_{ij}(\bar{\omega}), \bar{x}_{ij}(\bar{\omega}), \cdot \cdot \cdot)$$

where each F_{kl} is a polynomial with complex coefficients. We have

$$F_{1l}(\cdot \cdot \cdot, x_{ij}(\bar{\omega}_\sigma), \bar{x}_{ij}(\bar{\omega}_\sigma), \cdot \cdot \cdot) = \delta_{1l} \qquad (\sigma \varepsilon \mathfrak{G})$$

On the other hand, we have $x_{ij}(\bar{\omega}_\sigma) = f(i, j; \mathsf{P}_0)(\sigma)$ for all $\sigma \varepsilon \mathfrak{G}$ and $x_{ij}(\bar{\omega}) = \bar{\omega}(f(i, j; \mathsf{P}_0))$ for all $\bar{\omega} \varepsilon \mathfrak{G}_1$. Since $\bar{\omega} = \bar{\omega}'$, we have $\bar{x}_{ij}(\bar{\omega}) = \bar{\omega}(\bar{f}(i, j; \mathsf{P}_0))$. Since $\bar{\omega}$ is a homomorphism, the relations

$$F_{1l}(\cdot \cdot \cdot, f(i, j; \mathsf{P}_0), \bar{f}(i, j; \mathsf{P}_0), \cdot \cdot \cdot) = \delta_{1l}$$

imply $F_{1l}(\cdot \cdot \cdot, x_{ij}(\bar{\omega}), \bar{x}_{ij}(\bar{\omega}), \cdot \cdot \cdot) = \delta_{1l}$, which proves that

$$\gamma\Lambda(\bar{\omega})\gamma^{-1} = \begin{pmatrix} 1 & 0 \ldots 0 \\ * & * \ldots * \\ * & * \ldots * \end{pmatrix} \qquad (\bar{\omega} \varepsilon \mathfrak{G}_1)$$

and hence that Λ contains the unit representation of $\mathfrak{G}_1$. Proposition 1 is thereby proved.

In the case where $\mathfrak{G}$ has at least one faithful representation, the representation P_0 which was used in the proof is faithful. It then follows immediately from Proposition 1 that $\mathfrak{G}_1$ is isomorphic with $\mathfrak{G}$ (as a topological group).

Proposition 2. *If the compact Lie group $\mathfrak{G}$ admits a faithful representation, the group $\mathfrak{M}(G)$ is homeomorphic to the product of $\mathfrak{G}$ and R^n, where n is the dimension of $\mathfrak{G}$.*

We make use again of the representation P_0 of $\mathfrak{G}$ which was introduced in the proof of Lemma 1. Under our assumptions, P_0 is faithful. The corresponding representation $\tilde{\mathsf{P}}_0$ of $\mathfrak{M}(\mathfrak{G})$ maps topologically this last group onto a linear group $\mathfrak{H}$, and maps the elements of $\mathfrak{G}_1$ onto unitary matrices. Furthermore, the argument used in the proof of Lemma 1 shows immediately that the following statements hold true: if τ is any matrix in $\mathfrak{H}$, then τ^* also belongs to $\mathfrak{H}$; if τ is unitary, then τ represents an element of $\mathfrak{G}_1$. On the other hand, it follows from Proposition 3, §VIII, p. 198 that $\mathfrak{H}$ is an algebraic group. Proposition 2 will therefore follow from

Lemma 2. *Let $\mathfrak{H}$ be an algebraic subgroup of $GL(d; C)$ such that the condition $\tau \varepsilon \mathfrak{H}$ implies $\tau^* \varepsilon \mathfrak{H}$. If τ is any matrix in $\mathfrak{H}$, and if $\tau = \sigma\rho$ is the representation of τ as the product of an unitary matrix σ and a positive definite hermitian matrix ρ, then we have $\sigma \varepsilon \mathfrak{H}$, $\rho \varepsilon \mathfrak{H}$. The group $\mathfrak{H}$ is homeomorphic to the product of $\mathfrak{H} \cap U(d)$ and R^n, where n is the dimension of $\mathfrak{H} \cap U(d)$.*

We can find a matrix μ such that $\rho_1 = \mu\rho\mu^{-1}$ is a diagonal matrix; the coefficients of the main diagonal of μ are positive real numbers. We may write $\rho_1 = \exp \alpha_1$, with

$$\alpha_1 = \begin{vmatrix} a_1 & 0 \dots\dots 0 \\ 0 & a_2 \dots 0 \\ \dots\dots & \dots\dots \\ 0 & 0 \dots\dots a_d \end{vmatrix}$$

where $a_1, \cdots, a_d$ are real numbers.

The group $\mathfrak{H}_1 = \mu\mathfrak{H}\mu^{-1}$ is obviously also algebraic. We have $\rho^2 = (\tau^*)^{-1}\tau \varepsilon \mathfrak{H}$, whence $\rho_1^2 \varepsilon \mathfrak{H}_1$ and $\rho_1^{2k} \varepsilon \mathfrak{H}_1$ for every integer k.

Let $F(\cdots x_{ij} \cdots) = 0$ be any one of the algebraic equations which define $\mathfrak{H}_1$, and let $F'(x_1, \cdots, x_n)$ be the polynomial deduced from F by the substitution $x_{ij} \to 0$ if $i \neq j$, $x_{ii} \to x_i$. We have

$$(1) \qquad F'(e^{2ka_1}, \cdots, e^{2ka_d}) = 0 \qquad (k = 0, \pm 1, \pm 2, \cdots)$$

We shall deduce from (1) that $F'(e^{ta_1}, \cdots, e^{ta_d})$ is identically zero. In fact, we would otherwise have $F'(e^{ta_1}, \cdots, e^{ta_d}) = \Sigma b_m e^{tA_m}$, where each A_m is a real exponent and $b_m \neq 0$. Assuming that $A_1 > A_2 > \cdots$, we should have

$$|b_1 e^{2kA_1}| > |\Sigma_{m>1} b_m e^{2kA_m}|$$

for $|k|$ sufficiently large and k of the same sign as A_1: this would be in contradiction with (1).

In particular, we have $F'(e^{a_1}, \cdots, e^{a_d}) = 0$, whence $\rho_1 \varepsilon \mathfrak{H}_1$, $\rho \varepsilon \mathfrak{H}$, and $\sigma = \tau\rho^{-1} \varepsilon \mathfrak{H}$, which proves the first part of Lemma 2.

Since σ and ρ are continuous functions of τ, $\mathfrak{H}$ is clearly homeomorphic with the product of $\mathfrak{H} \cap U(d)$ and of the set H of all positive definite hermitian matrices contained in $\mathfrak{H}$. The proof of the first part of Lemma 2 shows that a matrix $\rho \varepsilon H$ is of the form $\exp \alpha$, where α is a hermitian matrix which has the property that $\exp t\alpha \varepsilon \mathfrak{H}$ for all *real or complex t*. The matrix $\sqrt{-1}\,\alpha$ is skew hermitian, whence $\exp t\sqrt{-1}\,\alpha \varepsilon \mathfrak{H} \cap U(d)$ if t is real. This means that $\sqrt{-1}\,\alpha$ belongs to the Lie algebra $\mathfrak{g}$ of $\mathfrak{H} \cap U(d)$. Conversely, let β be any matrix such that $\sqrt{-1}\,\beta \varepsilon \mathfrak{g}$; then, we have $\exp t\beta \varepsilon \mathfrak{H}$ when t is purely imaginary. Let $G(\cdots x_{ij} \cdots) = 0$ be any one of the algebraic equations which define $\mathfrak{H}$. If we substitute for the x_{ij}'s the coefficients of $\exp t\beta$, $G(\cdots x_{ij} \cdots)$ becomes an entire function of t which vanishes identically when t is purely imaginary. This function therefore vanishes identically, whence $\exp t\beta \varepsilon \mathfrak{H}$. The matrix β is hermitian because $\sqrt{-1}\,\beta$ belongs to the Lie algebra of $U(d)$; it follows that $\exp t\beta$ is positive definite hermitian if t is real, whence $\exp t\beta \varepsilon H$. We know that the mapping $\sqrt{-1}\,\beta \rightarrow \exp \beta$ maps $\mathfrak{g}$ topologically onto H, which proves the second part of Lemma 2.

It follows immediately from Proposition 2 that $\mathfrak{M}(\mathfrak{G})$ is a group of dimension $2n$. Moreover, it follows from the proof of Lemma 2 that $\mathfrak{g}$ and $\sqrt{-1}\,\mathfrak{g}$ have no element $\neq 0$ in common. Therefore, $\mathfrak{g} + \sqrt{-1}\,\mathfrak{g}$ is a vector space of dimension $2n$. Since it is obviously contained in the Lie algebra of $\mathfrak{H}$, it must coincide with it. This proves:

Proposition 3. *Let $\mathfrak{G}$ be a compact Lie group which admits a faithful representation, and let $\{M_1, \cdots, M_n\}$ be a base of the Lie algebra of $\mathfrak{G}$. Let $[M_i, M_j] = \Sigma_k c_{ijk} M_k$ be the corresponding equations of structure. The Lie algebra of the associated algebraic group $\mathfrak{M}(\mathfrak{G})$ has a base $\{M_1, \cdots, M_n, M'_1, \cdots, M'_n\}$ such that*

$$[M_i, M_j] = \Sigma_k c_{ijk} M_k, \qquad [M'_i, M'_j] = -\Sigma_k c_{ijk} M_k,$$
$$[M_i, M'_j] = \Sigma_k c_{ijk} M'_k$$

§X. EXAMPLES

We shall determine the associated algebraic groups of the linear groups which were introduced in Chapter I.

The associated algebraic group of $U(n)$ is obviously $GL(n, C)$.

Let us now consider the group $SU(n)$. The identity mapping P_0 of $SU(n)$ into $GL(n, C)$ is a representation of $SU(n)$. It follows from Proposition 3, §VII, p. 190 that the coefficients of $\mathsf{P}_0 \dot{+} \bar{\mathsf{P}}_0$ form a system of generators of the representative ring $\mathfrak{o}$ of $SU(n)$. If $\sigma \varepsilon SU(n)$, we have $\bar{\mathsf{P}}_0(\sigma) = ({}^t\mathsf{P}_0(\sigma))^{-1}$ and $|\mathsf{P}_0(\sigma)| = 1$; it follows that the coefficients of $\bar{\mathsf{P}}_0(\sigma)$ can be expressed as polynomials in the coefficients of $\mathsf{P}_0(\sigma)$. Therefore, the coefficients of P_0 alone form a system of generators of $\mathfrak{o}$. Let $\tilde{\mathsf{P}}_0$ be the representation of the associated algebraic group of $SU(n)$ which extends P_0 and let $\mathfrak{H}$ be the image of the associated group under $\tilde{\mathsf{P}}_0$. Since $SU(n) \subset SL(n, C)$, it follows from Proposition 3, §VIII, p. 198 that $\mathfrak{H} \subset SL(n, C)$. We have $\dim \mathfrak{H} = 2\,(\dim SU(n)) = 2(n^2 - 1) = \dim SL(n, C)$, which proves that $\mathfrak{H}$ is the component of the unit matrix in $SL(n, C)$. By

Lemma 2, §IX, p. 201, we know that $SL(n, C)$ is homeomorphic with $(SU(n)) \times R^{n^2-1}$; therefore we have $\mathfrak{H} = SL(n, C)$. Since $\tilde{\mathsf{P}}_0$ is a faithful representation, we may say that the associated group of $SU(n)$ is $SL(n, C)$.

The identity mapping of $O(n)$ into $GL(n, C)$ is a faithful representation of $O(n)$ by *real* matrices. It follows that the coefficients of a matrix in $O(n)$, considered as functions on $O(n)$, form a system of generators of the representative ring of $O(n)$. It can then be seen easily that the associated algebraic group of $O(n)$ is $O(n, C)$ and that the associated algebraic group of $SO(n)$ is $O(n, C) \cap SL(n, C)$, which is a subgroup of index 2 in $O(n, C)$.

The group $Sp(n)$ is the group of unitary matrices σ of degree $2n$ such that ${}^t\sigma J \sigma = J$, where

$$J = \begin{pmatrix} 0 & \epsilon_n \\ -\epsilon_n & 0 \end{pmatrix}$$

ϵ_n being the unit matrix of degree n. It follows that the condition $\sigma \varepsilon Sp(n)$ implies $\boxed{\sigma}^2 = 1$. On the other hand, we know that $Sp(n)$ is connected; it follows immediately that $\sigma \varepsilon Sp(n)$ implies $\boxed{\sigma} = 1$. Following the same procedure as in the case of $SU(n)$, we see that the coefficients of a matrix of $Sp(n)$, considered as functions on $Sp(n)$, form a system of generators of the representative ring of $Sp(n)$. The associated algebraic group of $Sp(n)$ may therefore be identified with a subgroup of $Sp(n, C)$. If τ is a matrix in $Sp(n, C)$, we have ${}^t\tau J \tau = J$, whence $({}^t\bar{\tau})^* \bar{J}^* \bar{\tau}^* = \bar{J}^*$; but $\bar{J}^* = J$ and $({}^t\bar{\tau})^* = {}^t(\bar{\tau}^*)$, whence $\bar{\tau}^* \varepsilon Sp(n, C)$. By Lemma 2, §IX, p. 201 we conclude that the associated algebraic group of $Sp(n)$ is $Sp(n, C)$. At the same time we prove that $Sp(n, C)$ is homeomorphic with $Sp(n) \times R^{2n^2+n}$, which proves, in particular, that $Sp(n, C)$ is connected and simply connected.

§XI. The Main Approximation Theorem

We have already seen (Lemma 1, §VII, p. 190) that, if $\mathfrak{G}$ is a compact group of matrices, then any continuous function on $\mathfrak{G}$ may be approximated as closely as we wish by a function belonging to the representative ring $\mathfrak{o}$ of $\mathfrak{G}$. As we announced earlier, this result holds also for arbitrary compact Lie groups. We shall now prove this fundamental result, which is due to H. Peter and H. Weyl.

Theorem 3 (Peter-Weyl). *Let $\mathfrak{G}$ be a compact Lie group and let f be a continuous function on $\mathfrak{G}$. If a is any number > 0, there exists in the representative ring of $\mathfrak{G}$ a function g such that $|f(\sigma) - g(\sigma)| \leqslant a$ for all $\sigma \varepsilon \mathfrak{G}$.*

We shall denote by $\mathfrak{F}$ the space of all complex valued continuous functions on $\mathfrak{G}$. Since $\mathfrak{G}$ is compact, every function $f \varepsilon \mathfrak{F}$ is bounded. We shall denote the maximum of the absolute value of f by $M(f)$. We have clearly

$$M(af) = |a|M(f) \qquad (a \varepsilon C)$$
$$M(f + g) \leqslant M(f) + M(g)$$

We can make $\mathfrak{F}$ a metric space in which the distance of two functions f and g is $M(f - g)$. A subset Φ of $\mathfrak{F}$ is said to be bounded if there exists a number A such that $M(f) \leqslant A$ for all $f \varepsilon \Phi$.

The *oscillation* of a function $f \varepsilon \mathfrak{F}$ on a subset E of $\mathfrak{G}$ is the least upper bound of the numbers $|f(\sigma) - f(\tau)|$ for all $\sigma, \tau \varepsilon E$.

A subset Φ of $\mathfrak{F}$ is said to be a set of *equicontinuous* functions if, for every number $a > 0$, there exists a neighbourhood U_a of the neutral element in $\mathfrak{G}$ such that, f being any function in Φ and σ being any element of $\mathfrak{G}$, the oscillation of f on the set σU_a is $\leqslant a$.

Lemma 1. *Let Φ be a bounded set of equicontinuous functions, and let a be a number > 0. If a subset Φ_1 of Φ is such that $M(f - g) > a$ for all pairs (f, g) of distinct elements of Φ_1, we may conclude that Φ_1 is finite.*

Let A be a number such that $M(f) \leqslant A$ for all $f \varepsilon \Phi$. If $\sigma_0 \varepsilon \mathfrak{G}$, the set of numbers $f(\sigma_0)$ (for all $f \varepsilon \Phi$) is contained in the compact region of the complex z-plane which is defined by the inequality $|z| \leqslant A$. It follows immediately that there exists a finite subset ψ_{σ_0} of Φ with the following property: to every $f \varepsilon \Phi$ there corresponds a function $f_1 \varepsilon \psi_{\sigma_0}$ such that the inequality $|f(\sigma_0) - f_1(\sigma_0)| < a/6$ holds true. Let σ be any point of $\sigma_0 U_{a/6}$; we have

$$|f(\sigma) - f_1(\sigma)| \leqslant |f(\sigma) - f(\sigma_0)| + |f(\sigma_0) - f_1(\sigma_0)| + |f_1(\sigma_0) - f_1(\sigma)| < \frac{a}{2}.$$

On the other hand, to every pair (f, g) of functions of Φ_1 we may associate a point $\sigma = \sigma(f, g)$ such that $|f(\sigma) - g(\sigma)| \geqslant a$. If ψ_{σ_0} contains m functions, there cannot exist more than m^2 pairs $(f, g) \varepsilon \Phi_1 \times \Phi_1$ such that $\sigma(f, g) \varepsilon \sigma_0 U_{a/6}$, because otherwise there would exist one of these pairs, say (f, g), such that $|f(\sigma) - f_1(\sigma)| < a/2$, $|g(\sigma) - f_1(\sigma)| < a/2$ with the same $f_1 \varepsilon \psi_{\sigma_0}$, and this would imply $|f(\sigma) - g(\sigma)| < a$.

Since $\mathfrak{G}$ is compact, it can be covered by a finite number of sets $V_1, \cdots, V_N$ each of which is of the form $\sigma_0 U_{a/6}$. Since there are only a finite number of pairs $(f, g) \varepsilon \Phi_1 \times \Phi_1$ such that $\sigma(f, g) \varepsilon V_i$ $(1 \leqslant i \leqslant N)$, we conclude that Φ_1 is finite.

Lemma 2. *If a sequence of functions belongs to a bounded set of equicontinuous functions, it is possible to extract from it a subsequence which converges uniformly on* $\mathfrak{G}$.

In fact, let μ be any integer > 0. We can construct a finite subset Φ_μ of Φ with the following properties: 1) if $f, g \varepsilon \Phi_\mu$, we have $M(f - g) \geqslant 1/\mu$; 2) if f is any function in Φ, there exists a function $g \varepsilon \Phi_\mu$ such that $M(f - g) < 1/\mu$.[1] Let now (f_m) be any sequence of functions in Φ. We denote by M_0 the set of all integers > 0. We shall define by induction on μ an infinite subset M_μ of M_0. Suppose that $\mu \geqslant 0$ and that M_μ has already been defined. Then, since Φ_μ is finite, there exists a function $g_{\mu+1} \varepsilon \Phi_{\mu+1}$ such that the inequality $M(f_m - g_{\mu+1}) < 1/(\mu + 1)$ holds for infinitely many integers $m \varepsilon M_\mu$; $M_{\mu+1}$ will be the set of these integers. We select in each M_μ an integer $m_\mu \geqslant \mu$. Since $M_\nu \subset M_\mu$ for $\nu > \mu$, we have $M(f_{m_\nu} - g_\mu) < 1/\mu$, $M(f_{m_\mu} - g_\mu) < 1/\mu$, whence $M(f_{m_\nu} - f_{m_\mu}) < 2/\mu$. It follows immediately that the sequence (f_{m_μ}) converges uniformly on $\mathfrak{G}$.

We shall now introduce in $\mathfrak{F}$ an operation which is a generalization of the scalar multiplication defined in Chapter 1, §III, p. 9. If f and g are any functions in $\mathfrak{F}$, we set

$$f \cdot g = \int_{\mathfrak{G}} f(\sigma)\bar{g}(\sigma)\, d\sigma$$

where the integral is the invariant integral defined in §VIII, Chapter V, p. 167, normalized by the condition $\int_{\mathfrak{G}} 1\, d\sigma = 1$.

The following properties are obvious:

1) For g fixed, $f \cdot g$ is linear in f, i.e. we have $(a_1 f_1 + a_2 f_2) \cdot g = a_1(f_1 \cdot g) + a_2(f_2 \cdot g)(a_1, a_2 \varepsilon C)$.

2) We have $g \cdot f = \overline{f \cdot g}$

3) If $f \neq 0$, we have $f \cdot f > 0$.

We shall let $\|f\|$ denote the number $(f \cdot f)^{\frac{1}{2}}$. If a, b are any real numbers, we have $(af + bg) \cdot (af + bg) = a^2(f \cdot f) + 2ab\Re(f \cdot g) + b^2(g \cdot g)$, where $\Re$ indicates the taking of the real part. Since this expression is $\geqslant 0$ for all real a, b, we have

$$(\Re(f \cdot g))^2 \leqslant (f \cdot f)(g \cdot g) \qquad \text{i.e.} \qquad |\Re(f \cdot g)| \leqslant \|f\| \cdot \|g\|$$

Let ϑ be a real number such that $\exp(\sqrt{-1}\,\vartheta)(f \cdot g) = |f \cdot g|$. If we replace f by $(\exp \sqrt{-1}\,\vartheta)f$ in our inequality, we get the *Schwarz*

[1] Let f_1 be any function in $\mathfrak{F}$. If $f_1, \cdots, f_r$ are already defined, and if there exists a function f such that $M(f - f_i) \geqslant 1/\mu$ for $1 \leqslant i \leqslant r$, we select such a function and call it f_{r+1}. This inductive process cannot be continued indefinitely (because of Lemma 1). If it stops with f_s, we may take $\Phi_\mu = \{f_1, \cdots, f_s\}$.

inequality:

$$|f \cdot g| \leqslant ||f||\,||g||$$

On the other hand, we have $||f + g||^2 = f \cdot f + 2\Re(f \cdot g) + g \cdot g \leqslant ||f||^2 + 2||f||\,||g|| + ||g||^2 = (||f|| + ||g||)^2$, which proves *Minkowski's inequality:*

$$||f + g|| \leqslant ||f|| + ||g||.$$

Now, let $k(\sigma, \tau)$ be a complex valued continuous function defined in $\mathfrak{G} \times \mathfrak{G}$ which satisfies the condition

$$k(\sigma, \tau) = \overline{k(\tau, \sigma)}.$$

If f is any function in $\mathfrak{F}$, we shall denote by Kf the function which is defined by the formula

$$Kf(\sigma) = \int_{\mathfrak{G}} k(\sigma, \tau) f(\tau)\, d\tau$$

We shall prove that Kf belongs to $\mathfrak{F}$. Let a be any number > 0. There exists a neighbourhood U_a of the neutral element ϵ in $\mathfrak{G}$ such that the inequality $|k(\sigma\rho, \tau) - k(\sigma, \tau)| < a$ holds whenever $\rho \varepsilon U_a$. In fact, there would otherwise exist three sequences (σ_m), (ρ_m), (τ_m) of elements in $\mathfrak{G}$ such that $\lim_{m\to\infty} \rho_m = \epsilon$, $|k(\sigma_m\rho_m, \tau_m) - k(\sigma_m, \tau_m)| \geqslant a$. Since $\mathfrak{G}$ is compact, the sequences (σ_m) and (τ_m) would have subsequences converging respectively to elements σ, τ and we would have $|k(\sigma\epsilon, \tau) - k(\sigma, \tau)| \geqslant a$, which is impossible. If ρ belongs to U_a, we have

$$(1) \qquad |Kf(\sigma\rho) - Kf(\sigma)| \leqslant a \int_{\mathfrak{G}} |f(\tau)|\, d\tau = a(1 \cdot |f|) \leqslant a||f||$$

which proves the continuity of the function Kf. Moreover, if A denotes an upper bound for the values taken by k on $\mathfrak{G} \times \mathfrak{G}$, we also have

$$(2) \qquad |Kf(\sigma)| \leqslant A \int_{\mathfrak{G}} |f(\tau)|\, d\tau \leqslant A||f||$$

These two inequalities prove

Lemma 3. *The operator K maps the set of all functions $f \epsilon \mathfrak{F}$ such that $||f|| \leqslant 1$ onto a bounded set of equicontinuous functions.*

Another property of the operator K is expressed in the formula

$$(3) \qquad Kf \cdot g = f \cdot Kg$$

which is entirely similar to the formula which expresses the hermitian character of a matrix. In order to prove (3), we observe that

$$Kf \cdot g = \int_{\mathfrak{G}} \left(\int_{\mathfrak{G}} k(\sigma, \tau) f(\tau)\, d\tau\right) \bar{g}(\sigma)\, d\sigma = \int_{\mathfrak{G} \times \mathfrak{G}} k(\sigma, \tau) f(\tau) \bar{g}(\sigma)\, d\sigma\, d\tau$$
$$= \int_{\mathfrak{G}} f(\tau) \left(\int_{\mathfrak{G}} \bar{k}(\tau, \sigma) \bar{g}(\sigma)\, d\sigma\right) d\tau = f \cdot Kg$$

We shall say that a number c is an "*eigenvalue*" of the function k if there exists in $\mathfrak{F}$ a function $\varphi \neq 0$ such that $K\varphi = c\varphi$. Any such function is called an *eigenfunction* belonging to the value c. Our next step will be to prove the existence of eigenvalues.

It follows immediately from (2) that $||Kf|| \leqslant A||f||$. Hence the numbers $||Kf||$ remain bounded when f varies in the set of functions f such that $||f|| = 1$. We shall denote by $||k||$ the least upper bound of $||Kf||$ for $||f|| = 1$. It is clear that $||Kf|| \leqslant ||k||\,||f||$ for any function $f \varepsilon \mathfrak{F}$.

Lemma 4. *The number $||k||$ is the least upper bound of the values taken by $Kf \cdot f$ when $||f|| = 1$.*

If $||f|| = 1$, we have $|Kf \cdot f| \leqslant ||Kf||\,||f|| \leqslant ||k||$. In order to prove the converse, let c be the least upper bound of $|Kf \cdot f|$ for $||f|| = 1$. It follows immediately that $|Kf \cdot f| \leqslant c||f||^2$ for any f. Let f and g be two functions such that $||f|| = 1$, $||g|| = 1$. We have

$$
\begin{aligned}
K(f+g) \cdot (f+g) &= Kf \cdot f + Kg \cdot g + Kf \cdot g + Kg \cdot f \\
&= Kf \cdot f + Kg \cdot g + 2\Re(Kf \cdot g) \leqslant c||f+g||^2 \\
(K(f-g)) \cdot (f-g) &= Kf \cdot f + Kg \cdot g - 2\Re(Kf \cdot g) \geqslant -c||f-g||^2
\end{aligned}
$$

whence $4\Re(Kf \cdot g) \leqslant c(||f+g||^2 + ||f-g||^2) = 2c(||f||^2 + ||g||^2) = 4c$ If we assume $Kf \neq 0$ and take $g = ||Kf||^{-1}Kf$, we obtain $||Kf|| \leqslant c$. This last inequality holds also if $Kf = 0$. We have therefore proved that $||k|| = c$.

Lemma 5. *One at least of the numbers $||k||$, $-||k||$ is an eigenvalue of the function k.*

We can find a sequence (f_m) of functions in $\mathfrak{F}$ such that $||f_m|| = 1$ and such that $Kf_m \cdot f_m$ tends towards one of the numbers $||k||$, $-||k||$. We set $c = \lim_{m\to\infty} Kf_m \cdot f_m$.

Replacing if necessary the sequence (f_m) by a subsequence, we may assume without loss of generality that the sequence (Kf_m) converges uniformly on $\mathfrak{G}$ to a function φ, which clearly belongs to $\mathfrak{F}$ (cf. Lemmas 2, 3). We therefore have $\lim_{m\to\infty} M(\varphi - Kf_m) = 0$.

We observe now that the operation $f \cdot g$ is continuous with respect to the metric defined by M. In fact, we have $||f||^2 = \int_{\mathfrak{G}} f(\tau)\bar{f}(\tau)\, d\tau \leqslant (M(f))^2$, whence $|(f_1 - f_2) \cdot (g_1 - g_2)| \leqslant ||f_1 - f_2||\,||g_1 - g_2|| \leqslant M(f_1 - f_2)M(g_1 - g_2)$.

Since $Kf_m \cdot f_m$ is real, we have

$$||Kf_m - cf_m||^2 = ||Kf_m||^2 + c^2||f_m||^2 - 2c(Kf_m \cdot f_m)$$

and the right side tends to $||\varphi||^2 - c^2$ as m increases indefinitely. It follows that $||\varphi|| \geqslant |c|$, whence $\varphi \neq 0$ provided $c \neq 0$. On the other hand, we have $||Kf_m||^2 \leqslant ||k||^2 \leqslant c^2$, and therefore the right side of our formula is $\leqslant 2c^2 - 2c(Kf_m \cdot f_m)$, a quantity which tends to 0 with $1/m$. We conclude that $\lim_{m\to\infty} ||Kf_m - cf_m|| = 0$ and therefore also $\lim_{m\to\infty} ||K(Kf_m) - cKf_m|| = 0$. Since $\lim_{m\to\infty} ||Kf_m - \varphi|| = 0$ we have $\lim_{m\to\infty} M(K(Kf_m) - K\varphi) = 0$, whence

$$||K\varphi - c\varphi|| = \lim_{m\to\infty} ||K(Kf_m) - cKf_m|| = 0$$

which proves that $K\varphi = c\varphi$. Lemma 5 is therefore proved if $c \neq 0$. If $c = 0$, we have $||k|| = 0$, $Kf = 0$ for every f and Lemma 5 is then trivial.

We shall say that two functions φ and ψ belonging to $\mathfrak{F}$ are *orthogonal* to each other if $\varphi \cdot \psi$ is equal to 0.

Let Φ be the set of eigenfunctions of k which belong to eigenvalues $\neq 0$. We can find a subset Φ^* of Φ with the following properties: *a*) if φ and ψ are in Φ^* and $\varphi \neq \psi$, then $\varphi \cdot \psi = 0$; *b*) we have $||\varphi|| = 1$ for every $\varphi \varepsilon \Phi^*$; *c*) Φ^* is maximal with respect to the properties *a*), *b*) (i.e. it is impossible to imbed Φ^* in a properly larger subset of Φ for which *a*), *b*) hold).[1]

Lemma 6. *Let a be a number* > 0. *There are only a finite number of functions in* Φ^* *which belong to eigenvalues greater than a in absolute value.*

In fact, let $\varphi_1, \cdots, \varphi_h$ be functions of Φ^* such that $K\varphi_i = c_i\varphi_i$, $|c_i| > a$ $(1 \leqslant i \leqslant h)$. We have $K(\varphi_i - \varphi_j) = c_i\varphi_i - c_j\varphi_j$ whence $||K(\varphi_i - \varphi_j)||^2 = c_i^2 + c_j^2 > 2a^2$. We know that $M(K(\varphi_i - \varphi_j)) \geqslant ||K(\varphi_i - \varphi_j)|| > a\sqrt{2}$. Lemma 6 then follows from Lemmas 1 and 3.

Lemma 7. *Any function f in* Φ *is a linear combination of a finite number of functions in* Φ^*.

Let c be the eigenvalue to which f belongs, and let $\varphi_1, \cdots, \varphi_h$ be the functions of Φ^* belonging to c. We set $f' = f - \Sigma_{i=1}^h (f \cdot \varphi_i)\varphi_i$. Since $Kf = cf$, $K\varphi_i = c\varphi_i$, we have also $Kf' = cf'$. Moreover, f' is orthogonal to $\varphi_1, \cdots, \varphi_h$. Let ψ be any function of Φ^* distinct from $\varphi_1, \cdots, \varphi_h$; then ψ belongs to an eigenvalue $d \neq c$. We have $f' \cdot \psi = 1/d(f' \cdot K\psi) = 1/d(Kf' \cdot \psi) = c/d(f' \cdot \psi)$ whence $f' \cdot \psi = 0$. We see that f' is orthogonal to every function in Φ^*. If we had $f' \neq 0$, the set composed of Φ^* and of the function $||f'||^{-1}f'$ would still have the properties *a*), *b*) above, which is impossible. Therefore $f' = 0$, which proves Lemma 7.

It follows from Lemma 6 that Φ^* is a countable set. We arrange the elements of Φ^* in a sequence (φ_μ) $(1 \leqslant \mu < \infty$ or $1 \leqslant \mu \leqslant \mu_0$

[1] This follows immediately from Zorn's lemma.

according as to whether Φ^* is infinite or finite), and we denote by c_μ the eigenvalue to which φ_μ belongs. We observe that, if $\Sigma_\mu c_\mu \varphi_\mu$ is a *finite* linear combination of functions of Φ^*, we have $||\Sigma c_\mu \varphi_\mu|| = \Sigma|c_\mu|^2$.

Lemma 8 (Bessel's Inequality). *If f is any function in $\mathfrak{F}$, the series $\Sigma_\mu |f \cdot \varphi_\mu|^2$ is convergent, and its sum is $\leqslant ||f||^2$.*

We set $g_i = f - \Sigma_{\mu \leqslant i}(f \cdot \varphi_\mu)\varphi_\mu$, where i is any positive integer if Φ^* is infinite, and is at most equal to the number of elements of Φ^* if Φ^* is finite. We have $g_i \cdot \varphi_\mu = 0$ $(1 \leqslant \mu \leqslant i)$, from which it follows easily that

$$||f||^2 = ||g_i||^2 + \Sigma_{\mu \leqslant i}|f \cdot \varphi_\mu|^2$$

which proves Lemma 8.

Lemma 9. *If $f \varepsilon \mathfrak{F}$, the series $\Sigma_\mu (Kf \cdot \varphi_\mu)\varphi_\mu$ converges uniformly on $\mathfrak{G}$ to the function Kf.*

We have

$$\Sigma_i^j(Kf \cdot \varphi_\mu)\varphi_\mu = \Sigma_i^j(Kf \cdot \varphi_\mu)\frac{K\varphi_\mu}{c_\mu} = \Sigma_i^j(f \cdot K\varphi_\mu)\frac{K\varphi_\mu}{c_\mu}$$

$$= \Sigma_i^j(f \cdot \varphi_\mu)K\varphi_\mu = K(\Sigma_i^j(f \cdot \varphi_\mu)\varphi_\mu)$$

whence, making use of (2),

$$M(\Sigma_i^j(Kf \cdot \varphi_\mu)\varphi_\mu) \leqslant A||\Sigma_i^j(f \cdot \varphi_\mu)\varphi_\mu|| = A(\Sigma_i^j|f \cdot \varphi_\mu|^2)$$

and it follows from Lemma 8 that the right side tends to 0 as i and j increase indefinitely (in the case where Φ^* is infinite). Therefore the given series does converge uniformly. There remains to prove that its sum is Kf.

We set $k_n(\sigma, \tau) = k(\sigma, \tau) - \Sigma_{\mu=1}^n c_\mu \varphi_\mu(\sigma)\bar{\varphi}_\mu(\tau)$. Since each c_μ is real, we have $k_n(\sigma, \tau) = \bar{k}_n(\tau, \sigma)$. If ψ is any function in $\mathfrak{F}$, we have $K_n\psi = K\psi - \Sigma_1^n c_\mu(\psi \cdot \varphi_\mu)\varphi_\mu$ (where K_n is defined for k_n as K was defined for k). It follows that $K_n\psi \cdot \varphi_\mu = K\psi \cdot \varphi_\mu - c_\mu(\psi \cdot \varphi_\mu)$ $= \psi \cdot K\varphi_\mu - c_\mu(\psi \cdot \varphi_\mu) = 0$ if $1 \leqslant \mu \leqslant n$. In particular, if ψ is an eigenfunction of K_n belonging to an eigenvalue $d \neq 0$, we have $\psi \cdot \varphi_\mu = 0$ $(1 \leqslant \mu \leqslant n)$, $K\psi = K_n\psi = d\psi$ and ψ is an eigenfunction of K belonging to d. It follows that ψ is a linear combination of the functions φ_ν for which $c_\nu = d$. We have $\psi = \Sigma_{c_\nu = d} b_\nu \varphi_\nu$ and $b_\nu = \psi \cdot \varphi_\nu$, whence $b_\nu = 0$ if $\mu \leqslant n$. Let a be a number > 0; if n is choosen to be larger than all indices μ for which $|c_\mu| \geqslant a$ (there are only a finite number of these indices), we may conclude that $|d| < a$, whence, by Lemma 5, $||K_n|| \leqslant a$. It follows that $\lim_{n \to \infty} ||K_n f|| = 0$ (if Φ^* is infinite; if Φ^* is finite, $K_n f = 0$ if n is the number of elements in Φ^*). But $K_n f = Kf - \Sigma_{\mu=1}^n(Kf \cdot \varphi_\mu)\varphi_\mu$, whence $||Kf - \Sigma_\mu(Kf \cdot \varphi_\mu)\varphi_\mu|| = 0$, which completes the proof of Lemma 9.

Lemma 10. *If the function $k(\sigma, \tau)$ is of the form $\chi(\sigma^{-1}\tau)$, where χ is a continuous function on $\mathfrak{G}$ such that $\chi(\sigma^{-1}) = \bar{\chi}(\sigma)$, then every eigenfunction of k belonging to an eigenvalue $c \neq 0$ is a representative function on $\mathfrak{G}$.*

If f is any function in $\mathfrak{F}$ and $\rho \varepsilon \mathfrak{G}$, we denote by f^ρ the function defined by $f^\rho(\sigma) = f(\rho\sigma)$. Assume that f is an eigenfunction of k belonging to an eigenvalue $c \neq 0$. We have

$$\begin{aligned} cf^\rho(\sigma) &= \textstyle\int_{\mathfrak{G}} \chi(\sigma^{-1}\rho^{-1}\tau) f(\tau)\, d\tau \\ &= \textstyle\int_{\mathfrak{G}} \chi(\sigma^{-1}\tau') f(\rho\tau')\, d\tau' = \int_{\mathfrak{G}} \chi(\sigma^{-1}\tau') f^\rho(\tau')\, d\tau' \end{aligned}$$

in virtue of the invariant character of our integration process. Hence f^ρ is again an eigenfunction belonging to c. We may assume that the functions of the set Φ^* which belong to c are $\varphi_1, \cdots, \varphi_h$. By Lemma 7, we have

$$\varphi_i^\rho = \Sigma_{j=1}^h g_{ij}(\rho)\varphi_j \qquad g_{ij}(\rho) = \varphi_i^\rho \cdot \varphi_j$$

Let $\mathsf{P}(\rho)$ be the matrix $(g_{ij}(\rho))$. Since $\varphi_i^{\rho_1\rho_2} = (\varphi_i^{\rho_2})^{\rho_1}$ we have $\mathsf{P}(\rho_1\rho_2) = \mathsf{P}(\rho_1)\mathsf{P}(\rho_2)$. On the other hand, since each φ_i is continuous on the compact group $\mathfrak{G}$, it is also uniformly continuous on $\mathfrak{G}$; it follows that, for every $a > 0$, there exists a neighbourhood U_a of the neutral element ϵ in $\mathfrak{G}$ such that $M(\varphi_i^\rho - \varphi_i) < a$ whenever $\rho \varepsilon U_a$. It follows that $\lim_{\rho\to\epsilon} g_{ij}(\rho) = \lim_{\rho\to\epsilon} \varphi_i^\rho \cdot \varphi_j = \delta_{ij} = g_{ij}(\epsilon)$. The mapping $\rho \to \mathsf{P}(\rho)$ of $\mathfrak{G}$ into the set of matrices of degree h is therefore a continuous representation of $\mathfrak{G}$, and the functions g_{ij} are representative functions. But we have $\varphi_i(\rho) = \varphi_i^\rho(\epsilon) = \Sigma_{j=1}^h g_{ij}(\rho)\varphi_j(\epsilon)$, which shows that each φ_i is likewise a representative function. Lemma 10 is thereby proved.

Lemma 11. *Let χ be a continuous function on $\mathfrak{G}$ such that $\chi(\sigma^{-1}) = \bar{\chi}(\sigma)$, and let a be a number > 0. Then there exists a representative function g of $\mathfrak{G}$ such that $M(\chi - g) < a$.*

Let K be the operator associated with the function $k(\sigma, \tau) = \bar{\chi}(\sigma^{-1}\tau)$. We determine a neighbourhood V of the neutral element such that the inequality $|\bar{\chi}(\sigma^{-1}\tau) - \bar{\chi}(\sigma^{-1})| < a/2$ holds for all $\tau \varepsilon V$. We can construct a continuous function f_1 with real non negative values such that $f_1(\epsilon) \neq 0$, $f_1(\sigma) = 0$ if σ does not belong to V. The number $J = \int_{\mathfrak{G}} f_1(\tau)\, d\tau$ is $\neq 0$; we set $f = J^{-1}f_1$, whence $\int_{\mathfrak{G}} f(\tau)\, d\tau = 1$. We have

$$|\chi(\sigma) - Kf(\sigma)| = |\bar{\chi}(\sigma^{-1}) - Kf(\sigma)| = |\textstyle\int_{\mathfrak{G}} (\bar{\chi}(\sigma^{-1}) - \bar{\chi}(\sigma^{-1}\tau))f(\tau)\, d\tau| < \dfrac{a}{2}$$

Making uses of Lemmas 9 and 10, we see that there exists a representative function g such that $M(Kf - g) < a/2$. It follows that $M(\chi - g) < a$, which proves Lemma 11.

Let now f be an arbitrary continuous function on $\mathfrak{G}$. We set $\chi_1(\sigma) = f(\sigma) + \bar{f}(\sigma)$, $\chi_2(\sigma) = \sqrt{-1}\,(f(\sigma) - \bar{f}(\sigma))$, whence $\chi_i(\sigma^{-1}) = \bar{\chi}_i(\sigma)$ $(i = 1, 2)$ and $f = \frac{1}{2}(\chi_1 - \sqrt{-1}\,\chi_2)$. Since χ_1 and χ_2 may be approximated as closely as we want by representative functions, the same holds for f. Theorem 3 is thereby proved.

§XII. First Applications of the Main Approximation Theorem

Theorem 4. *A compact Lie group admits at least one faithful representation.*

Let $\mathfrak{G}$ be a compact Lie group. Making use of Proposition 6, §VII, p. 193, we see that, in order to prove that $\mathfrak{G}$ admits a faithful representation, it is sufficient to prove that, if σ is any element of $\mathfrak{G}$ distinct from the neutral element ϵ, there exists a representation P of $\mathfrak{G}$ such that $\mathsf{P}(\sigma)$ is not a unit matrix. Let f be a continuous function on $\mathfrak{G}$ such that $f(\sigma) \neq f(\epsilon)$. Since f can be approximated as closely as we want by representative functions, we see that there exists a representative function which takes distinct values at σ and ϵ, and our assertion follows immediately from this fact.

Proposition 1. *If $\mathfrak{H}$ is a closed subgroup of a compact Lie group $\mathfrak{G}$, any irreducible representation of $\mathfrak{H}$ is contained in the trace on $\mathfrak{H}$ of some representation of $\mathfrak{G}$.*

This follows at once from Theorem 4 above and from Proposition 4, §VII, p. 191.

Theorem 5 (Theorem of Tannaka). *Let $\mathfrak{R}$ be the set of all representations of a compact Lie group $\mathfrak{G}$. Let $\mathfrak{G}_1$ be the set of all representations ζ of $\mathfrak{R}$ which satisfy the supplementary conditions $\zeta(\bar{\mathsf{P}}) = \overline{\zeta(\mathsf{P})}$, where P is any element of $\mathfrak{R}$ and where $\bar{\mathsf{P}}$ is the imaginary conjugate representation of P. If we define a multiplication in $\mathfrak{G}_1$ by the formula $\zeta_1\zeta_2(\mathsf{P}) = \zeta_1(\mathsf{P})\zeta_2(\mathsf{P})$, $\mathfrak{G}_1$ becomes a group. Let σ be any element of $\mathfrak{G}$ and define the representation ζ_σ of $\mathfrak{R}$ by $\zeta_\sigma(\mathsf{P}) = \mathsf{P}(\sigma)$. Then the mapping $\sigma \to \zeta_\sigma$ is an isomorphism of $\mathfrak{G}$ with $\mathfrak{G}_1$.*

This follows immediately from Proposition 1, §IX, p. 199 and from Theorem 4 above.

Corollary. *Let $\mathfrak{G}$ be a compact Lie group of dimension n. Then the associated algebraic group of $\mathfrak{G}$ is homeomorphic to the product of $\mathfrak{G}$ and R^n.*

This follows immediately from Proposition 2, §IX, p. 200 and from Theorem 4 above.

Proposition 2. *Let f be a continuous function on a compact Lie group $\mathfrak{G}$ such that $f(\sigma) = f(\tau\sigma\tau^{-1})$ for any σ, $\tau\varepsilon\mathfrak{G}$. If a is any number > 0, there exists a function f_1 which is a linear combination of characters of irreducible representations of $\mathfrak{G}$ such that $|f(\sigma) - f_1(\sigma)| \leqslant a$ for all $\sigma\varepsilon\mathfrak{G}$.*

Let P be any irreducible unitary representation of $\mathfrak{G}$; denote by $g_{ij}(\sigma)$ the coefficients of $\mathsf{P}(\sigma)$. We have

$$g_{ij}(\tau\sigma\tau^{-1}) = \Sigma_{kl} g_{ik}(\tau) g_{kl}(\sigma) g_{lj}(\tau^{-1})$$

and $g_{lj}(\tau^{-1}) = \bar{g}_{lj}(\tau)$. Making use of the orthogonality relations, we obtain

$$\int_{\mathfrak{G}} g_{ij}(\tau\sigma\tau^{-1})\,d\tau = 0 \qquad \text{if} \qquad i \neq j$$
$$\int_{\mathfrak{G}} g_{ii}(\tau\sigma\tau^{-1})\,d\tau = d^{-1}\chi(\sigma)$$

where d and χ are respectively the degree and the character of the representation P.

We know from the general approximation theorem that there exists a function f_2 of the representative ring of $\mathfrak{G}$ such that $|f(\sigma) - f_2(\sigma)| \leqslant a$ for all $\sigma\varepsilon\mathfrak{G}$. It follows that $|\int_{\mathfrak{G}} f(\tau\sigma\tau^{-1})\,d\tau - \int_{\mathfrak{G}} f_2(\tau\sigma\tau^{-1})\,d\tau| \leqslant a$. Since $f(\sigma) = f(\tau\sigma\tau^{-1})$, we have $\int_{\mathfrak{G}} f(\tau\sigma\tau^{-1})\,d\tau = f(\sigma)$. On the other hand, f_2 is a linear combination of coefficients of irreducible unitary representations of $\mathfrak{G}$ and therefore the function $f_1(\sigma) = \int_{\mathfrak{G}} f_2(\tau\sigma\tau^{-1})\,d\tau$ is a linear combination of characters of irreducible representations of $\mathfrak{G}$. Proposition 2 is thereby proved.

§XIII. COMPACT ABELIAN GROUPS

Proposition 1. *A compact connected abelian Lie group $\mathfrak{G}$ of dimension n is isomorphic (as a topological group) with the n-dimensional torus T^n.*

In fact, let $\mathfrak{g}$ be the Lie algebra of $\mathfrak{G}$. Since $\mathfrak{G}$ is abelian, we have $[X, Y] = 0$ for any elements X and Y of $\mathfrak{g}$. It follows that $\mathfrak{g}$ coincides with the Lie algebra of R^n. Since R^n is simply connected, the universal covering group of $\mathfrak{G}$ is R^n. But it is well known that a compact connected group which is locally isomorphic with R^n is isomorphic with T^n.

Let $\mathfrak{x}$ be a real number modulo 1 (i.e. a residue class of the additive group of real numbers modulo the group of integers), and let x be a real number whose residue class modulo 1 is $\mathfrak{x}$. Since the value of $\exp(2\pi\sqrt{-1}\,x)$ depends only upon $\mathfrak{x}$, we may set

$$\exp(2\pi\sqrt{-1}\,\mathfrak{x}) = \exp(2\pi\sqrt{-1}\,x)$$

The mapping $\mathfrak{x} \to \exp(2\pi\sqrt{-1}\,\mathfrak{x})$ is clearly a representation of T^1.

Any element $\sigma \varepsilon T^n$ may be represented in the form $(\mathfrak{x}_1, \cdots, \mathfrak{x}_n)$, with $\mathfrak{x}_i \varepsilon T^1$ $(1 \leqslant i \leqslant n)$. If we set $\mathsf{P}_i(\sigma) = \exp(2\pi\sqrt{-1}\,\mathfrak{x}_i)$, each P_i is a representation of T^n. Moreover, $\mathsf{P} = \mathsf{P}_1 \dotplus \cdots \dotplus \mathsf{P}_n$ is a faithful representation of T^n. If $m_1, \cdots, m_n$ are any integers, the mapping $\sigma \rightarrow \exp(2\pi\sqrt{-1}\,\Sigma_i m_i x_i)$ is a representation of T^n which we shall denote by $\mathsf{P}_1^{m_1} \cdots \mathsf{P}_n^{m_n}$. We have $\mathsf{P}_i^{-1} = \bar{\mathsf{P}}_i$ $(1 \leqslant i \leqslant n)$; if $m_1, \cdots, m_n$ are all positive, $\mathsf{P}_1^{m_1} \cdots \mathsf{P}_n^{m_n}$ is the Kronecker product of m_1 times the representation $\mathsf{P}_1, \cdots, m_n$ times the representation P_n.

We have $\bar{\mathsf{P}} = \mathsf{P}_1^{-1} \dotplus \cdots \dotplus \mathsf{P}_n^{-1}$; by Proposition 3, §VII, p. 190, the set $\{\mathsf{P}, \bar{\mathsf{P}}\}$ contains sufficiently many representations of T^n. It follows that every irreducible representation of T^n is contained in some representation obtained by Kronecker multiplication of P and $\bar{\mathsf{P}}$, each taken a suitable number of times. It follows immediately that every irreducible representation of T^n is of the form $\mathsf{P}_1^{m_1} \cdots \mathsf{P}_n^{m_n}$ with suitable integers $m_1, \cdots, m_n$.

In the case of T^n, the irreducible representations are of degree 1. Therefore, the Kronecker product of two irreducible representations is likewise irreducible. It follows that the irreducible representations form a group under the operation of Kronecker multiplication, the inverse operation in this group being the passage to the imaginary conjugate representation. This group $\mathfrak{R}$ is the product of n times the additive group of integers by itself. It is easy to see directly that T^n is isomorphic with the group of all irreducible unitary representations of $\mathfrak{R}$: this is a particular case of the famous duality theorem of Pontrjagin. This fact can also be deduced from Theorem 5, §XII, p. 211; the latter theorem is therefore a far reaching generalization of Pontrjagin's theorem.

Let us finally observe that the application to T^n of the theorem of Peter-Weyl yields the following well known approximation theorem:

Let $f(x_1, \cdots, x_n)$ be a continuous function of n real variables which is periodic of period 1 *with respect to each variable. Given any number $a > 0$, there exists a trigonometric polynomial* (i.e. *a function of the form $g(x_1, \cdots, x_n) = \Sigma c(m_1, \cdots, m_n) \exp(2\pi\sqrt{-1}\,\Sigma_i m_i x_i)$*) *such that $|f(x_1, \cdots, x_n) - g(x_1, \cdots, x_n)| \leqslant a$ for all values of $x_1, \cdots, x_n$.*

INDEX

Adjoint representation, 123
alternate functions, 141
analytic groups, 100
associated algebraic group, 197
automorphism group, 135

Bessel's inequality, 209

canonical coordinates, 118
Cartan, differential forms of, 146
character of a representation, 186
Clifford numbers, 61
compact abelian groups, 212
coordinate systems, 70
countability axioms, 94
covering group, 53
covering spaces, 40

derived algebra, 125
derived group, 125
differentials, 78

exponential mapping, 115
exponential of a matrix, 5

Grassmann algebra, 145

hermitian product, 9
homogeneous spaces, 29
homomorphism
  analytic, 111
  local, 48

ideals of Lie algebras, 114
infinitesimal transformations, 82
integration of differential forms, 161
invariant integration on a group, 167
involutive distribution, 86
isomorphism, local, 37

Kronecker product, 179

Lie algebras, 101
Lie groups, 129
  compact, representations of, 171 ff.

linear groups
  general, 3, 101
  special, 4, 119

manifolds
  analytic, 68–98
  integral, 87
  oriented, 158
  products of, 75
matrices, 2
  hermitian, 11
  skew-hermitian, 8
  skew-symmetric, 8
  unitary, 4, 10
Maurer-Cartan, forms of, 152
monodromy, principle of, 46
multilinear functions, 139

orthogonal groups, 4, 32, 36, 60, 65, 203
  complex, 4
  special, 119, 203

Peter-Weyl, theorem of, 203
Pfaffian forms, 146
Poincaré group, 52

quaternions, 16

representative ring, 188

Schur's lemma, 182
spinor group, 65
star representation, 178
submanifolds, 85
symplectic geometry, 18
symplectic groups, 21, 33, 36, 60, 203
  complex, 23, 203

tangent vectors, 76
Tannaka, theorem of, 211
topological groups, 26
  connectedness of, 35
  local characterization of, 28
  products of, 27

unitary groups, 4, 33, 36, 60, 119, 202
  special, 119, 203